Cooperative Intelligent Transport Systems

SCIENCES

Networks and Communications, Field Director – Guy Pujolle

Network Management and Control, Subject Head – Francine Krief

Cooperative Intelligent Transport Systems

Control and Management

Coordinated by

Léo Mendiboure

ISTE

WILEY

First published 2024 in Great Britain and the United States by ISTE Ltd and John Wiley & Sons, Inc.

ISTE Ltd
27-37 St George's Road
London SW19 4EU
UK

www.iste.co.uk

John Wiley & Sons, Inc.
111 River Street
Hoboken, NJ 07030
USA

www.wiley.com

Library of Congress Control Number: 2024936174

British Library Cataloguing-in-Publication Data
A CIP record for this book is available from the British Library
ISBN 978-1-78945-180-1

ERC code:
PE6 Computer Science and Informatics
PE6_2 Distributed systems, parallel computing, sensor networks, cyber-physical systems
PE6_7 Artificial intelligence, intelligent systems, natural language processing
PE7 Systems and Communication Engineering
PE7_8 Networks, e.g. communication networks and nodes, Internet of Things, sensor networks, networks of robots

Contents

Preface

Léo MENDIBOURE
ERENA, COSYS, Université Gustave Eiffel,
Marne-la-Vallée, France

Automated vehicles could eventually reduce greenhouse gas emissions from transport and improve road safety and traffic flow. However, its safe implementation will require a high-performance communication system that enables vehicles to obtain information from their neighbors and from the infrastructure, such as lane changes, the presence of obstacles, diversions and extended perception.

Cooperative Intelligent Transport Systems (C-ITS), designed to enable these exchanges, will therefore play an essential role in the advent of the automated and connected vehicle. However, their deployment in a highly constrained and mobile environment could prove problematic in terms of guaranteeing quality of service (QoS), as well as the reliability and security of exchanges.

In response to these problems, this book presents new solutions for managing and controlling performance and security for C-ITS. After two introductory chapters presenting the concept of local interactions and the current development of use cases for C-ITS, this book will explore various ways of optimizing the control and management of C-ITS: hybridization of access technologies (cellular, ITS-G5), use of new tools (e.g. artificial intelligence), etc.

May 2024

PART 1

Introduction to Cooperative Intelligent Transport Systems

1

Local Interactions for Cooperative ITS: Opportunities and Constraints

Jean-Marie BONNIN[1] and Christophe COUTURIER[2]

[1] *IMT Atlantique, Rennes, France*

[2] *YoGoKo, Rennes, France*

1.1. Introduction

Since the advent of wireless communication and its integration into consumer devices, the concept of intelligent environment or pervasive application has emerged. The ability to communicate with all objects in our immediate environment makes it possible to take information or trigger actions. Information collection feeds a context that applications take into account to adapt their behavior to the situation.

For this type of application, direct interaction with objects in the environment greatly facilitates matters, since it is not necessary to rely on a precise location and database to associate information (or objects) with this location. If we need to know the room temperature, all that is needed is to discover a temperature sensor and query it directly. Acquiring the same information when a server is in charge of collecting and exposing the building's temperature data firstly implies discovering the server that has the information at its disposal, then dialoging with it to retrieve

Cooperative Intelligent Transport Systems,
coordinated by Léo MENDIBOURE.

the temperature of the room in which the sensor is located, and finding consequently a way to determine that the location is necessary. The machinery to be put in place is much more complex and yet it seems more intuitive, as the majority of the industry has been built on this model.

The difficulty when it comes to building services on direct (we will also use the term "local") interactions is that this implies standardizing the method of communication, the frequency (or frequencies) used and the message format. For road or city applications, it is therefore necessary to bring many actors to agreement, and to impose choices on the entire ecosystem.

Direct interactions are widely used today for service discovery; for example, Wi-Fi devices continuously scan all frequencies used in the 2.4 GHz and 5 GHz bands to determine if there is an access point available in the environment. The presence of such an access point in no way indicates that the terminal will know how to connect to it, and even in the case where it is able to connect, whether it will be able to obtain a service (an Internet connection). The other technology widely used on consumer terminals is Bluetooth. Again, part of the terminals expose their presence by regularly sending messages at a determined frequency. All Bluetooth devices in proximity are able to see these messages and determine whether or not they know the correspondent. They can then either establish a connection to perform the service (e.g. hands-free kit) by taking advantage of the keying material previously established during pairing, or ask to perform a pairing, which requires the user's intervention.

It should be noted that even when the two correspondents know each other, whether via Wi-Fi or Bluetooth, the discovery and connection establishment time frame is far too long for services with significant time constraints. We will return to this when we examine how the specificities of ITS-G5 make it possible to significantly reduce the time required to exchange information for road safety-related services.

In the second part of this chapter, we will present the concept of ephemeral local interactions, giving examples of services based entirely (or partially) on this type of interaction. We will describe how the first services that will be deployed in the context of cooperative ITS (awareness) are based on this type of interaction and the advantages/constraints of this approach. Lastly, before concluding, we will explore the place infrastructure holds in the implementation of services, based on ephemeral local interactions.

1.2. Ephemeral local interactions: concept and examples

1.2.1. *Examples of services using ephemeral local interactions*

Once it has been established that the different devices in interaction use the same communication technology on a subset of frequencies well known to all, it is necessary to specify the type of interaction targeted. Indeed, we will focus more specifically on interactions where no connection is established. When two devices are in proximity, they can "see" each other because of their technology community; they have at their disposal information that is spontaneously sent by their peers without having to go through the time-consuming establishment of a connection. When the communication technology has a fairly short range, simply being in communication and seeing a device gives an indication of co-spatiality that can form an integral part of the service. Therefore, when a telephone receives an advertisement on one of the three Bluetooth Low-Energy (BLE) channels, it knows that it is in close proximity to the tag whose identity is transmitted in the message, in addition to the information contained in the message itself. In a supermarket, the reception of its advertisements makes it possible to locate the mobile as long as the service provider has the precise placement of the tags in the store at its disposal. However, tags can also directly send information that may be used by the smartphone itself, such as a price, promotions or a link to the page describing a product.

Within the framework of the *TousAntiCovid* backtracking application and other mechanisms for identifying at-risk contacts, developed in the context of the Covid-19 pandemic, this co-spatiality property was used to identify transmission risks (Roca 2022). The complexity of the application comes mainly from the need to protect the privacy of users, while ensuring contact identification that is as accurate as possible. It was therefore necessary to avoid storing the list of contacts but to transmit within the advertisement messages the information required to make it possible to determine a posteriori whether their smartphone had been in contact with that of a contaminated person.

The case of contactless payment applications is rather different, since it instead involves establishing as secure a connection as possible to carry out a monetary transaction. It is therefore absolutely necessary to ensure that we are faced with the right device, and to prove that a valid transaction has taken place. However, the co-spatiality property is used to ensure that the payment card with which the transaction is carried out is in immediate proximity to the payment terminal. NFC (Near Field Contact) technology has been specifically adapted to reduce range and impose "near-contact". The operation of radio transmissions makes this work quite complex because of the propagation of waves in the frequency bands used. It poses security problems, since it makes it possible, for example, to use relays. It then becomes necessary to go beyond controlling the transmission power to limit the range and to very finely control the transmission time, which also depends on the distance; this makes it possible, when it is excessive, to detect an attempt to relay the signal.

Prior to the emergence of the Bluetooth Low-Energy (BLE) version, applications used RFID technology, which has the great advantage of being able to install devices in the environment at very low cost, capable of "responding" to a request and sending previously configured data. This is generally a simple unique identifier, relatively similar to that of eBeacons in BLE. These RFID tags also have the property of being passive most of the time and of using the energy of the reader, which lights them up to wake up and respond. They therefore do not require a battery but are, however, inactive as long as they are not lit up. Readers also need to consume a fairly significant amount of energy to power the tags remotely.

In the different examples that we have seen, local interactions are essentially used to transmit an identifier, which makes it possible to establish our position by referring to prior knowledge of the position of the various devices. Richer applications make it possible to transmit information that the correspondent can directly use and that is most often linked to the position of the sender (a URL describing a product). This somewhat removes the need to maintain a geographic information system (GIS). In the case of BLE, this information is transmitted regularly, whether or not there is a correspondent to listen to it and to do something with it. The information broadcast in this way forms part of the environment and enriches it. The outlines of what we will call "ephemeral local interactions" are given below.

1.2.2. *Characteristics of ephemeral local interactions*

The first characteristic of local interactions is that they are established in the event of a *close contact* supported by a short- or medium-range wireless communication (BLE, NFC, RFID, ITS-G5, etc.).

The examples presented above highlight the *opportunistic* nature of these contacts. The objects considered evolve within a very large scope. They interact, sometimes ephemerally, with many other objects that they do not know beforehand. As stated above, from the outset, this excludes communication technologies requiring a form of pairing (e.g. Bluetooth) or a connection to a network (e.g. Wi-Fi or cellular networks). Indeed, beyond the fact that the necessary establishment time would often be prohibitive with regards to the applications envisaged, it is simply impossible for objects to memorize specific association parameters for each of these relationships; it would also be even more complicated to memorize the association parameters of all potential interactions beforehand.

As a result, the most suitable technologies to support local interactions are those that allow messages to be exchanged directly *without prior configuration*. It is of course still necessary to have knowledge of low-level parameters such as the type of technology, the operating frequency, the modulation parameters or security elements, as appropriate. On the contrary, the fact that the sender has no prior knowledge about the recipients (and their addresses) generally requires the use of communications in broadcast (sometimes multicast) mode, rather than specifically targeting a correspondent (unicast).

It should be noted that once the initial service discovery phase has been completed, it is possible to establish "traditional" connections in order to deepen exchanges with certain special objects. BLE enables, for example, discovery in opportunistic mode and then switching to dedicated channels in BLE or Bluetooth for more substantial exchanges. The same type of example can be envisaged in V2X, where the detection of a vehicle (or group of vehicles) on the control channel (CCH) can lead to the establishment of special relationships on a service channel (SCH), for example, to process exchanges within a platoon or to carry out a financial transaction (toll).

The characteristics of ephemeral local interactions are therefore as follows:

– The information senders and receivers (which can be the same, or two devices with different functionalities) do not know each other beforehand and – often due to mobility – change over time.

– They require prior knowledge of a technology and the channels (frequencies) used to broadcast the information. This therefore presupposes regulation or standardization.

– There is a spontaneous broadcasting of information – without prior contact – in a predetermined, and therefore most often standardized, format. This information is visible to all devices within communication range and listening.

It follows from these basic characteristics that:

– It is difficult to operate on multiple channels, since this presupposes listening to several channels successively, and significantly complicates the encounter between the sent data and a device that is potentially interested.

– This leads to a risk of bandwidth overload, since the load cannot be distributed over several channels as is done in current technologies. Therefore, applications using ephemeral local interactions must be limited to fairly simple data (we will not transmit a 4k video stream).

– Moreover, since an extension of the range is sought, more robust data encoding is often used, since it is understandable with a lower signal-to-noise ratio and therefore at a longer distance. This reduces the available throughput, so in ITS-G5 an encoding of 6 Mb/s is used, while the technology would allow 27 Mb/s to be reached.

– As generalized diffusion (broadcast) or selective diffusion (multicast) is used in most cases, we cannot have an acknowledgment system. Indeed, if the different receivers had to acknowledge each broadcast message, the responses would need to be spread out over a long period of time to avoid collisions. Moreover, this would not be very useful, given that since the sender does not know the list of recipients beforehand, it could not determine that there have been losses;

– The securing of exchanges is quite complicated insofar as the assumption is made that there is no prior exchange of information between the protagonists and that they do not know each other beforehand. Setting up cryptographic material to authenticate the sender or to sign (or encrypt) the content is therefore difficult as there is no trusted third party that can be reached at all times. This is all the more difficult because all or part of the devices are mobile and associated with users; therefore, the use of permanent identifiers is prohibited as this would make it possible to track the user's journey. We will see how different technologies protect themselves from this risk and how the world of cooperative ITS has addressed the need to secure the exchange of sensitive data.

1.2.3. *Advantages of ephemeral local interactions*

The use of ephemeral local interactions does not make it possible to maintain the usual mode of operation of applications and services; however, it offers advantages that are of great interest.

Therefore, ephemeral local interactions do not require prior knowledge of the protagonists. This mode of interaction is often used in an initial discovery phase before returning to a more conventional mode and establishing a connection. Therefore, a Wi-Fi access point will very regularly send announcements (beacons) that enable the stations to discover its existence. The presence of an access point does not make it possible to determine whether or not the latter can offer a service to the mobile terminal, which, in order to do so, will have to attempt to connect to it and establish a secure session. However, simply receiving beacons in itself enables information to be obtained that is useful to the mobile terminal (Chandra et al. 2007), since this is how smartphones obtain their positioning, through fingerprinting techniques and Geographic Information Systems maintained by service providers. As soon as part of this database is stored locally on the mobile terminal, the latter no longer needs the infrastructure in order to calculate its position.

Bluetooth technologies also use beacons that are transmitted on specific channels listened to by mobile devices (Bluetooth Special Interest Group (SIG) 2016). This enables a terminal to discover the presence of a Bluetooth device, and to connect to it if it has already performed the pairing phase. Here, we are therefore not in the operating principle of ephemeral local interactions, but the diverted use of its beacons to detect the presence of mobile terminals or vehicles (through their hands-free kit) corresponds closely to this logic. Therefore, it is possible, from the roadside, to detect vehicles and the unique identifier of their Bluetooth device, which makes it possible to establish input/output matrices by listening to highway access ramps (Barceló 2013; Boudabous 2021).

One of the main benefits of services based on ephemeral local interactions is that they do not require the prior deployment of a network infrastructure. It is therefore possible to rapidly deploy services without relying on an infrastructure, and to withstand permanent or accidental absences on the network infrastructure. The direct or device-to-device (D2D) mode of 3GPP networks on which cellular versions of V2X technologies are based have also found their first application in the field of emergency tactical networks, which must be able to be set up rapidly, when the telecommunication infrastructure has been harmed by a natural disaster. Depending on the application, we will see that it may occasionally be necessary to rely on an infrastructure to manage the securing of exchanges but without requiring it to be permanently available or involved in the normal operation of the service.

Depending on the needs and the communication technologies employed, using ephemeral local interactions enables very fast interactions that depend solely on the message sending frequency and the traffic capacity of the communication technology used. Therefore, the transmission delay will be higher when using BLE beacons (Yang et al. 2020), which must be able to operate for several months/years on a button battery, and which, as a result, will only send a "beacon" every minute. The maximum detection time of a beacon will therefore be of the order of a minute if we take into account the potential losses and the periods when the receiver is not listening to the correct channel. In the case of ITS G5, CAMs (Cooperative Awareness Messages (ETSI 2014)) are sent by default every 100 ms, which, even considering the potential losses, enables very short latencies.

1.2.4. *Suitability of communication technologies for this type of interaction*

The communication technologies used must enable an interaction model that is consistent with ephemeral local interactions. Therefore, it is very much possible to build above IP a service that functions through the regular broadcasting of messages and the use of this information to provide a service without any further dialog with the sender. However, while this presupposes prior connection to a network (e.g. Wi-Fi or Cellular), a large part of the advantages described previously disappear. The underlying communication technologies therefore need to be aligned with the interaction model and the targeted properties.

Adapted technologies generally use a shared multiple-access channel, which means that to access the communication medium, it is not necessary to obtain prior authorization and that everyone can transmit on the channel. There is therefore competition for access to the radio resource, which is for exclusive use (only one sender at a time). A device seeking to send a frame will listen and wait until the channel is free. There is a risk – and mechanisms to reduce it – that two devices will start sending at the same time. When the messages of two senders overlap in time, there is a collision. Some of the potential receivers do not receive either of the two messages, while the other potential receivers decode one message or the other according to the capturing-effect principle (Roberts 1975; Gezer et al. 2010). However, the senders cannot hear the other transmission and do not notice the collision. Regardless of the organization of the radio channel, as soon as there is no device in charge of organizing the resource (such as a base station in the 4/5G networks), and no prior exchange to decide who is allowed to use the resource (RTS/CTS mode in Wi-Fi), the nodes use the resource and then manage the effects of the collisions. When the transmission is directed towards a specific receiver, it is possible to use an acknowledgment and retransmission system (this is the case for Wi-Fi); however, when it is broadcast, losses cannot be detected by the sender, and this must be taken into account in the very construction of the service.

In the case of ITS-G5, a device listens to the channel and transmits only if it is free for a minimum period of time, thus limiting the occurrence of collisions. This type of self-organized system (CSMA: Carrier Sense Multiple Access) works very well up to a certain level of channel occupancy (approximately 60% of the bandwidth) (Bianchi 2000). It is therefore appropriate to introduce mechanisms to limit the network load. Mechanisms are thus used to vary the message frequency according to the usage scenarios and the load of the radio access network. Of course, reducing the sending frequency increases the average time between two receptions and the time needed to discover a new node; this is why it is important to take into account the usage scenario, such as vehicle speed in the context of ITS. Indeed, the lower the velocity of the vehicle, the less its position changes between two announcements.

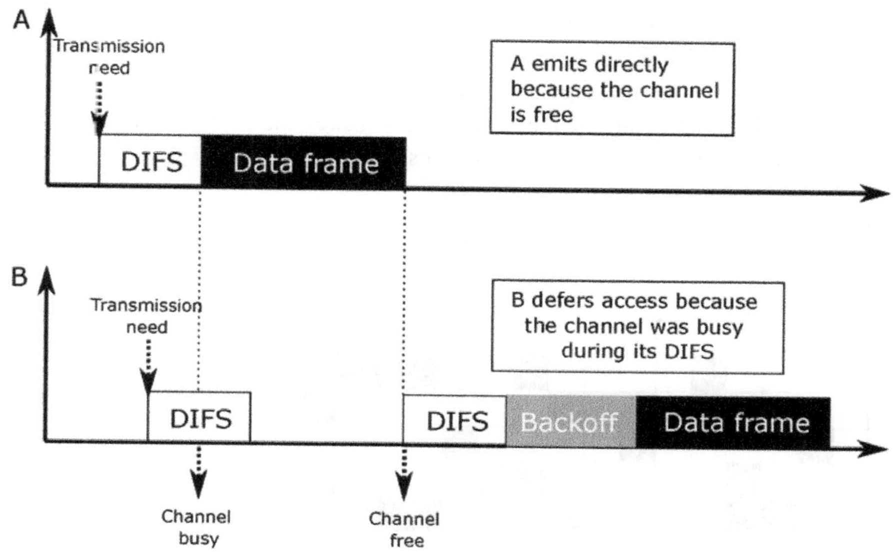

Figure 1.1. *Operation of Wi-Fi CSMA (IEEE 2009)*

In the case of BLE, the very limited bandwidth (typically 1 Mb/s under conventional conditions) will quickly impose constraints on the announcement sending frequency, even if, by design, the messages are limited in size. Since the range is also quite small, this technology could be used for low-velocity applications (pedestrians and VRUs (Vulnerable Road Users) more generally).

In the case of LTE-V2X (Garcia-Roger et al. 2020), the technology developed relatively recently has slightly different properties. Indeed, the radio resource is organized in a frame, with concurrent access to each slot, so that there are several transmissions in parallel on the channel. To limit collisions, due to the very short duration of a slot, it is not possible to listen to decide whether to transmit or not. When the channel is organized by a base station – this is one of the LTE-V2X modes – a station asks to be allowed to transmit before obtaining the resource from the base station. While there can be a collision on the request, the transmission of useful packets is no longer subject to collision. When the channel is self-organized, all the stations share the temporal structure of the frame and must therefore be strictly synchronized. They use semi-persistent scheduling (SPS). This involves listening to the channel and taking advantage of the repetitive nature of regular announcement messages (CAMs) by noting the slots that are occupied in the recent period. For each slot, the station determines the probability that it will be occupied by observing the power level received in that slot for a given duration (1 second by default). It will then select one of the "free" slots for its own transmissions. This selection will then be reviewed regularly. This mechanism works quite well for regular and permanent messages. However, when a collision occurs, it is not detected and lasts until one of the two senders makes the selection again, which becomes critical when the real-time constraints are high.

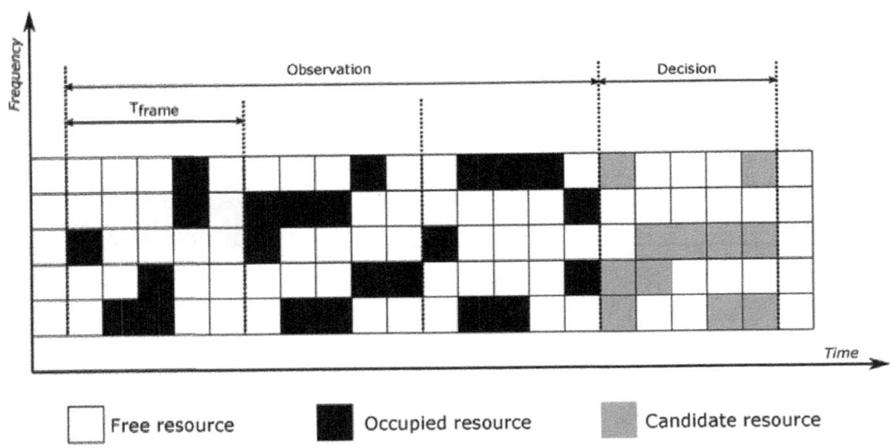

Figure 1.2. *Structure of frames and operation of LTE-V2X semi-persistent scheduling (Haider and Hwang 2019)*

Due to the direct communications between the devices concerned, direct interactions allow very short connection times, given that the announcement messages are frequent and that the receivers constantly listen to the correct channel. Mechanisms to time-multiplex sending and reception extend the time frames and the available bandwidth automatically. When everyone listens to the same channel, the connection time, i.e. the average time required to receive the first message that the receiver is able to decode, is directly related to the message sending period and the distribution of losses on the channel. Indeed, if the losses are not distributed uniformly, the calculation of this average time depends mainly on the maximum duration of the consecutive message loss sequences and on the probability of loss. Losses can be due to collisions, other forms of signal interference or masking. The different forms of interference may be considered to be random. On the contrary, as already mentioned above, collisions are positively correlated with the load submitted to the network, hence the importance of controlling the amount of traffic submitted to the network.

1.3. Local interactions serving cooperative ITS

In this section, we will focus more specifically on the use of ephemeral local interactions in the context of cooperative ITS.

1.3.1. *Cooperative ITS services*

The scope of cooperative ITS (C-ITS) applications is particularly broad. It is traditionally broken down according to the maturity of the technology required. So-called "Day 1" applications are deployable with the technologies currently available. These include the transmission of alerts (slowing down, approaching a priority vehicle, accidents, road works, etc.), signage (on-board display or traffic-light phase) or of presence (position, speed, direction, etc.). In contrast, "Day 2" applications require performance and standardization levels that have not yet been achieved. This covers, for example, driving in convoy ("platooning"), remote driving or vision sharing ("see through"). "Day 1.5" applications are at an intermediate stage: they are feasible for particular cases but their level of standardization does not allow them to be immediately generalized on a large scale. This is the case, for example, with the protection of Vulnerable Road Users (VRU), parking space management or dynamic routing.

As diverse as they are, these applications can be based on the standardization proposed by the "Facilities" layer of the ITS-Station stack. This messaging layer is a kind of middleware between communication layers and applications. This layer is regularly enriched with new functionalities. At this stage, we can cite the following messages as examples:

– CAMs (Cooperative Awareness Messages) (ETSI 2014) are sent regularly (typically every 100 ms) by vehicles to signal their position, speed, direction and physical characteristics. Other vehicles use this information to add the sender to their perception of the environment.

– DENMs (Decentralized Event Notification Messages) (ETSI 2014) are used to signal one-off events such as accidents, construction sites, slowdowns, etc.

– SPAT/SPATEMs (Signal Phase and Time/Extended Messages – ETSI TS 103 301) transmit the traffic-light phase status. They are typically associated with

MAP/MAPEMs (Map Data/Extended Messages) that describe the geometry of roads and intersections.

– IVI (In-Vehicle Information – ETSI TS 103 301) dynamically relays the signage information for on-board display. They replace or complement conventional road signage and are more easily exploitable by the vehicle's automations.

– CPMs (Cooperative Perception Messages – ETSI TR 103 562) enable actors (vehicles and infrastructure) to dynamically exchange information on their perception of the environment (obstacles, vehicles, pedestrians, etc.).

These standardized messages offer an impressive toolkit to accelerate the development and penetration of applications into the market. However, before they start, application designers must analyze the consequences of the choice of message type on the architecture of their solution. Will the application be dependent on an infrastructure or not? Can this infrastructure be decentralized (RSU) or will it require a central server? The answer to these questions will have a strong impact on the cost and ease of deployment of the solution, as well as on its performance (for example, reaction time).

1.3.2. *Benefit of ephemeral local interactions for cooperative ITS*

Here, local interactions offer numerous advantages. They are simple to deploy (as their configuration is very lightweight), require little or no infrastructure (e.g. a light connected to an RSU) and offer very good responsiveness (of the order of several ms to several 100 ms). In addition, limiting wave propagation makes it possible to natively reduce diffusion around areas of interest of the different messages.

The benefit is obviously for collaborative perception applications (based on CAM or CPM messages) or for signage (IVI, SPAT, etc.). In the first case, the speed of transmission is decisive. Also, in all cases, the geographical limitation of broadcasting and the ability to exchange information without the need for prior association constitutes a decisive advantage.

The transmission of warnings (typically by DENM message) represents an interesting case. A network access priority and a sending frequency is associated with the message, based on the emergency associated with the event at the origin of the warning. Therefore, DENM messages relating to an extreme emergency (e.g. in the event of a collision) are sent with maximum priority and at a high frequency to ensure that all approaching vehicles dispose of the information early enough to react (in a few ms).

DENM messages are also used to signal road works or other hazards present on the roadway, in which case, the time constraint is generally more relaxed. For example, it is generally not indispensable to signal the presence of road works in under a second. Moreover, this type of information is sometimes transmitted in V2N (Vehicle to Network) via cellular networks. However, use is often made of V2X – and therefore of local interactions – for these messages too. The reason is twofold: the natural geographical limitation of broadcasting, as well as the fact that the vehicles already have a V2X receiver for other types of messages. In this context, imposing a second receiver only for DENM messages would be counterproductive. Moreover, since it may be necessary to cover a geographical area larger than the range of a V2X transmission, ETSI has provided for the possibility of addressing a geographical area directly in the message header, which is named "geocasting" (ETSI 2009). The messages are thus relayed hop by hop to reach the area of interest without requiring any equipment in the road-side infrastructure.

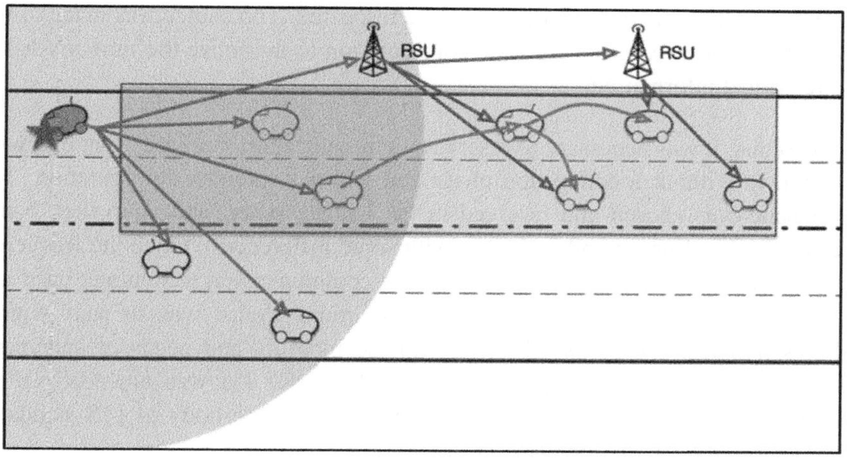

Figure 1.3. *Different modes of DENM message broadcasting. For a color version of this figure, see www.iste.co.uk/mendiboure/transportsystems.zip*

The example of DENMs is also interesting in terms of architecture. It is appropriate to equip a mobile construction site with a mobile RSU to send DENMs, signaling the construction site to approaching vehicles. However, its range will be insufficient to announce the construction site 5 km upstream. For this, the RSU can send the messages in geocasting, but these will not reach their target if there is no vehicle to relay the messages. It is then possible to make use of RSUs that will relay messages as vehicles would do, and without needing to be connected to a network infrastructure; however, this requires RSUs to be specifically deployed if the road is not yet equipped. Lastly, it is possible to make use of a centralized ITS infrastructure and a non-dedicated network infrastructure (e.g. a server that can be reached by a cellular network) so that its message is transferred directly to the vehicles or to the RSUs, which will have to broadcast it downstream. In this case, when no technology meets all the needs, it is useful to combine traditional communications with opportunistic communications.

1.3.3. *V2X communication technologies*

During the work on WAVE architecture (Eichler 2007; Gräfling et al. 2010) in the United States, several types of services were considered, some of which relied on a connected infrastructure. However, covering all roads with a communication infrastructure before the deployment of services is not a reasonable alternative – especially since even if it had been possible, the communication networks of the time did not allow the time constraints to be respected. The choice was made to use direct interactions, requiring the least configuration to minimize the time needed to acquire critical information.

To do this, it was necessary to reserve a communication frequency that everyone can listen to without a connection phase and without complex configuration. The 5 GHz band was chosen and reserved in the United States, and equivalent bands were reserved in Europe and in Japan (with some differences). Once the frequency was selected, it was necessary to agree on the communication technology used and on the organization of radio resources. The initial choice was to start with a well-mastered technology by adapting it to the mobility and needs of ephemeral local interactions. For this purpose, Wi-Fi (IEEE 802.11) was amended (IEEE 802.11p). In Europe, this technology is integrated into the body of ITS standards developed by ETSI under the name ITS-G5.

It should be noted that Wi-Fi, due to its fully distributed channel access control mode, is well suited to ephemeral direct interactions; indeed, by definition, all those who listen to a channel hear all of the messages and decide based on their perception of the channel, whether or not they can send a Wi-Fi frame. Natively, broadcast is used, which makes it possible to mobilize all stations capable of decoding the message. However, in Wi-Fi, even if all of the stations perceive the frames, the communications are relayed by the access point, requiring the prior deployment of an infrastructure, even if the latter does not need to be connected to the Internet to relay the frames between local stations. Among the adaptations made to Wi-Fi, the choice was made to introduce a new operating mode without infrastructure in Wi-Fi: OCB (Outside the Context of a BSS) mode. The other adaptations concern improvements to radio in order to improve the reliability of transmissions in contexts with high-speed differentials (Doppler effect), and with a lot of interference due to fading (urban environment).

For more than 15 years now, numerous projects (SCORE-F, SCOOP@F (Aniss 2016), InDiD, C-Roads and Intercor) have been deploying, experimenting and testing the limits of ITS-G5 in different contexts and for a wide variety of services, so its operating scope is well known.

It should be noted that many other technologies have been (and are still being) considered for direct communications between two vehicles, in particular, to accommodate much higher throughputs (See Through) or even shorter time frames (Platooning and Cooperative Maneuvers). For example, we can mention highly directive technologies such as millimeter waves or visible light communication. For basic services related to security and the perception of other users, as well as the discovery of available services, vehicular Wi-Fi (ITS-G5 in Europe) has long been the only alternative really considered. Today, the situation is a little more complex; two families of incompatible technologies are in the running.

Indeed, cellular network actors have proposed to adapt a variation of LTE's D2D (Device-to-Device) mode to V2X communications (Garcia-Roger et al. 2020; ESTI 2018). Even more recently, 3GPP published standards to offer more advanced services through the adaptation of the new 5G radio interface to V2X (NR-V2X) (ETSI 2020; Storck and Duarte-Figueiredo 2020). The possibility of implementing high value-added and potentially marketable services is one of the challenges of the battle between the two technology families. It should be noted that V2X cellular communications have an infrastructure-free mode that does not rely on the base station to control access and optimize the radio resource. However, they can also use an infrastructure to offer services and, in particular, with the 5G network architecture that promises guaranteed interaction times of the order of milliseconds, which would make it possible, for example, to organize complex maneuvers from servers in proximity to traffic (Mobile Edge Computing).

Technological differences and the debate regarding the choice of technologies is not what interests us in this chapter. It will be necessary, however, to quickly choose a technology, even if it is not perfect, because this is a prerequisite for the deployment of the first services, based on direct interactions. Indeed, everyone must speak at least one common language, if only to discover the presence of others and exchange, without prior complex interactions, messages related to emergency situations.

Other technologies enable direct interactions; for example, they target access control services (wireless car-key) but it requires prior pairing or complex configuration to manage identifications. Note that it is increasingly possible to use a smartphone and standard communication technology (Bluetooth, BLE or Wi-Fi) to implement these services. In any case, these are not the forms of interaction that interest us.

1.3.4. *Properties of C-ITS services built on local interactions*

1.3.4.1. *Latency and reduced responsiveness*

The different C-ITS services do not require the same level of service, but for all, a fairly low latency (less than 100 ms) is sought. It is in fact even shorter for advanced services such as convoy maneuvers (platooning) or cooperative maneuvers. The maximum latency depends on several factors; first and foremost on the communication technology used and how it controls access to the radio resource. Latency also depends on the network load and the state of the radio channel. How the load impacts transmission performance depends directly on the technology used. Therefore, in ITS-G5, the average channel access time increases with the load submitted to the network; however, it is above all the losses due to collisions that end up making the time frames too long. Indeed, the messages are generally transmitted at a certain frequency (10 times per second for CAM messages), which makes it possible to support frame losses while not jeopardizing the service. The losses, which also increase with the load submitted to the network, therefore amount to adding time between messages received successfully.

It should be remembered that due to transmissions in broadcast, it is impossible to use an acknowledgment mechanism and therefore to make transmissions more reliable by retransmitting lost messages. To increase the message transmission rate, LTE-V2X prefers prevention rather than cure, systematically transmitting critical messages twice. While this consumes twice the radio resource, it greatly increases the probability of reception.

The two competing technologies, ITS-G5 and LTE-V2X, use very different mechanisms to control access to the radio resource, and this results in different load behaviors. In both cases, the collision rate increases with the network load, but in the case of ITS-G5, each message competes for access to the channel and can therefore be subject to collisions; in the case of LTE-V2X, the SPS algorithm protects the recurring traffic already established, because the vehicles seeking to send know which resources are occupied. Initial access to the channel is faster in ITS-G5, however, with messages very often delivered in a few milliseconds. In the case of

LTE-V2X, sending a new message requires a certain time frame for choosing the resource (Resource Block) and a collision, the probability of which depends on the network load; this can take place on this resource. In this case, messages are not delivered throughout the duration of the collision (the 100 ms re-selection window), which is a real problem for the most urgent warning messages.

While it is difficult to reduce collision delays in a fully distributed mechanism, it is easier to prioritize access to the radio resource for the most urgent messages. Therefore, ITS-G5 allocates a different priority to the message according to their service class, using the Wi-Fi EDCA (Enhanced Distributed Channel Access) mechanism (ETSI 2019).

1.3.4.2. *Frugality of exchanges and congestion mitigation*

Due to the capacity limits of the radio channel and the number of devices with the potential to send, exchanges must be limited to what is strictly necessary. In addition, it is necessary to favor short messages that free the channel up faster. This is especially true in ITS-G5 because owing to the channel access method (CSMA), a message in the process of being sent cannot be interrupted and prevents vehicles having more urgent messages from accessing the channel. In order to reduce the size of messages and avoid protocol overload due to the TCP/IP stack, the messages have been installed directly above the connection layer with ad hoc protocols[1]. For the same reasons, certificates used to verify the legitimacy of the sender are transmitted only once per second.

Despite these precautions, the vehicle density may be such that it is necessary to prevent channel congestion by limiting the load submitted to the network. In order to evaluate the network load, and because of the lack of centralized coordination, each vehicle can only have, at its disposal, information that it is able to observe by itself, such as: the proportion of time the channel is occupied and the number of vehicles observable in the vicinity. When it measures that the network is loaded, it can play on time by reducing the number of messages and space by limiting the transmission power to reduce the "cost" of a transmission in terms of spectrum occupancy. Some of the mechanisms apply at the radio technology level, while others involve the upper layers (called cross-layer control) to limit the number of messages submitted to the network. This reduction takes into account an assessment – by necessity, simplified – of the usefulness or criticality of the messages. Therefore, the frequency of CAM messages can be reduced to 1 Hz when the vehicle is slow or stationary.

1. The choice of doing without IP in order to limit the protocol overhead is questionable, given that header compression mechanisms exist. When used in the IoT, they are very efficient and reduce the header size to a few bytes (Gomez et al. 2020).

1.3.4.3. *Facilitated privacy*

If the contents of the messages are not directly private in nature, the information transmitted by a vehicle makes it possible to trace the latter and thus reconstruct its journey. More generally speaking, each time a permanent or semi-permanent identifier can be associated with a positioning, there is a risk of allowing tracing.

It is therefore indispensable to mask unique and permanent identifiers. Indeed, radio technologies use unique identifiers (MAC address) that are allocated when manufacturing the radio interfaces and are used as source addresses in the frames. The upper layers also use identifiers (certificate, IP address, etc.) in the different layers of the protocol stack used. For several years now, it has been commonly accepted that the different identifiers used must no longer be able to be associated with a particular terminal. To avoid this, regularly changing random identifiers are used (BT or Wi-Fi). However, care must be taken to change all the identifiers present in a message's different headers at the same time, to avoid two successive identifiers from being associated too easily. Indeed, if the MAC address changes but the identifiers used in the upper layers do not change, an observer will have no trouble attributing the two successive MAC identifiers to the same terminal.

It should be noted that the cost to reconstruct a vehicle's movements is quite significant, since it must be possible to observe the vehicle (listen to messages) at different points of the journey. An infrastructure covering a highway system could thus very easily reconstruct journeys and trip times, as can be done with Bluetooth scanners (Klinkusoom et al. 2014). The fact of masking the identifiers using pseudonyms is relevant in the sense that it greatly complicates the work of the observer seeking to reconstruct the journeys; it is nevertheless not impossible but the effort required to reconstruct journeys depends directly on the pseudonym change rate (Kountché et al. 2017).

In the case of cooperative ITS, the problem is made more complex because of the need to secure exchanges and to ensure a form of control of the legitimacy of the sender to send certain data, and therefore a form of authentication. Authentication and anonymity are generally very difficult to combine, especially when the communications are purely local and therefore without a permanently accessible, trusted third party.

1.3.4.4. *Independence with respect to the functioning of the infrastructure*

It is clear that it will be necessary to have a communication infrastructure to manage deployments and maintain roadside devices. However, once configured, the operation of basic services will depend little on this infrastructure. This makes it possible to imagine very progressive deployments: it is possible to deploy roadside devices only in locations that are hazardous or whose operation can be optimized (complex intersections) and to have devices with only a weak or partial connection to the Internet, the relationship to the management system being used only for device monitoring and configuration. There are therefore few performance constraints weighing on the interconnection network and an already existing non-dedicated network (cellular network) can be used.

This property is significant, given that the economic model to render the communicating infrastructure profitable remains to be found. This is especially true of Day-1 services that are related to road safety aspects and for which a business model based on users' propensity to pay is not easy to find.

1.3.4.5. *Facilitated emergence of new actors*

Due to a naturally open design, the implementation of new services does not depend on the volition of a communication infrastructure operator, since this can be implemented based on direct exchanges.

However, due to security requirements, it may well be necessary to obtain authorizations to be allowed to use messages on frequencies reserved for ITS. Furthermore, signing messages using certificates creates a dependency on the certificate management infrastructure (Haidar et al. 2017). It will therefore be necessary to open the use of these security tools up to other areas (energy, home automation, logistics, etc.) to enable the emergence of new services.

1.3.4.6. *New opportunities in the exploitation of locally produced information (AoA, TOF, RSSI, etc.)*

Local interactions teach us something about the protagonists in interactions, and this information can be used to consolidate or, on the contrary, to question exchanged information. Therefore, for example, the simple fact of being in communication with a BLE beacon teaches us that we are a few meters away from it. The level of the received signal is in itself difficult to use to calculate a distance, but observing the power level of the messages of several protagonists gives us information on their relative positioning. Apart from co-presence (i.e. being within the range of a sender), little use is made of this information today; however, it could, for example, serve to perform a consistency check between the information received and what is observed.

With the evolution of radio technologies, a new type of information is set to become available in the coming years, once active antennas become widespread: the angle of arrival (AoA). Associated with the power level, this information will make it possible to build up an understanding – albeit partial and approximate – of the positioning relative to ourselves of communicating objects within our environment.

1.3.5. Limitations and constraints of implementing services built on local interactions

Local and ephemeral interactions make it possible to do a lot at low cost, but they also suffer from limitations that complicate their implementation and impose hybrid modes in which there is less use made of a communication infrastructure; however, the latter remains just as necessary. The benefit then lies in reducing the constraints weighing on the communication infrastructure in order to use pre-existing, non-dedicated infrastructures.

1.3.5.1. *The need to standardize communication technologies and data exchange*

The first "obvious limitation" is that the implementation of service based on this type of interaction assumes that all devices use the same technology on a reduced set of frequency bands known to all. When the interaction time has to be kept very short (services related to road safety), it is even preferable to reduce the meeting time to a minimum and therefore have a single frequency that everyone listens to. The more sophisticated services can use other channels, or even other technologies, once the appointment has been made on the common channel.

Beyond communication technology and its intrinsic limitations, the fact that everyone listens to the same channel at all times implies that the content of the messages is intelligible to all those who may be concerned. In the case of cooperative ITS, all vehicles and road users, as well as the communicating elements of road infrastructure are concerned. An additional difficulty comes from the fact that messages can contain information intended for users, and they therefore have to be understandable in several languages. The standardization groups have developed message formats that adapt to the uses of the different countries and are translatable through a dictionary mechanism in the different languages.

Lastly, it was necessary to provide a minimum subset of standard messages for vehicles coming from other geographical areas and circulating on European roads through harmonization work between the standardization bodies dedicated to cooperative ITS (e.g. WAVE in the USA, ETSI/CEN ITS in Europe, etc.).

1.3.5.2. *Management of security and trust in distributed mode*

When services are connected, the mechanisms to ensure the security of exchanges is now well mastered and rely on the same solutions as those implemented for conventional Internet services. In particular, it is possible to use the certificate mechanism, which relies on a trusted authority in charge of guaranteeing the identity of a correspondent. The problem worsens when the operation is purely local without the possibility of verifying the identity of a correspondent with a trusted third party – for example, via a public-key infrastructure (PKI).

In the case of cooperative ITS, the issue of security is extremely important as messages may feed automated processes (autonomous driving or advanced driver-assist) that could take full or partial control of a vehicle (emergency braking). Therefore, the simple fact of sending a DENM message to signal emergency braking on a highway at a busy time could create a significant traffic jam. It is therefore necessary to give the message receiver the ability to verify, one way or another, that its content has not been modified and that the sender is legitimate. Verifying the identity of a sender with whom there has been no prior relationship most often relies on a trusted third party. However, as seen, it is not possible to make the assumption that a trusted third party will always be reachable; in any case, the time frames would have been far too lengthy to perform this verification before taking the message content into account.

In the context of cooperative ITS, ETSI has opted to use a mechanism based on certificates signed by a well-known authority. It is thus possible to have confidence in the certificate without first checking with the authority that signed it. However, this does not entirely solve the problem, since this certificate contains an identity and we saw that permanent identities had to be hidden to avoid exposing users' journeys. Therefore, in the same way that pseudonyms are used instead of the permanent identifiers of the different network protocols, pseudonymous certificates are used instead of certificates containing the immutable identity of vehicles and roadside devices. These certificates are produced by a dedicated PKI. Vehicles request a pool of pseudonymous certificates (several hundred) and renew it from the PKI when the current pool is exhausted. These are then used to sign messages and are changed very regularly to avoid the tracing of journeys. In addition to the identity of the referring authority, pseudonymous certificates contain the validity period (lifetime) and a temporary identity. They also include information regarding the role of the sender, thus enabling an emergency vehicle or a road-side unit (traffic lights) to send messages that a simple vehicle will not be allowed to broadcast.

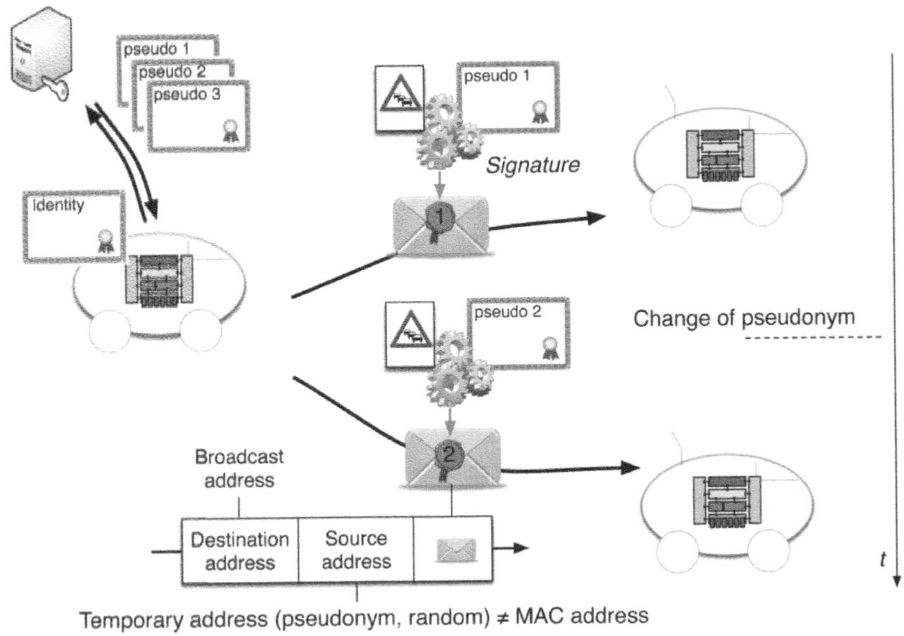

Figure 1.4. *Functioning of pseudonymous certificates. For a color version of this figure, see www.iste.co.uk/mendiboure/transportsystems.zip*

It is worth retaining the concept of a pseudonymous certificate used for a short period of time to protect location privacy, and that they contain a brief description of the actions/messages that a device may legitimately perform. The trusted third party exists but its mobilization is disassociated from the emergency on the field.

The fact that the public-key architecture is necessary implies the need to have a permanent identity provided by the authorities in charge of cooperative ITS for all devices capable of emitting on the reserved frequency band. This introduces significant rigidity compared to the general context of services based on ephemeral local interactions and is a burden for the deployment of new services. While listening to messages is still possible, signature verification presupposes significant resources and dedicated hardware, whenever it is necessary to process a certain message volume. The sent messages will need to be signed in order to be considered valid by the receivers, which presupposes the acquisition of a permanent certificate and a temporary certificate from the public-key architecture dedicated to cooperative ITS.

1.3.5.3. *Interoperability management*

As we have already mentioned, the problem of interoperability between devices that do not have the same technologies is difficult to address. From the point of view of protocols relating to communication, format or semantics of messages, methods exist to mitigate its impacts and it is always possible to update in-vehicle systems to support a new message format and/or a new version of a protocol. For this purpose, the use in networks is to clearly indicate to the receiver which version of the protocol or format of the messages the transmission uses and possibly to provide an extension mechanism that is readable only by those who recognize its identifier. This is very important in a context where standards are very regularly updated and enriched with new functionalities. Moreover, they are not all stabilized yet. Unfortunately, the first versions of the protocols proposed by ETSI did not have these mechanisms.

However, it is important to bear in mind the objective of the first protocols, which was to provide a basic information (awareness) service, in which both vehicles and roadside devices could participate. For these basic services, the objective is to minimize the load and therefore the size and complexity of messages. The benefit of having options that only a few receivers can understand could legitimately be questioned. The more sophisticated services then pass through a dedicated session, thanks to the appointment made on the common channel. This makes it much easier to manage service versions.

Regardless, once deployments have begun, it becomes very complicated to deploy an update that is incompatible with the existing one. It is necessary to provide/allow the possibility of updating the devices already deployed (over the air), and to organize coexistence between different versions of the standards for an indefinite period. This problem of updating vehicle embedded software in deployed objects is a critical subject beyond cooperative ITS. A large part of the communicating objects (IoT) deployed today are rarely, if ever, updated beyond the initial deployment period. Security vulnerabilities are therefore present long after the vulnerabilities have been discovered and documented.

The deployment of several communication technologies incompatible with one another in the same frequency band is very difficult to manage for V2X communications. It is necessary, on the one hand, to organize the coexistence, i.e. sharing of the radio resource; and, on the other hand, to manage the fact that not all vehicles or roadside devices will see all the messages. The problem arises both between two families of technologies and between two generations in the same family. This debate is all the more critical today because there are two families competing, operating in very different ways. However, it would have arisen at one time or another regardless, if only upon the deployment of a new generation of V2X technology, which will inevitably occur at some point. It is completely unrealistic to modify all the devices over a very short period of time, especially considering that some of the devices scattered about in nature will not be connected. It is therefore necessary to anticipate the ways of managing the cohabitation between the different generations.

Sharing the radio resource: organizing sharing of the same radio resource (the same ITS channel) amounts to organizing usage over time. Even if this worked well, it would only solve part of the problem, since devices with one of the technologies would still be deaf to messages broadcast using another technology. In addition, it would have the major defect of reducing the bandwidth available for each of the technologies, especially since the organization of the resource would consume part of the time. Several proposals have been made, and work is underway for ITS-G5 to "hear" the transmissions of C-V2X and take them into account when it decides to emit, and vice versa. However, the semi-persistent scheduling used in C-V2X would be rendered inoperative, since CSMA technology is unable to take it into account.

Managing heterogeneity: the simplest solution would no doubt be to reserve different channels for each of the technologies, but the messages should probably be broadcast on both channels by devices with both technologies and possibly relayed by these devices when the source has only one of them at its disposal. The organization of this relaying poses similar problems to multi-hop broadcasting in self-organized (ad-hoc) networks and is not trivial to solve. It is indeed necessary to choose who relays the message when several devices are capable of doing so in order to maximize the number of receivers, while not over-occupying the channel with multiple copies of the same message.

1.4. Role of infrastructure in cooperative ITS services

1.4.1. *Infrastructures dedicated to cooperative ITS*

If we consider the need to deploy infrastructures dedicated to certain services – in the following, we will discuss mainly cooperative ITS services and more generally services related to the mobility of goods and people – it is important to consider the distribution of the cost of deployment and operation of its infrastructures. While at the start of WAVE architecture standardization in the United States, it was envisaged for all roads to be covered with a dedicated infrastructure (based on the IEEE 802.11p amendment), this idea did not survive the various financial crises. In any case, it is unrealistic and does not solve any of the engineering problems related to the necessary support of the discontinuity of coverage; if only because it is impossible to achieve full coverage before deploying the first services.

It is therefore essential that the various services operate opportunistically to take advantage of existing infrastructure when available, whilst otherwise maintaining autonomous operation. For a large part of the services developed within the framework of cooperative ITS, the V2X mode without infrastructure involving only vehicles is therefore the favored mode. Augmented modes with more functionalities or better performance are considered when the infrastructure is available. Hybrid modes using Internet connectivity (often cellular) are also envisaged for information services with low time constraints.

The presence of active devices in the infrastructure does not necessarily mean that these devices are interconnected, nor that they are connected to the Internet, and even less that they offer vehicles the possibility of interacting with servers, whether they are centralized (cloud) or remote as close to usage as possible (edge computing). However, these are configurations that each offer different possibilities:

– Independent infrastructure devices: in this configuration, it is possible to simply deploy devices that will serve to support certain services and that do not need to be connected to the Internet (or only for remote maintenance matters). The device will use V2X technologies and behave like a vehicle to listen to and transmit information. As an example of this type of device, we can cite the traffic lights that announce the phase of the lights and the configuration of the intersection to approaching vehicles. Another example is the replacement of the current passive signage and a simple small device can announce a priority to the right or a speed limit. Lastly, these devices can play a very important role in extending the scope of DENM messages by relaying them. For more advanced uses, a micro-server placed in an intersection or on a highway entrance ramp, can be in charge of facilitating communications and possibly organizing traffic. The possibilities offered by this type of deployment are already very numerous and the deployment can be very gradual, since it is possible to instrument only the places that require it (hazardous intersections), and vehicles using the information that is available to make traffic more efficient.

– Connected infrastructure devices: the services offered can be much more dynamic here; for example, it is possible to announce coming hazards or deviations. Announcements made can replace variable-message signs. This type of infrastructure is also of great interest in offering much more advanced services, such as platooning assistance, if it is possible to follow vehicle convoys along their journey. Lastly, thanks to roadside stations, it is possible to collect messages transmitted by vehicles and transmit them to central servers that will be able to calculate journey times, identify distributed-attack attempts and relay warning messages sent locally by the vehicles on a larger scale. Some of the observations that are made by cameras today could thus be made through messages regularly transmitted by vehicles.

– Infrastructure devices relaying the traffic to central servers: in this configuration, we leave the world of ephemeral local interactions in the sense that interactions do not remain local. Vehicles learn by listening to roadside stations that they are able to offer Internet connectivity. A vehicle needing to download a list of pseudonymous certificates, for example, can establish a session through the connectivity offered by the roadside station. The difficulty lies in maintaining connectivity long enough to establish a secure session with a server and to exchange data. It is thus much simpler to use conventional cellular communications to achieve this type of service or to wait to be static for a while.

1.4.2. *Towards an active infrastructure*

When we talk about V2X communication technologies, we think first of all of inter-vehicular communications, but it also includes communications between vehicles and the nearby infrastructure: RSUs (Road-Side Units). The latter can relay or directly host a very large number of services intended for both TOAs (Transport Organizing Authorities) and infrastructure managers to collect information, vehicles to offer information services (traffic-light phase) and collaborative perception or support services, such as the organization of traffic in a complex intersection. They can also serve other users, such as cyclists or pedestrians, and form a link between different user categories. We will then speak of active infrastructure, and in the following, we give an example of two service categories.

1.4.2.1. *Collaborative perception*

An active infrastructure is very useful for increasing the perception capacity of vehicles in complex environments. Therefore, surveillance cameras positioned in an intersection can announce a description of the intersection, including moving objects and pedestrians. Consequently, a vehicle will be able to perceive pedestrians that are masked from it and take them into account in its trajectory and speed decisions. The question of confidence in the information announced is crucial here and will need to be addressed; indeed, an autonomous vehicle could only pass at full speed through a blind intersection if the infrastructure tells it that the lane is clear, insofar as the vehicle trusts it.

1.4.2.2. *Heterogeneity support*

This same active infrastructure could take charge of relaying information between two incompatible technologies, provided that it has the various communication technologies at its disposal. In the same way, it can draw on its ability to perceive non-communicating objects (pedestrians, cyclists, as well as non-equipped cars) to instead send CAM messages, which will enable equipped vehicles to take them into account (Tsukada et al. 2020). The active infrastructure becomes a means of facilitating deployments by reintegrating non-communicating objects into the C-ITS mechanisms.

1.5. Conclusion and prospects

Services based on ephemeral local interactions are not yet frequent if potential partner discovery mechanisms present in many wireless technologies are excluded. Once the discovery is made, direct interactions are used to establish a connection and offer a service (connection to a hands-free kit in BT).

In this chapter, we reviewed the properties of services relying on this type of interaction and demonstrated the benefit of relying on them in the field of cooperative ITS.

One of the difficulties is then to put in place standards – this is largely done at ETSI, ISO and CEN – to define a language and a mode of interaction common to all objects that can cooperate. Since this mode of interaction is relatively new, it is important to experiment on a large scale what has been done for several years now with major European initiatives. The projects SCOOP@F (Aniss 2016), INDID, C-road, etc., aim to precisely organize the ecosystem under the aegis of the French Ministry of Transport to evaluate the functioning of the services that will be deployed in the coming years. This pooling of resources within a national experimental platform is necessary in that devices, whether in-vehicle, deployed along roads or in central systems, will have to cooperate with each other. The tests, in particular, of interoperability, that could be performed by a group of actors would necessarily have too little coverage to be representative of the complexity of real deployment.

Moreover, the question of whether it is appropriate to deploy the vehicles or the infrastructure first has delayed the process for a long time, the former waiting for the services offered by the infrastructure to become available in order to justify the additional cost related to the implementation of OBUs (On-Board Units) in vehicles. While the simplest approach would probably be for a regulation to require all new vehicles to have V2X communications, another approach consists of gradually deploying infrastructure devices first for the needs of infrastructure managers and then gradually offering new services to equipped vehicles. Existing services (traffic-light control) can then be modified to use V2X communications rather than proprietary solutions. They would immediately benefit from all protocols and messages already defined to further improve the service. For example, SPAT/MAP messages can be used to regulate approach speed and prevent a bus from coming to a complete stop.

The progressive enrichment of the infrastructure with RSUs (Road-Side Units) equipped with V2X technology enables the environment to be enriched with data that may be used by a wide variety of services. These same devices can also broadcast the same information using other media to reach other classes of users: for example, smartphones. This will make it possible to apply the same basic technologies, operating principles and exchange protocols to other fields, and thus reduce the costs of deploying new innovative services.

1.6. References

Aniss, H. (2016). Overview of an ITS project: SCOOP@ F. In *Communication Technologies for Vehicles: 10th International Workshop, Nets4Cars/Nets4Trains/Nets4Aircraft 2016,* San Sebastián, 6–7 June, 131–135. Springer, Cham.

Barceló, J., Montero, L., Bullejos, M., Serch, O., Carmona, C. (2013). A Kalman filter approach for exploiting bluetooth traffic data when estimating time-dependent OD matrices. *Journal of Intelligent Transportation Systems*, 17(2), 123–141.

Bianchi, G. (2000). Performance analysis of the IEEE 802.11 distributed coordination function. *IEEE Journal on Selected Areas in Communications*, 18(3), 535–547.

Bluetooth Special Interest Group (SIG) (2016). Specifications and documents [Online]. Available at: https://www.bluetooth.com/specifications/bluetooth-core-specification [Accessed July 2021].

Boudabous, S. (2021). Vehicular traffic analysis based on Bluetooth sensors traces. Doctoral Dissertation, Institut polytechnique de Paris.

Chandra, R., Padhye, J., Ravindranath, L., Wolman, A. (2007). Beacon-stuffing: Wi-fi without associations. In *Eighth IEEE Workshop on Mobile Computing Systems and Applications*, 53–57.

Eichler, S. (2007). Performance evaluation of the IEEE 802.11 p WAVE communication standard. In *2007 IEEE 66th Vehicular Technology Conference*, 2199–2203.

ETSI (2009). TR 102 638 V1.1.1; Intelligent Transport Systems (ITS); Vehicular communications; Basic set of applications; Definitions. ETSI Technical Report, Sophia Antipolis.

ETSI (2011). TR 102 863 V1.1.1; Intelligent Transport Systems (ITS); Vehicular communications; Basic set of applications; Local dynamic map (LDM); Rationale for and guidance on standardization. ETSI Technical Report, Sophia Antipolis.

ETSI (2014). EN 302 637-2 V1.3.1; Intelligent Transport Systems (ITS); Vehicular communications; Basic set of applications; Part 2: Specification of cooperative awareness basic service. ETSI Standard, Sophia Antipolis.

ETSI (2018). TS 103 574 – V1.1.1; Intelligent Transport Systems (ITS); Congestion control mechanisms for the C-V2X PC5 interface; Access layer part. ETSI Standard, Sophia Antipolis.

ETSI (2019). EN 302 663 – V1.3.1; Intelligent Transport Systems (ITS); ITS-G5 Access layer specification for Intelligent Transport Systems operating in the 5 GHz frequency band. ETSI Standard, Sophia Antipolis.

ETSI (2021). TS 124 587 – V16.0.9; 5G; Vehicle-to-Everything (V2X) services in 5G System (5GS); Stage 3 (3GPP TS 24.587 version 17.8.0 Release 17). ETSI Standard, Sophia Antipolis.

Garcia-Roger, D., González, E.E., Martín-Sacristán, D., Monserrat, J.F. (2020). V2X support in 3GPP specifications: From 4G to 5G and beyond. *IEEE Access*, 8, 190946–190963.

Gezer, C., Buratti, C., Verdone, R. (2010). Capture effect in IEEE 802.15. 4 networks: Modelling and experimentation. In *2010 IEEE 5th International Symposium on Wireless Pervasive Computing*, 204–209.

Gomez, C., Minaburo, A., Toutain, L., Barthel, D., Zuniga, J.C. (2020). IPv6 over LPWANs: Connecting low-power WANs to the Internet (of Things). *IEEE Wireless Communications*, 27(1), 206–213. doi: 10.1109/mwc.001.1900215.

Gräfling, S., Mähönen, P., Riihijärvi, J. (2010). Performance evaluation of IEEE 1609 WAVE and IEEE 802.11 p for vehicular communications. In *2010 Second International Conference on Ubiquitous and Future Networks (ICUFN)*, 344–348.

Haider, A. and Hwang, S.H. (2019). Adaptive transmit power control algorithm for sensing-based semi-persistent scheduling in C-V2X mode 4 communication. *Electronics*, 8(8), 846.

Haidar, F., Kaiser, A., Lonc, B. (2017). On the performance evaluation of vehicular PKI protocol for V2X communications security. In *2017 IEEE 86th Vehicular Technology Conference (VTC-Fall)*, 1–5.

IEEE 802 LAN/MAN Standards Committee (2009). IEEE Standard for Information Technology-Telecommunication and information exchange between systems-Local and metropolitan area networks-Specific requirements Part 11: Wireless LAN Medium Access Control (MAC) and Physical Layer (PHY) Specifications Amendment 1: Radio Resource Measurement of Wireless LANs [Online]. Available at: http://standards.ieee.org/getieee802/download/802.11n-2009.pdf.

Klinkusoom, L., Aswakul, C., Janpugdee, P. (2014). Bluetooth sensors for vehicular traffic monitoring. In *The 29th International Technical Conference on Circuit/Systems Computers and Communications*, Phuket.

Kountché, D.A., Bonnin, J.M., Labiod, H. (2017). The problem of privacy in cooperative intelligent transportation systems (C-ITS). In *2017 IEEE Conference on Computer Communications Workshops (INFOCOM WKSHPS)*, 482–486.

Roberts, L.G. (1975). ALOHA packet system with and without slots and capture. *ACM SIGCOMM Computer Communication Review*, 5(2), 28–42.

Roca, V. (2022). TousAntiCovid : gros plan sur les deux protocoles de traçage numérique de l'application. In *Médecine et intelligence artificielle*, Nordlinger, B., Villani, C., de Fresnoye, O. (eds). CNRS Editions, Paris.

Storck, C.R. and Duarte-Figueiredo, F. (2020). A survey of 5G technology evolution, standards, and infrastructure associated with vehicle-to-everything communications by internet of vehicles. *IEEE Access*, 8, 117593–117614.

Tsukada, M., Oi, T., Kitazawa, M., Esaki, H. (2020). Networked roadside perception units for autonomous driving. *Sensors*, 20(18), 5320.

Yang, J., Poellabauer, C., Mitra, P., Neubecker, C. (2020). Beyond beaconing: Emerging applications and challenges of BLE. *Ad Hoc Networks*, 97, 102015.

2

Evolution of Use Cases for Intelligent Transport Systems

Sassi MAALOUL[1,2], Hasnaâ ANISS[1], Marion BERBINEAU[3] and Léo MENDIBOURE[1]

[1] *ERENA, COSYS, Gustave Eiffel University, Marne-la-Vallée, France*

[2] *JUNIA, École d'ingénieurs, Lille, France*

[3] *COSYS-LEOST, Gustave Eiffel University, IFSTTAR, Villeneuve d'Ascq, France*

2.1. Introduction

In recent years, Intelligent Transport Systems have been designed to improve road safety, optimize traffic and enhance the experience of road users. These are systems that rely on vehicular communication architectures in order to exchange relevant data and information related to the road. Their proper functioning requires information exchanges between the vehicles and other communicating entities such as other vehicles, infrastructure and other road users, such as pedestrians. In addition, ITS services are defined in several use cases such as emergency brake warnings, collision warnings, green light optimal speed advisories (GLOSA), traffic information, etc. Nevertheless, each use case represents a specific scenario for exchanging information concerning a potential event between two communicating entities.

Within this context, this chapter presents the evolution of use cases, which has progressed thanks to the evolution of vehicular communication architecture and enablement of improvements to road safety, mainly for connected and autonomous vehicles (CAV).

Cooperative Intelligent Transport Systems,
coordinated by Léo MENDIBOURE.

2.2. Vehicular communication technologies

Emerging communication technologies are currently one of the main building blocks of the Intelligent Transport Systems research field. Indeed, this ecosystem of communication technologies is necessary to provide better support for aspects such as road safety, traffic efficiency and user comfort.

These communication technologies also contribute to the performance of vehicular networks by enabling different entities of the ecosystem to communicate effectively through various modes. The Infrastructure-to-Vehicle (I2V), Vehicle-to-Infrastructure (V2I), Vehicle-to-Vehicle (V2V) and Vehicle-to-Network (V2N) communication modes, collectively referred to as Vehicle-to-Everything (V2X), aim to enable data exchange between vehicles and their environment.

These systems' primary objective is to improve road safety by making drivers aware of potential hazards in the vicinity. With alert notifications, drivers can be helped in avoiding collisions and sudden threats from different actors on the roads.

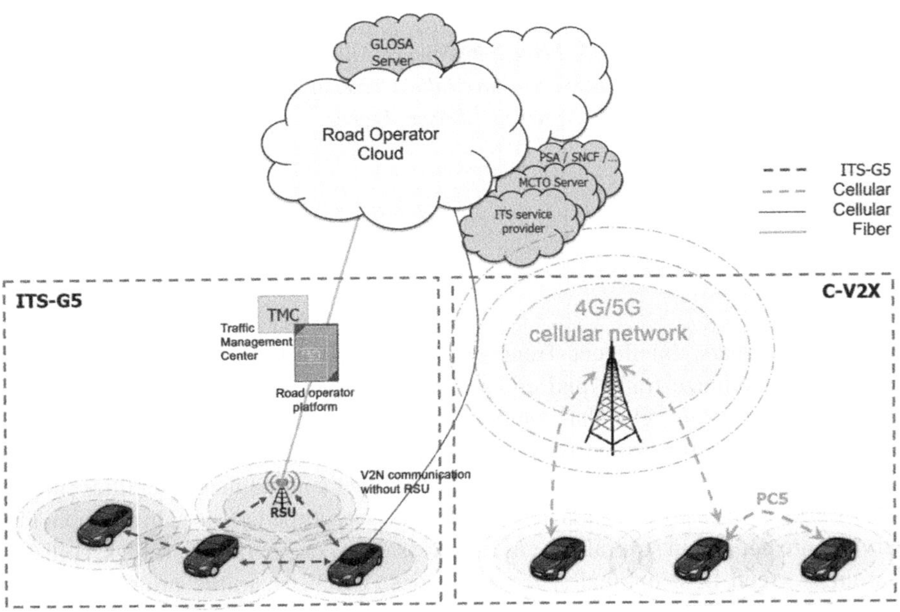

Figure 2.1. *Vehicular network architecture. For a color version of this figure, see www.iste.co.uk/mendiboure/transportsystems.zip*

Several wireless technologies are currently available to support communications for these vehicular networks. The two most promising technologies intended for ITS services are ITS-G5 and C-V2X, as shown in Figure 2.1. ITS-G5 is the

European standard for vehicular communications, proposed by ETSI (European Telecommunications Standards Institute). 3GPP offers LTE-V2X or C-V2X (Cellular Vehicle-to-Everything) as an extension of LTE (Long-Term Evolution) cellular technology for vehicular communications.

2.2.1. *ITS-G5/IEEE 802.11p technology*

ITS-G5 was presented by the ETSI, based on Standard IEEE 802.11p (2012) (ETSI 2013). It is considered a promising technology to support V2X communications through OCB (Outside the Context of BSS) mode. It operates in the frequency band (5.9 GHz), from which it derived its name, ITS-G5. The latter is explicitly designed to provide direct (short-range) V2V, I2V and V2I communications, with the aim of providing various critical and non-critical services such as road safety.

ITS-G5 forms a distributed network between vehicles, or between vehicles and infrastructure: road-side units (RSUs) deployed by the road manager. It is characterized by the integration of the DCC (Distributed Congestion Control) functionality for channel control and the use of channel access priorities through the EDCA (Enhanced Distributed Coordination Access) mechanism to ensure the quality of service required by ITS services.

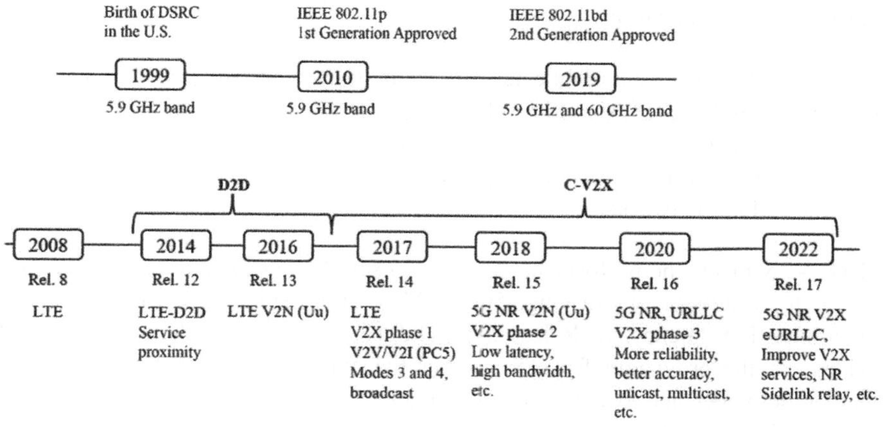

Figure 2.2. *Evolution of ITS technologies*

Over the past few years, IEEE 802.11 technology has continued to evolve to better support the requirements of ITS applications (Hongyuan et al. 2018; Naik et al. 2019). As shown in Figure 2.2, the IEEE has approved a new Standard, IEEE 802.11bd (NGV: Next Generation V2X). Work on this standard is defined to better meet the needs of new applications, anticipating new requirements in terms of throughput, reliability, efficiency and range. Different use cases are covered by this standard, such as simultaneous operation on several channels, applications supported by infrastructure, autonomous driver assist, ITS applications of aerial vehicles (drones), train-to-train or vehicle-to-train communications, etc.

2.2.2. *The 3GPP standard: C-V2X*

LTE-V2X, or Cellular V2X (C-V2X), is a communication technology standardized by 3GPP in Releases 14 and 15 (3GPP 2017; Molina et al. 2017) to extend the LTE standard to vehicular applications. C-V2X delivers V2X services using two communication modes: mode 3 and mode 4. Mode 3 uses the Uu cellular interface to take advantage of the existing mobile network infrastructure for V2N (Vehicle-to-Network) long-range communications. It provides services with a high data rate and no latency or reliability requirements. In this mode, the cellular network manages the radio resources used by vehicles.

Mode 4 uses the PC5 (Proximity Communication 5) interface for direct, short-range communications in the 5.9 GHz band. In this mode, vehicles autonomously select radio resources through a semi-persistent scheduling (SPS) mechanism. When vehicles are not under cellular coverage, mode 4 is used to ensure the different communications.

The evolution of current mobile network technology to the fifth generation 5G-NR (New Radio) (3GPP 2018) should result in a significant increase in network capacity, a decrease in latency, an improvement in reliability and availability, more accurate positioning and high-definition mapping, which will improve the services provided by traditional cellular networks.

LTE-V2X meets the performance needs of a set of services, but in advanced use cases such as autonomous driving, 5G could be more suitable, as it is supposed to guarantee low latency. In Releases 16 and 17, as shown in Figure 2.2, 3GPP provides specific provisions for ITS systems that will improve the performance of vehicular networks and benefit from the many advantages of 5G. In this sense, integrating 5G technology as a successor to LTE technology is designed to provide complementary capabilities and smoother network exchanges while maintaining backward compatibility. Various use cases related to CAV are covered by 5G-V2X.

A brief comparison of the performance indicators of these technologies is presented in Table 2.1.

	ITS-G5	C-V2X	
Launch	2012	2016	2018
Standard	IEEE 802.11p	3GPP LTE	3GPP 5G
Communication mode	Direct V2V/V2I	Direct V2V/V2I and indirect V2N	
Operation	OCB mode	Assisted and unassisted mode	
Range	Short	Short (PC5)/long	
Throughput	6 Mbps	50 Mbps	10 Gbps (V2N)/100Mbps (V2V)
Latency	Low	Low	Ultra-low
Reliability	++	+++	++++
Maturity	+++	++	+

Table 2.1. *Comparison of ITS technologies*

2.2.3. *Deployment of ITS technologies*

The United States Federal Communications Commission (FCC) allocated 75 MHz of spectrum in the 5.9 GHz band for industrial use for ITS systems. In 2008, the European Telecommunications Standards Institute (ETSI) allocated 30 MHz of spectrum in the 5.9 GHz band for ITS services in Europe. In November 2020, the FCC reallocated the entire spectrum (EC 2020) where a 40 MHz band was dedicated to the use of ITS systems. Currently, ETSI is proposing studies on the feasibility of the coexistence of two ITS technologies in the same frequency channel (ETSI 2021a, 2021b). These studies are based on time-division multiplexing to overcome collisions and imbalances in channel access (excessive occupation by one technology).

Many European car manufacturers consider ITS-G5/IEEE 802.11p to be the most mature short-range communication technology for ITS services. In addition, LTE-V2X is another communication technology introduced by 3GPP that combines long- and short-range capabilities. These two technologies have divided the industrial community into two opposing positions, one favoring IEEE 802.11p/ITS-G5 technology, and the other favoring LTE-V2X within the context of 5G standardization. One possible solution for coexistence lies in frequency channel sharing or in a deployment strategy. For example, ITS-G5 can be activated in locations where C-V2X signals are weak (rural/remote areas). In this sense, ETSI studies support the hypothesis of the simultaneous operation and deployment of the two technologies. This hypothesis allows vehicular networks to better meet the growth and performance needs of ITS services.

2.3. Evolution of use cases

A use case represents a scenario for information exchange between a vehicle and its environment (infrastructure manager, other vehicles, road user, etc.) concerning a potentially hazardous event in the vicinity. The required performance such as latency, reliability and throughput depends mainly on the event and the significance of the information exchanged in order to prevent and avoid an accident. Indeed, the evolution of vehicular network communication technologies is leading to improvements in ITS system connectivity, enabling better performance. Moreover, these improvements lead to the progression and development of current use cases and the emergence of other use cases, more demanding in terms of performance.

2.3.1. *Classification of use cases*

The following section presents a number of innovative use cases that have emerged in recent years by standardizing organizations (Gao et al. 2017; 3GPP 2018; 5GAA 2019; Fernandez et al. 2019). V2X services are divided into several groups, each comprising several use cases. It should be noted, however, that a use case may belong to more than one of these groups.

2.3.1.1. *Use cases: warning and road safety*

The use cases of this group serve to improve road safety. Consequently, the information exchanged is generally prioritized on the communication channels in order to guarantee the safety of the vehicle and its driver on the road. The performance requirements in this regard are mainly low latency, high reliability and broadcasting of messages at high-frequency periodicity. These applications require these characteristics so that the data can reach its destination in time. These use cases are very helpful for drivers and road operators in order to improve road safety and ITS systems. They are also applicable to (semi-)autonomous vehicles: they can assist drivers, one of the notable examples being live-view camera assistance for human drivers.

Examples of this group of use cases include:

- emergency brake warning;
- lane change warning;
- emergency vehicle approaching;
- vulnerable road user warning;
- stationary vehicle warning;
- overtaking vehicle warning;
- collision warning;
- road works warning.

2.3.1.2. *Use cases: traffic management and efficiency*

Use cases related to traffic efficiency are less demanding on network performance. Yet, they remain crucial since they inform the driver of the traffic situation and provide added value to providers of infrastructures where the vehicles will be operated. They require low-frequency message periodicity, good reliability especially for traffic management messages and necessitate less performance in latency terms. These use cases are very useful for drivers and companies (transport, delivery, etc.) since they have potential value with respect to consumption and routing optimization and can offer environmental advantages.

The use cases include:

- green light optimal speed advisory (GLOSA);
- traffic jam information;
- routing advise;
- smart routing;
- cooperative flexible lane change;
- in-vehicle signage;
- level crossing;
- electronic toll collect;
- traffic information;
- parking, park and ride, multimodality;
- multimodal cargo transport optimization.

2.3.1.3. *Use cases: comfort and entertainment*

This group includes use cases that offer value and convenience to drivers. These are use cases that may not be mandatory from a road safety point of view, but that offer significant value to the drivers or passengers of the vehicle. Entertainment use cases are generally lowest in priority and do not specifically require performance in terms of reliability and latency, but are the most bandwidth-intensive. From a business strategy perspective, these are use cases that would be attractive to vehicle drivers/passengers and could be provided by car manufacturers to improve the efficiency of vehicle maintenance and monitoring.

The use cases include:

- infotainment;
- autonomous smart parking;
- connected vehicle;
- mobile high data rate;
- payment services;
- software updates;
- remote support.

2.3.1.4. *Use cases: connected and autonomous driving*

Developments in vehicular technologies are mainly aimed at meeting the demand of the automotive sector for autonomous driving systems. Autonomous-driving vehicles will enable drivers and passengers to access various services (watching television, listening to music, road maps, routing, parking, traffic, news, etc.) in real time while the vehicle is in autonomous driving mode.

These systems are evolving rapidly and are set to revolutionize the driving experience and make it safer. Within this context, these technologies are based on advanced driver-assist systems (ADAS), which use sensors, cameras and radars to improve vehicle safety. In addition, V2X communication should complement and extend the capabilities of these ADAS systems and offer additional advantages such as improved energy efficiency, reduced vehicle emissions, etc.

This group concerns the use cases relevant to autonomous driving with a high level of automation (according to the SAE (Society of Automotive Engineers)), as well as the use cases concerning platooning. These use cases require high network performance. Indeed, in the case of platooning, it is indispensable for vehicles to use ultra-low-latency networks. The same applies to all other autonomous driving use cases such as remote driving, collision anticipation, etc. Certain safety use cases require cooperative interaction between vehicles in order to be efficient and safe. These use cases are of interest to all road transport actors (drivers, transport or delivery companies, car manufacturers, road operators, etc.). They could also help strengthen road safety, make more efficient use of road infrastructure and offer environmental advantages.

Examples of this group of use cases include:

– use cases for advanced driving: cooperative driving for vehicle, cooperative collision avoidance, intersection safety information provisioning for urban driving, emergency trajectory alignment, cooperative lane change, etc.;

– use cases for remote driving: automated valet parking, charging stations management, public transport, last mile delivery;

– use cases related to sensors (extended sensors): sensor information sharing, collective perception of environment, video data sharing for automated driving, handling of dynamic maps, bird's eye view, etc.;

– use cases for platooning: information exchange within platoon, automated cooperative driving for short distance grouping, information sharing for limited automated platooning, changing driving mode, etc.

2.3.2. *Required performance*

Many use cases have been developed, from convenience services to complex services with strict requirements in terms of network latency and reliability. For each use case, a specific latency, a packet size and different data rates that the network must support are given. In this sense, not all services have the same potential requirements. For example, certain services require low latencies to function properly, while others require high network reliability (low packet losses). A direct relationship therefore exists between the potential requirements of the different services and the key performance indicators provided by the network.

Latency is an indicator of prime importance for V2X services, because a delay in transmitting or receiving messages can affect the proper functioning of the intended services and, in the worst case scenario, can lead to conflicting results. The size of the packets to be transmitted depends on the message type, whether via the V2V, V2I, V2P or V2N service. In addition to this, the network must guarantee an adequate connection to vehicles traveling at high speed. Table 2.2 presents the main requirements for V2X services as described in Standard 3GPP NR-V2X (3GPP 2018). These services are divided into four categories: "Vehicle platooning", "Advanced driving", "Extended sensors" and "Remote driving".

V2X service	Latency (ms)	Size (bytes)	Reliability (%)	Data rate (Mbps)	Range (m)
Vehicle platooning	10–500	50–6,000	90–99.99	50–65	80–350
Advanced driving	3–100	300–12,000	90–99.999	10–50	360–500
Extended sensors	3–100	1,600	90–99.999	10–1,000	50–1,000
Remote driving	5	–	99.999	DL:25, UL:1	–

Table 2.2. *The potential requirements for NR V2X (3GPP 2018)*

The requirements depend on the specificity of the use case. For example, for remote driving that makes it possible to control a vehicle remotely (operator), the maximum latency that the network must offer must be less than 5 ms. Other services aimed at enabling the implementation of extended sensors (data exchange of sensors or video information between vehicles, pedestrians, RSUs, servers, etc.) may require more bandwidth to transmit video information. The network reliability must be greater than 90% in all use cases.

2.3.3. *Example of use cases*

A use case can include several scenarios that differ depending on the road configuration (intersection, urban, rural or highway), the actors involved, the benefit of the service (road safety or traffic management), etc. Several use cases are briefly described below.

2.3.3.1. *Emergency brake warning*

– *Objective of use case.* This use case aims to alert the driver of an imminent and significant hazard so that they can act accordingly. Therefore, it can significantly reduce accidents, thanks to automatic detection.

– *Description of use case.* A vehicle automatically detects that the driver has applied emergency braking and broadcasts a warning message to other vehicles via V2V communications. This detection occurs by automatically analyzing its various in-vehicle systems. Surrounding vehicles then receive the information and display it to their driver, who adapts their behavior by braking, reducing their speed, changing lanes, etc. In addition, the message related to the braking event contains other information on the vehicle responsible for the warning, such as positioning, speed, acceleration, etc.

The potential requirements of this use case are low latency and high reliability to transfer messages to drivers as rapidly as possible. The expected advantages of this use case are reducing the risk of accidents and avoiding numerous braking operations that could lead to traffic jams.

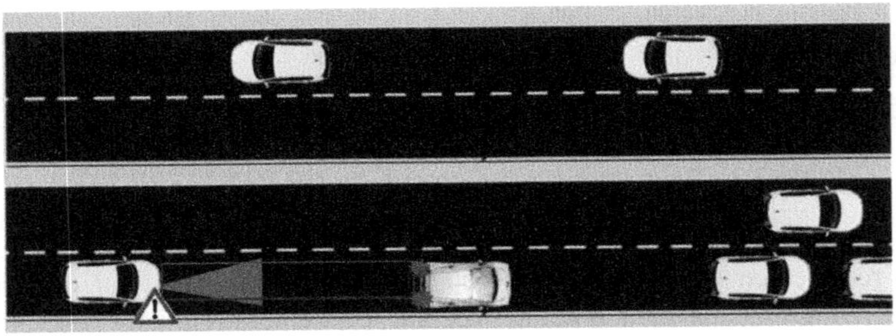

Figure 2.3. *Emergency brake warning. For a color version of this figure, see www.iste.co.uk/mendiboure/transportsystems.zip*

2.3.3.2. *Lane change warning*

– *Objective of use case.* The objective of this use case is to improve driver perception. This involves managing the merging of vehicles effectively, by avoiding sudden stops or forced lane changes for vehicles traveling on the main road.

– *Description of use case.* A vehicle traveling on a section of the main road continuously sends its position via CAMs (Cooperative Awareness Messages). The vehicle affected by the lane change receives the warning and displays it to the driver, who adapts/controls the driving speed to make room for the vehicle on the main road before changing lanes.

The warning must be displayed on the HMI (Human–Machine Interface) sufficiently early. The expected advantages of this use case are improved road safety for all road users, improved visibility of road users by enabling perception of other vehicles and improved traffic by reducing the number of interventions due to accidents.

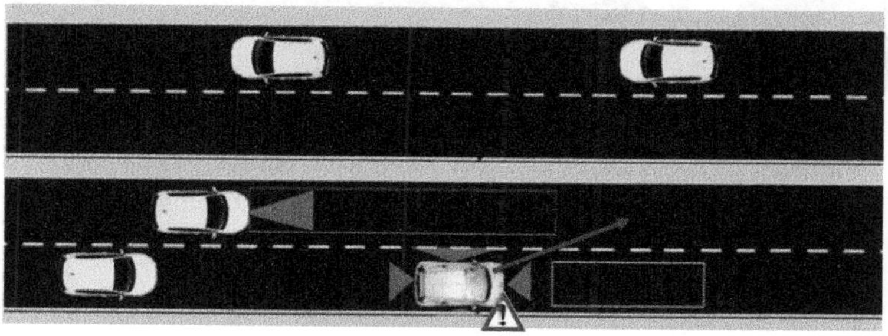

Figure 2.4. *Lane change warning. For a color version of this figure, see www.iste.co.uk/mendiboure/transportsystems.zip*

2.3.3.3. *Road works warning*

– *Objective of use case.* The objective of this use case is to inform road users of a road works area, in which they will encounter operators. A warning occurring sufficiently in advance to warn drivers must enable them to adapt their behavior: greater vigilance, adapting their speed and changing lanes (if necessary).

– *Description of use case.* The road manager programs the mobile road works and signals them in their traffic management system (TMS). A message is then broadcast by I2V communication to road users, or by V2V communication directly from the vehicle that protects the operatives conducting the road works. The message contains all of the elements needed to accurately describe the road works area (start/end position of the road works area, and duration). Moreover, further information can be added, such as the speed limit of each neutralized portion.

Vehicles approaching mobile road works receive the message, process it and display the information to the driver to allow them to adapt their speed and position on the roadway. The expected benefits of this use case are a reduction in the risk of accidents for road users and operators, and better traffic management.

The warning must be displayed on the HMI sufficiently early. Low latency and a good message repetition frequency are needed. If the road operator's vehicle is on site and broadcasts a message signaling a mobile construction site, the HMI may also have to process these messages. However, priority must be given to the information provided by the vehicle on site.

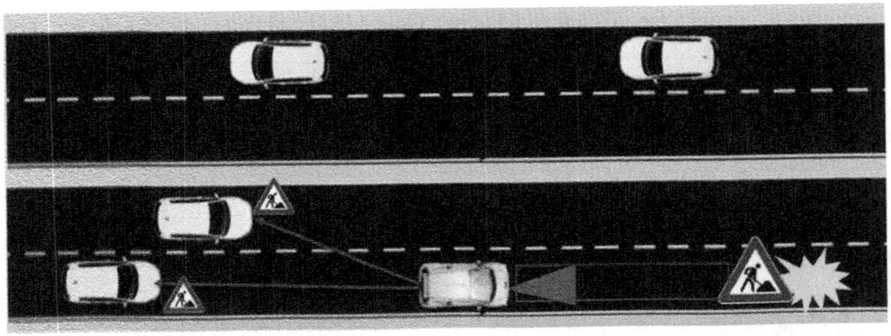

Figure 2.5. *Road works warning. For a color version of this figure, see www.iste.co.uk/mendiboure/transportsystems.zip*

2.3.3.4. *Traffic jam information*

– *Objective of use case.* The objective of this use case is to inform the driver of a traffic jam they will encounter. The latter can thus raise their vigilance, anticipate a possible increase in their journey time and react by modifying their routing or adapting their speed as they approach the traffic jam.

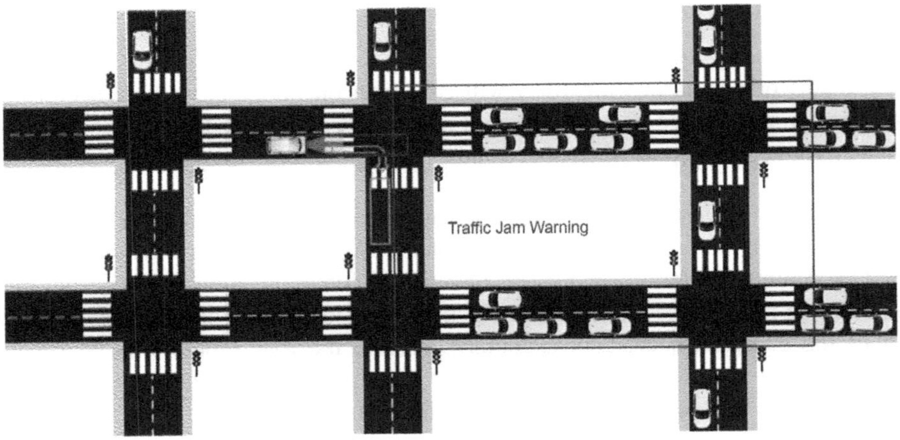

Figure 2.6. *Traffic jam information. For a color version of this figure, see www.iste.co.uk/mendiboure/transportsystems.zip*

– *Description of use case.* In the TCC (Traffic Control Center), the road operator obtains information on a traffic jam on their road network. This information can be collected by cameras, operatives, other vehicles that detected the traffic jam and signaled it to the road operators, etc. The road operator places the information in their TMS, confirms it with their position, the section/lanes concerned, the time it takes to get through the traffic jam and other useful information if available. The message is then broadcast to road users approaching the traffic jam area via I2V communications. Vehicles in the vicinity of the traffic jam area receive the information and display it to their drivers in order to adapt their routing and their behavior. Subsequently, the road operator can automatically update the information (the duration of the traffic jam, lanes concerned, etc.) and indicate the end of traffic jam area.

Traffic jams do not usually occur over a short period of time. Given the communication range, which is sufficiently large, warning messages are received before the driver is close to the traffic jam area. This leaves them more time to make a decision and adapt their routing and their behavior. On the contrary, the road positioning accuracy may be low depending on the available information sources used by the road operator and the information that can be configured in their TMS.

2.3.3.5. *Green light optimal speed advisory*

– *Objective of use case.* The objective of the use case is to give drivers advice to enable them to optimize their passage through traffic lights (maintaining their speed, slowing down, adopting a specific speed, time remaining until the green light, etc.).

– *Description of use case.* The traffic light controller is responsible for sending messages when a driver approaches a traffic light. The messages are broadcast via I2V communications from the infrastructure to approaching vehicles, informing them of the traffic light status. This information, along with the position, speed and distance of the vehicle from the light, allows it to determine the optimal approach speed (below the mandatory speed limit) and display it on the HMI. Next, the driver then either adapts their speed to pass at a permitted speed or prepares to brake, according to the advice displayed. The expected benefits of the use case are: smooth driving conditions, keeping to average speeds, safety, reduced consumption and lower CO_2 emissions.

This use case is well suited to urban traffic lights. The information is processed locally by the vehicle to determine the harmonized speed advice that depends on the data provided by the traffic light controller.

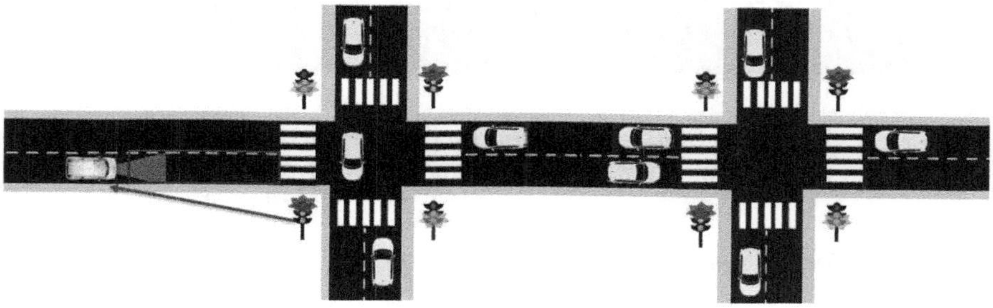

Figure 2.7. *Green light optimal speed advisory. For a color version of this figure, see www.iste.co.uk/mendiboure/transportsystems.zip*

2.3.3.6. *Extended sensors*

– *Objective of use case.* The use case objective is to warn vehicle drivers that there are pedestrians located on their trajectory who are crossing the road. The service aims to warn affected drivers of approaching vehicles when a collision risk is identified.

– *Description of use case.* The infrastructure monitors and collects data through cameras and sensors placed at intersections (commercial areas, schools, etc.). When a pedestrian crosses an intersection, the infrastructure detects them and broadcasts an I2V message to all surrounding vehicles. The message informs and warns us of the presence of a pedestrian crossing the road, and a collision risk is identified. Approaching vehicles receive the message and process it. If a risk is detected by the vehicle, a warning message is displayed on the driver's HMI, signaling the presence of a pedestrian on its trajectory. The driver then reacts to the message by adopting cautious behavior. This use case is particularly useful in particular when a pedestrian crosses the intersection with a green light and a vehicle approaches with a green light (right turn).

Messages are sent to drivers as rapidly as possible, with a high display priority. The expected advantages of this use case are as follows: improved safety, better vehicle/pedestrian interactions and a reduction in the number of accidents and conflicts in dense urban areas.

Figure 2.8. *Bird's eye view. For a color version of this figure, see www.iste.co.uk/mendiboure/transportsystems.zip*

2.3.3.7. *Collective perception of the environment*

– *Objective of use case.* The objective of this use case is to improve the perception of the vehicle environment thanks to sensor data exchanged in real time between vehicles. This information exchange leads to a collective perception of the environment, improving road safety and avoiding accidents.

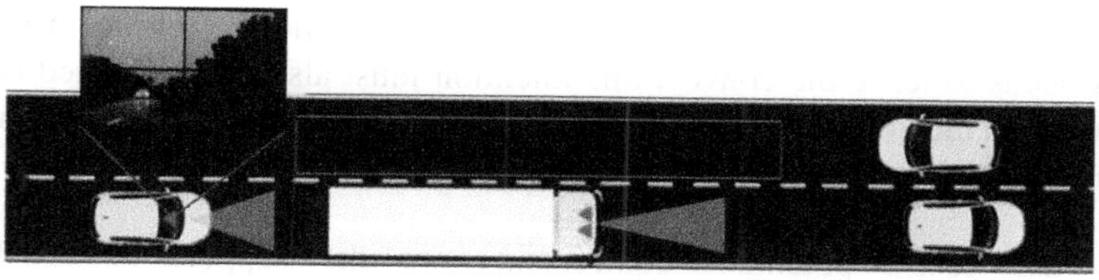

Figure 2.9. *See-through. For a color version of this figure, see www.iste.co.uk/mendiboure/transportsystems.zip*

– *Description of use case.* At the beginning of the overtaking maneuver, the vehicle behind the truck cannot perceive the vehicle in the opposite direction. The truck detects the approaching vehicle using its sensors located at the front of the truck and sends the corresponding information to the vehicle following it, via V2V communication. Then, the vehicle wishing to overtake analyzes the information received and warns the driver of the hazard. The driver interrupts their overtaking maneuver due to the warning. Depending on the scenario, sensor data must be sent either in low resolution as processed data, or in high resolution as raw data. The raw data is necessary for reasons of liability in the event of an accident and to obtain a map fusion and accurate positioning of object.

Information related to sensor data is shared to increase the collective perception of the other vehicles. The information generated by each vehicle must be transmitted to all neighboring vehicles within a given radius. The potential requirements of this use case are low end-to-end latency and high reliability for message transfer between vehicles. In addition, it must be capable of supporting a significant data rate for a short period of time in the event of an imminent collision.

2.3.3.8. *Management of vehicle convoys (platooning)*

– *Objective of use case.* The objective of this use case is platooning management. Platooning consists of operating a group of closely linked vehicles (convoy), such that the vehicles move like a train. To maintain distance between vehicles, they must share information on their status, such as speed, course and intentions (braking, acceleration, etc.).

– *Description of use case.* When vehicles are moving along the road, they can dynamically form a convoy. To do this, they must exchange their intentions, such as the benefit of forming a convoy and being either a leader or follower in the convoy. When the convoy is formed, several messages are exchanged for its management. For example, vehicles in the convoy must exchange information concerning the moment at which a particular road is to be taken, whether to brake or accelerate, preventing potential threats to safety, etc. At the same time, the convoy leader must receive real-time I2V messages, including information on road and traffic conditions, and share them with the convoy members. All convoy members can also share information within the group via V2V communications. When a vehicle reaches a destination where it needs to leave the convoy, this intention must also be exchanged between the vehicles of the convoy. This exchange of intentions may take place at any time while the convoy is active.

This use case's potential requirements are the support of low-latency communications, a high periodicity for broadcasting messages among a group of vehicles and the ability to add/remove vehicles. The use of platooning decreases the distances between vehicles, lowers the overall fuel consumption and reduces the number of drivers required.

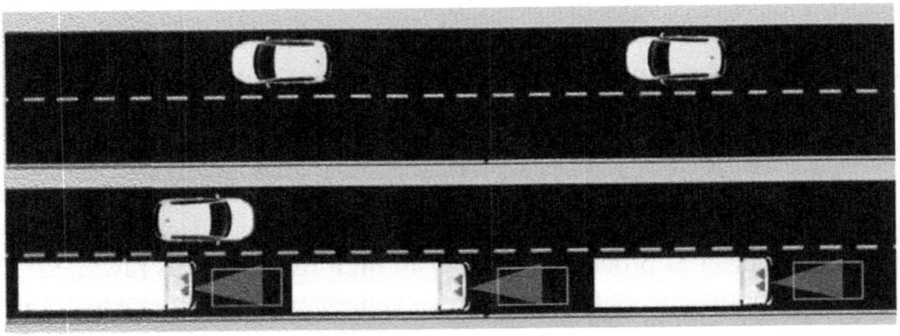

Figure 2.10. *Platoon management. For a color version of this figure, see www.iste.co.uk/mendiboure/transportsystems.zip*

2.4. Challenges and future services of V2X

Cellular technologies are evolving rapidly, and the sixth generation, 6G, is currently in the specification phase for commercial deployment around 2030. CAV are set to be one of the most promising applications to fully operate in 6G (Tang et al. 2020). Generally speaking, 6G should differ from 5G on several key performance indicators (Saad et al. 2020): data rate (50 times faster, 1 Tbps vs. 20 Gbps), service latency (10 times lower, 0.1 ms vs. 1 ms), positioning capacity (10 times more accurate, scale of several cm), reliability (100 times more reliable), wider coverage, energy efficiency (10 times better), etc. Moreover, 6G can increase the "intelligence" of vehicles and offer a high/full automation capability for these vehicles.

The research community has begun work on next-generation vehicle networks. The challenges consist of identifying the shortcomings of current vehicular networks and highlighting the performance of the next generation. The following section illustrates several shortcomings of current networks that 6G networks could help overcome, improving existing use cases and enabling the emergence of new use cases.

– *Advanced remote driving*. In this regard, new network capabilities can help remote drivers to connect to vehicle sensors in order to create an augmented and virtual reality view of surrounding environments that would improve driver perception. To increase safety, transmitting high-quality video and multi-sensor data is often necessary to help remote drivers capture rapid changes in traffic conditions around vehicles. However, transmitting very high-resolution or higher-quality videos can require up to several tens of Gbps. This extremely high throughput is still difficult to achieve for current communication technologies (even 5G).

– *The future transport and mobility model*. Booking a flying taxi will become a reality in the coming years. Major brands such as Uber and Hyundai and numerous startups have already spent a lot of money on developing air shuttle services and cargo drones. Therefore, many new functions such as Flying Parking Management, Cooperative Flight and Flight Sharing/Revocation require good Internet communications and even data sharing between flying vehicles, base stations and remote control centers. Although satellite Internet providers can provide limited connectivity, unfortunately there is no technology (even 5G) to ensure interoperability of communications between multiple network infrastructures (space/air/ground/sea).

2.5. Conclusion

The wireless communications field advances promise faster connections, lower latency, increased reliability, greater capacity and wider coverage. Integrating these advances into vehicular communication systems enables intelligent transportation systems to increase vehicle safety and autonomy in addition to road safety, while saving energy and reducing costs.

This chapter provides an overview of ITS technologies envisaged for intelligent transport systems. It goes on to present a summary of a potential grouping of use cases, with several examples and requirements. Lastly, we identify the limits of current vehicular networks and describe prospects (particularly in terms of services) that the future generation of vehicular networks could potentially open up.

2.6. References

3GPP (2017). TS 36.300 V14.3.0; Evolved Universal Terrestrial Radio Access (E-UTRA) and Evolved Universal Terrestrial Radio Access Network (E-UTRAN); Overall description; Stage 2. 3GPP Standard.

3GPP (2018). TR 22.886 V16.2.0; Study on enhancement of 3GPP support for 5G V2X services (Release 16). 3GPP Standard.

5GAA Germany (2019). C-V2X use cases: Methodology, examples and service level requirements. White paper [Online]. Available at: https://5gaa.org/wp-content/uploads/2019/07/5GAA_191906_WP_CV2X_UCs_v1.pdf.

ETSI (2013). EN 302 663 V1.2.1; Access layer specification for Intelligent Transport Systems operating in the 5 GHz frequency band. ETSI Standard.

ETSI (2021a). TR 103 667 V1.1.1; Study on Spectrum Sharing between ITS-G5 and LTE-V2X technologies in the 5855 MHz-5925 MHz band. ETSI Technical Report.

ETSI (2021b). TR 103 766 V1.1.1; Pre-standardization study on co-channel co-existence between IEEE and 3GPP based ITS technologies in the 5855 MHz-5925 MHz frequency band. ETSI Technical Report.

Fernandez, A.E., Fallgren, M., Brahmi, N. (2019). 5GCAR scenarios, use cases, requirements and KPIs. Technical Report, 5GCAR/D2.1, Germany.

Gao, L., Li, Y., Misener, J., Patil, S. (2017). C-V2X based basic safety related ITS spectrum requirement analysis. In *Proc. IEEE 86th Veh. Technol. Conf.*, 1–5.

Hongyuan, Z., Rui, C., Yan, Z., Liwen, C., Jinjing, J., Hui-Ling, L., Manish, K., Sudhir, S., James, L., Michael, M. et al. (2018). IEEE 802.11-18/0513r2: 802.11 for Next Generation V2X Communication. Working Report, IEEE NGV Meeting.

Molina-Masegosa, R. and Gozalvez, J. (2017). LTE-V for sidelink 5G V2X vehicular communications: A new 5G technology for short-range vehicle-to-everything communications. *IEEE Vehicular Technology Magazine*, 12(4), 30–39.

Naik, G., Choudhury, B., Park, J.M. (2019). IEEE 802.11 bd & 5G NR V2X: Evolution of radio access technologies for V2X communications. *IEEE Access*, 7, 70169–70184.

Official Journal of the European Union (2020). Commission Implementing Decision (EU) 2020/1426 of 7 October 2020 on the harmonised use of radio spectrum in the 5 875-5 935 MHz frequency band for safety-related applications of intelligent transport systems (ITS) and repealing Decision 2008/671/EC.

Saad, W., Bennis, M., Chen, M. (2020). A vision of 6G wireless systems: Applications, trends, technologies, and open research problems. *IEEE Network*, 34(3), 134–142.

Tang, F., Kawamoto, Y., Kato, N., Liu, J. (2020). Future intelligent and secure vehicular network toward 6G: Machine-learning approaches. *Proceedings of the IEEE*, 108(2), 292–307. doi: 10.1109/JPROC.2019.2954595.

PART 2

Optimization of Data Transmission for Cooperative Intelligent Transport Systems

3

Towards an Optimization of Data Transmission in Cooperative Intelligent Transport Systems

Mohamed BENZAGOUTA[1], Ramzi BOUTAHALA[2], Secil ERCAN[1], Sassi MAALOUL[1,5], Hasnaâ ANISS[1], Léo MENDIBOURE[1], Marwane AYAIDA[3] and Hacène FOUCHAL[4]

[1]*ERENA, COSYS, Gustave Eiffel University, Marne-la-Vallée, France*

[2]*Université de Reims, Champagne-Ardenne, France*

[3]*IEMN, Université Polytechnique Hauts-de-France, Valenciennes, France*

[4]*CReSTIC, Université de Reims, Champagne-Ardenne, France*

[5]*JUNIA, École d'ingénieurs, Lille, France*

3.1. Introduction

The number of vehicles on the roads is steadily increasing, leading to a rise in accident situations. In order to improve road safety and thus prevent, for example, accidents, traffic jams and pollution, Cooperative Intelligent Transport Systems (C-ITS) have been developed. These enable vehicles and road infrastructure to exchange information through what are known as C-ITS messages (Festag 2014). They also support the deployment of autonomous vehicles by providing them with a distant view of the road scene through information exchanges, particularly with the infrastructure.

For a color version of all figures in this chapter, see www.iste.co.uk/mendiboure/transportsystems.zip.

Cooperative Intelligent Transport Systems,
coordinated by Léo MENDIBOURE.

The basic operation of these systems relies on the regular exchange of different types of messages, including those to report the status of vehicles and their environments (Bauder et al. 2022): CAMs (Cooperative Awareness Messages) and CPMs (Cooperative Perception Messages), or those for notifying of critical events or situations such as DENMs (Decentralized Environmental Notification Messages) and IVIMs (In-Vehicle Information Messages). Two final message types describe intersections and junctions: MAPEMs (MAP Extended Messages); and give the phases of traffic lights turning to green (or red): SPATEMs (Signal, Phase and Time Extended Messages).

C-ITS messages are used by a set of services possessing their own broadcasting rules according to their relevance and criticality. Certain messages necessarily need to be repeated in order to assure the service that they are to provide. This repetition follows different rules, depending on the channel used: either an ad-hoc network of the ITS-G5 type, or an IP network on a cellular network of the C-V2X. When ITS-G5 networks are used, messages are relayed by the infrastructure and surrounding vehicles in order to reach areas that are far from the sending source. The GeoNetworking protocol plays the role of relay in this context, and widens the coverage of a given area. However, this relay mechanism is limited, with the maximum number of hops authorized in the ETSI (European Telecommunications Standards Institute) standard being 11.

Therefore, several mechanisms exist, such as the GeoNetworking protocol, to ensure the broadcasting of messages. These include the possibility of repeating the sending of messages several times, without measuring the potential additional costs involved. However, this message sending necessarily involves calculations on the transmitting devices, which results in an overconsumption of energy. Also to be kept in mind is the additional cost linked to mechanisms for securing exchanges that are based on computations (signature, signature verification, chain-of-trust verification, etc.) and the sending of relatively large packets.

At a time of digital and energy sobriety, it is legitimate to ask the question of the relevance of these innumerable repetitions of message sending and to measure the consequences of this approach.

In this chapter, we propose a technical study of the performance level of the different C-ITS services, based on experiments in urban areas. A quantitative and qualitative analysis of the observed results will allow a discussion on the causes of the proliferation of C-ITS messages and the recommendations and prospects for reducing digital and energy consumption.

3.2. Context

Cooperative Intelligent Transport Systems (C-ITS) offer vehicular communication mechanisms that enable the exchange of messages between road-traffic actors, in particular, vehicles and road infrastructure. The final objectives of such a system can be summarized as:

– improving road safety;

– improving road transport efficiency;

– managing traffic and reducing congestion;

– protecting the environment and reducing air pollution.

Given the requirements (quality of service) associated with C-ITS, these systems come under the scope of critical applications, i.e. applications that require very low latency and high reliability of communications. In addition, vehicles must also have real-time information at their disposal, relating to their environment. The availability of information and the multiplication of information sources are therefore an essential condition for the functioning of the vehicular communication system.

3.2.1. *C-ITS Services*

C-ITS is based on several services implemented in the Services (Facilities) layer of the protocol stack (Sun et al. 2016). Each service is linked to a specific message type. These services use Vehicle-to-Infrastructure (V2I) and/or Vehicle-to-Vehicle (V2V) communications. Certain services use only messages transmitted from the infrastructure to vehicles, while others use both communication schemes. As regards services that use both V2V and V2I, the most notable services are Cooperative Awareness (CA) and Decentralized Environmental Notifications (DENs):

– *CA* (Cooperative Awareness) (ETSI 2014) is a service that controls CAM transmission and reception. The CAM is a periodic message sent by vehicles and Road-Side Units (RSUs) at frequencies between 1 and 10 Hz in broadcast. The message is sent to all neighboring vehicles and RSUs and contains information on the original sender, such as its position, speed and other information. The objective is to create awareness between the vehicles of their surroundings in order to promote cooperation between them.

– *DEN*, or Decentralized Environmental Notification (ETSI 2019) is the service responsible for broadcasting DENM messages. A DENM is a message that indicates an event on the road, such as an accident or road works. When an event on the road is detected, the vehicle or RSU can transmit a DENM to the affected area. Unlike the CAM, a vehicle receiving a DENM can retransmit it to other nodes in the network. A DENM message can remain persistently broadcast as long as the event continues.

Where V2I communications are concerned, in addition to CA and DEN, there are services whose messages are solely generated by infrastructure. Messages are transmitted from the infrastructure to the vehicle and not vice versa. The following services can be mentioned (ETSI 2020b):

– *TLM* (Traffic-Light Maneuver) is a service that manages the generation, transmission and reception of SPATEM messages. Its objective is to control user access to intersections and conflict areas. It sends safety information to the vehicles present in an intersection and informs them of the real-time status of the traffic light and its future status, as well as the time margin between the two. The SPATEM message is then sent periodically to all participants for as long as they are in the intersection to circulate or cross (pedestrian).

– *RLT* (Road and Lane Topology) is the service responsible for generating, transmitting and receiving MAPEM messages. A MAPEM is a message containing a digital topological map of an area's geometry; this topology describes the lanes, pedestrian crossings, conflict areas and allowed maneuvers. A lane is made up of several connected landmarks. When these passage points connect with a conflict area, a set of allowed maneuvers is modeled as a connection between the two points that connect the two lanes with the area. The MAPEM is transmitted at the same time as a SPATEM, when the vehicle approaches an intersection.

– *IVI* (*In-Vehicle Information*) is a road signage service; it uses the IVIM message to provide information on physical or virtual road signs, such as contextual speeds or road warnings, as well as the presence of road works.

3.2.2. Communication standards

As regards radio communications, the two most commonly used standards for vehicular communications are ITS-G5 (and its equivalent, DSRC) and C-V2X (Mannoni et al. 2019). The first is a standard based on Wi-Fi technology and is mainly short-range. C-V2X, based on the cellular network, has a greater communication range and possesses the same mechanisms as cellular technologies, such as scheduling of radio channel occupations for transmission (Maaloul et al. 2022).

C-V2X (Cellular Vehicle-to-Everything) was standardized by 3GPP in Release 14 (ETSI 2021). In the latter, four types of communication in broadcast mode were specified: a) V2V, or Vehicle-to-Vehicle, b) V2I, or Vehicle-to-Infrastructure (RSU), c) V2N, or Vehicle-to-Network, when communication is made with a server from the cellular network, and d) V2P, or Vehicle to a vulnerable user (Pedestrian or cyclist).

For these communication schemes, two operating modes have been specified:

– direct communication between vehicles via the PC5 air interface, also known as "Sidelink". It refers to a type of direct communication between users without passing via the network;

– communications between vehicles via the Uu air interface, where each communication takes place through a first link with the network (Uplink) and then another link with the destination (Downlink).

Meanwhile, ITS-G5 (ETSI 2013) is Wi-Fi-based and operates on the 5.9 GHz band. It implements communication standard IEEE 802.11p and its extension, 802.11bd (Naik et al. 2019). It has a shorter communication range and only supports ad hoc links between vehicles, and also RSUs. The sender occupies the entirety of the channel available during transmissions.

Regarding the protocol stack, ITS-G5 differs from other standards with the presence of the GeoNet layer (ETSI 2020a). Therefore, in order to relay a message in a dynamic, mobile environment on an ITS-G5 network, geographical routing is used. The GeoNetworking protocol is an ETSI standard and enables packets to be relayed from the source to an area called the relevance area. In Taysi et al. (2012), authors provide a survey of the existing GeoRouting protocols and evaluate the compatibility of these protocols with the GeoNet architecture. After a detailed discussion on routing requirements for V2X communications, they select protocols that meet their routing properties. The purpose of Geonetworking is to continuously build a list of neighbors and then choose the most suitable neighbor to whom the message is to be sent for it to reach the destination. The idea is thus to perform a series of hops that will ultimately see the message transmitted to the final destination.

3.3. Experimental evaluation of the performance of the C-ITS message broadcasting system

Several pilot deployment projects have emerged in France since 2014. These projects have provided very large volumes of data, enabling detailed analyses to be conducted on the performance of these systems.

3.3.1. *C-Roads France project and COOPITS application*

The C-Roads France project (Lokaj et al. 2020) corresponds to a succession of pilot C-ITS deployment projects in France, aiming to develop and test innovative C-ITS solutions. Firstly, the Scoop@F project (Aniss 2016) initiated the work and implemented a cooperative system via Road-Side Units. The Scoop@F2 project then introduced new hybrid modes of communication. Lastly, the C-Roads France project defined use cases in urban areas and focused more on vehicle-to-vehicle (V2V) communications.

New services have been developed: 1) services in the urban environment (GLOSA (Green Light Optimal Speed Advisory), Traffic-Light Optimization, multimodal information, information on Park-and-Ride facilities, etc.) and 2) Traffic Information services (smart routing, information on truck parking areas, rest-area services, etc.).

Within the framework of the C-Roads France project, hybrid communication (multiple access technologies) has been specified. It is based on the ITS-G5 infrastructure dedicated to C-ITS services and the LTE mobile cellular network. The objective of this hybridization is to enable the services requested of the vehicles, for example, according to the availability of the communication link. Figure 3.1 shows the functional links between all actors of the functional architecture of the C-Roads France project.

The C-Roads France project has thus specified, developed and distributed a mobile application called COOPITS, which enables C-ITS messages to be broadcast directly to the smartphone via the cellular network (Bhattacharyya et al. 2022). As shown in Figure 3.1, this application forms part of the overall architecture of the C-Roads France project.

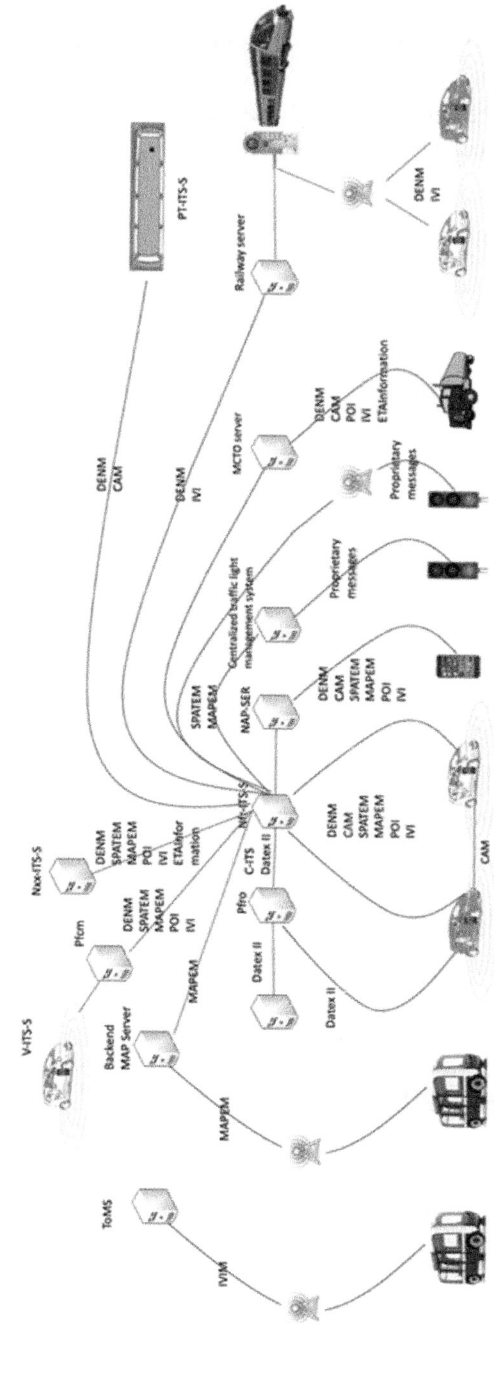

Figure 3.1. *Simplified view of SDVN architecture*

3.3.2. *Experimental environment and measurements*

COOPITS was launched on the Play Store in January 2021. The data was collected for 11 months, focusing on the Nouvelle–Aquitaine region, and more specifically, the DIRA (Interdepartmental Directorate of Atlantic Roads) network and Bordeaux Métropole (see Figures 3.2 and 3.3).

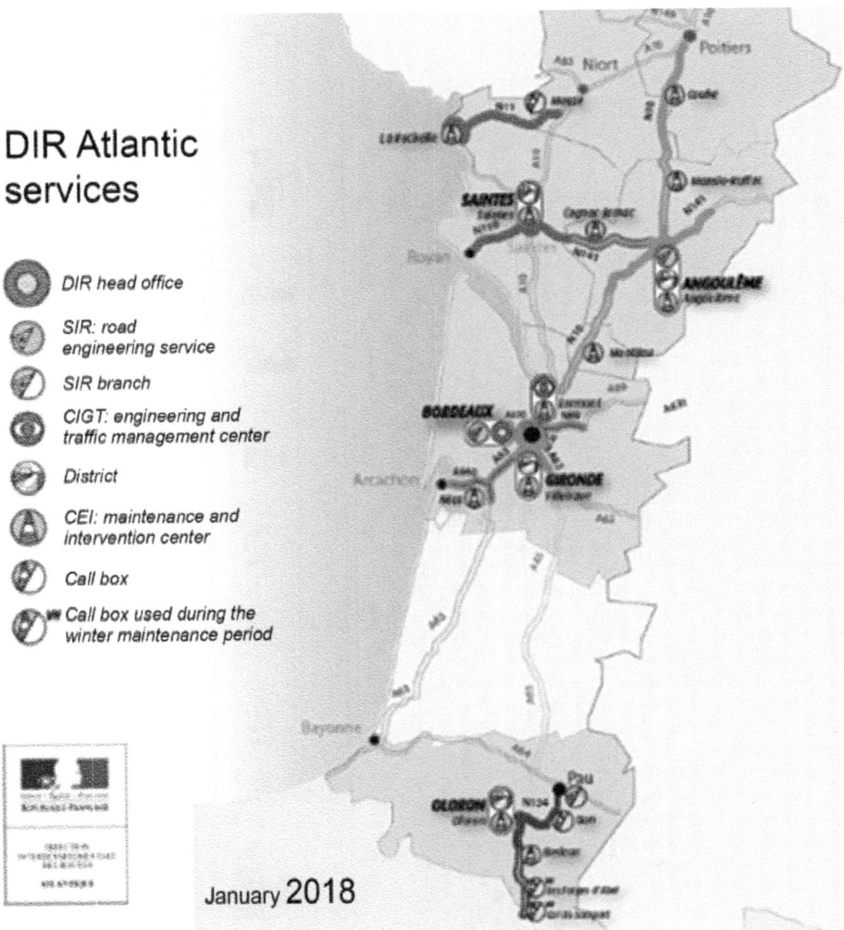

Figure 3.2. *DIRA network*

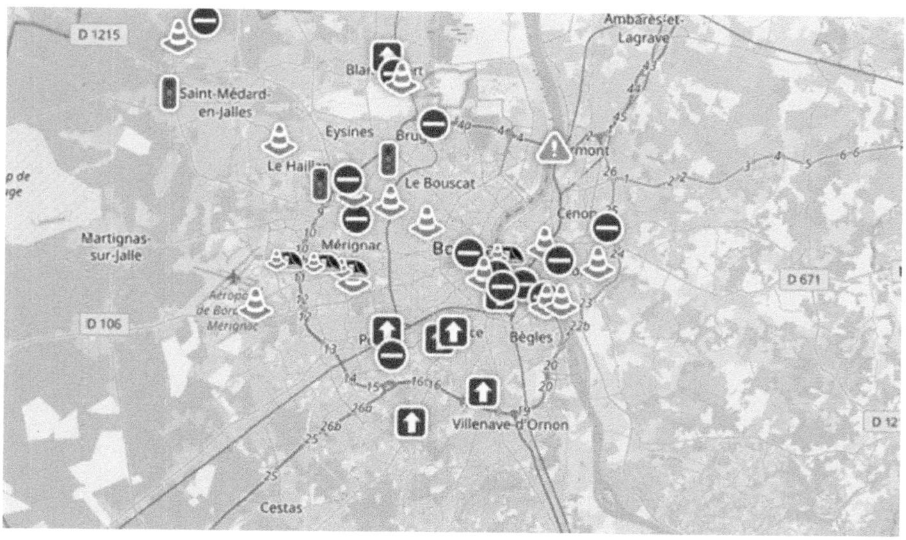

Figure 3.3. *Example of a real-time event on the Bordeaux Métropole network (see: https://sedeplacer.bordeaux-metropole.fr/)*

3.3.3. *Analysis of results*

The data collected from the logs of the COOPITS application was used to conduct several technical analyses: 1) network load induced by the messages; 2) number of messages received and sent by smartphones; 3) end-to-end latency; and 4) geographical distribution of the use cases. From a functional point of view, the balance between system behavior and the initial objective was evaluated, as well as the efficiency of the use case.

3.3.3.1. *Message broadcast area*

As shown in Figure 3.4, the location of the C-ITS messages indicates that some users activated the application outside the Bordeaux region. However, it was mainly used within the Bordeaux Métropole.

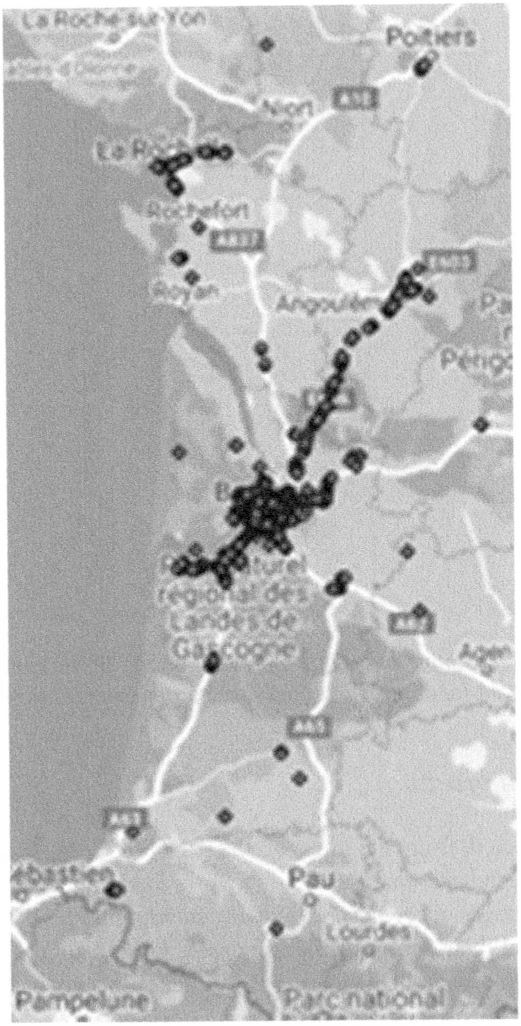

Figure 3.4. *Location of C-ITS messages*

3.3.3.2. *Message size*

In order to quantify the load induced on the network by C-ITS messages, the number of messages and their size were evaluated. Message size varies owing to the different types of information sent (CAM, DENM, SPATEM or IVI). Therefore, the approximate volume of exchanged messages is calculated from the average size of each message type (see Table 3.1).

Message type	Description of the message	Overall number of messages	Average message size (bytes)	Total message size (Mb)
CAM sent by smartphones	Status message (position, speed, etc.)	773,116	281	217.25
DENM sent by smartphones	Event-type message (obstacle or pedestrian warning)	249	163	0.04
DENM received by smartphones	Event-type message (road works, obstacle or pedestrian warning)	10,344	163	1.69
IVIM received by smartphones	VMS (Variable-Message Sign) type text information message	254,056	231	58.69
MAPEM received by smartphones	Message describing an intersection	523,072	281	146.98
SPATEM received by smartphones	Message indicating the traffic-light phase in an intersection	425,695	73	31.08
Total		1,986,532		455.73

Table 3.1. *The number of messages exchanged during the experiment*

Over the entire period, just under 2 billion messages were exchanged between smartphones and the infrastructure, of which just over half were received by users. This induced 455.73 Megabytes (Mb) of traffic. The impact on the network load seems to be small, since nearly 6,000 stations generate less than 500 Mb. However, this result needs to be viewed in terms of the number of stations and the collection duration. Scaling-up would lead to an exponential increase in the number of messages and therefore in the network load.

This result is also to be considered with respect to the number of messages presented on the HMI. What proportion of the messages received provokes a warning on the driver's HMI? To answer this, let us first look at the use cases that have been triggered by the user or the infrastructure.

3.3.3.3. *Day-1 use cases*

Day-1 use cases correspond to on-board signage events and driver information on unexpected and hazardous events (HLN: Hazardous Location Notification). We differentiate the event from the message because an event causes the production of several messages (repeat, update or cancel). Among the use cases involving the sending of DENMs from a smartphone (triggered by the user), accident warnings are the events most sent by users, although they represent only 15% of the messages received (Figure 3.5). However, the obstacle on the road is the most-received message (29%) and accounts for 27% of the events.

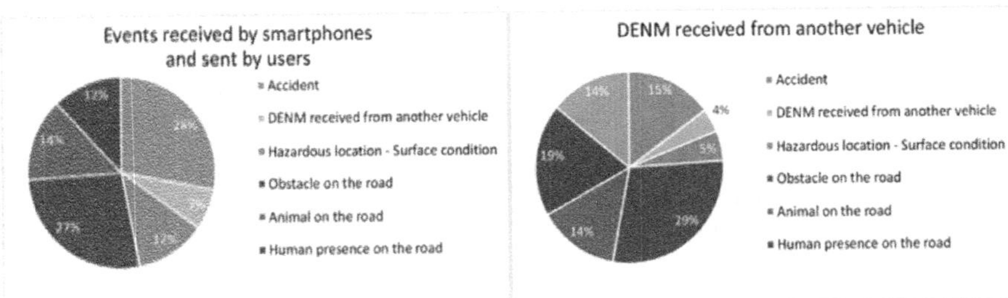

Figure 3.5. *DENMs and events sent by users*

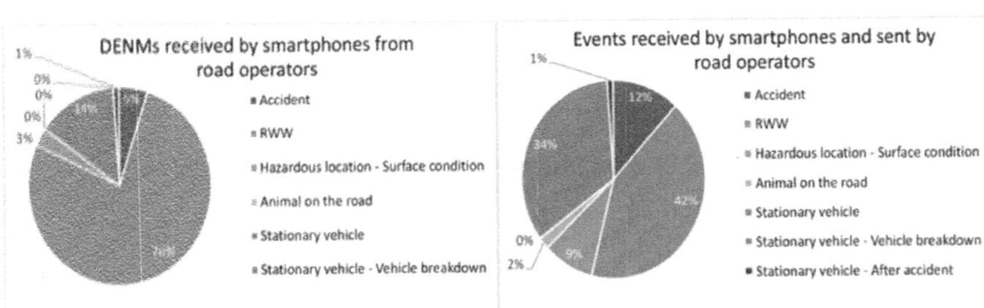

Figure 3.6. *DENMs and events sent by the road operator*

Moreover, 42% of events are road works warnings; this represents 76% of messages received by smartphone (Figure 3.6).

3.3.3.4. *Event display*

During the 11 months of experiments, 73 HLN events were communicated to the driver (Table 3.2) out of 10,344 messages received (Table 3.1), i.e. 0.7% of the messages received are relevant to the user and trigger a display.

A total of 805 pieces of variable-message sign (VMS) information were displayed, i.e. a ratio of 0.3% of efficient messages. When it comes to SPATEMs and MAPEMs, which enable realization of the GLOSA (Green Light Optimization Speed Advisory) use case, this figure falls to 0.17% of efficient messages. These very low display rates are a true indicator for the reduction effort that could be initiated in order to reduce the digital footprint of these broadcasts, while maintaining the same efficiency in terms of display and therefore prevention of the user in time to avoid accidents.

Number of HLN events displayed	73
Number of VMSs (Variable Message Signs) displayed	805
Number of GLOSAs displayed	1,662

Table 3.2. *Number of events displayed on smartphones*

3.3.3.5. *Messages in ITS-G5 versus in cellular*

A comparison of the number of messages exchanged in ITS-G5 and in cellular was carried out as part of an experiment involving vehicles equipped with hybrid units (ITS-G5 and cellular mode simultaneously). The same information was sent by the road operator over the short- and long-range networks. We can see from Table 3.3 that the number of DENM messages received is relatively similar for ITS-G5 and cellular. The message repetition modalities defined by the ETSI standards for ITS-G5 have in fact been transposed to cellular.

	DENM
ITS-G5	2,384
Cellular	2,966

Table 3.3. *The number of ITS-G5 and cellular messages received*

The following sections focus on the main causes of this significant transmission of messages, and aim to explain the low message efficiency rate. This is applicable to both cellular and ITS-G5 technology.

3.4. Discussion of the main causes

In this section, we detail the main causes contributing to the proliferation of messages that can lead to a lack of digital and energy efficiency in Cooperative Intelligent Transport Systems.

3.4.1. *Absence of adaptation to actual conditions*

In general, messages that broadcast one-off events (for example, DENMs) have very high repetition frequencies. For example, the "*RepetitionInterval*" field of the DENM is set between 200 ms and 1,000 ms in the European specification of the C-Roads Platform (see: https://www.c-roads.eu/platform.html). These choices have been made so that the message arrives in low latencies for new vehicles arriving in the area. However, these repetition rates can overload the bandwidth, and induce chain-reaction collisions and thus message losses. We can cite the example of a traffic jam warning that would be broadcast continuously in the form of a DENM by a set of vehicles blocked in this slowdown. Since the information is already known by all vehicles, it should therefore only be broadcast by the tail of the jam for new vehicles joining it, which is not taken into account in the algorithms used as standard, such as GeoNetworking. This lack of adaptation overloads the network and goes against energy sobriety.

3.4.2. *Duplication of non-scalable information*

Beyond situations in which the same events are broadcast continuously without justification (see the previous section), there are several scenarios in which vehicles receive the same information several times. The example that can be cited here concerns MAPEMs, which describe the topology of intersections. These messages are broadcast every second in ITS-G5 in accordance with the standard, even if there is no change. Indeed, these messages aim to describe the topology of the intersection. Even if an intersection's topology very rarely changes, the MAPEMs describing it are broadcast continuously. The objective is to inform vehicles crossing this intersection for the first time. This also implies that vehicles that are used to passing through this intersection receive several copies of this information at each crossing. We can thus see that there is significant room for optimization in this scenario without any deterioration of the service rendered.

3.4.3. *Broadcasting of information in wide geographical areas*

C-ITS messages are designed to be transmitted in dissemination areas that can be quite large. For example, in the Scoop@F project, it has been specified that messages should be broadcast in a circular region with a radius of 10 km centered on the event (which makes a surface area of 314 km^2). This creates a lot of unnecessary sending and sharing that overloads the network, without any certainty that the information is relevant to a user.

3.4.4. *High level of security in relation to the risks involved*

The security of exchanges between C-ITS devices (Road-Side Units and On-Board Units) uses authentication as a major principle. Therefore, a signature is added to the packet body of all messages sent. The certificate (which will serve to verify the signature) is included in the message sent in the majority of situations.

Several solutions for reducing packet size have been proposed and sometimes even adopted by ETSI standards. These solutions propose considering sending the certificate hash only, but this remains insufficient. In the following, we discuss the possibility of further reducing the size of messages exchanged when a level of trust is installed between a stable set of vehicles driving in convoy.

We propose a lightweight security concept for C-ITS, aimed at reducing the size of the bandwidth consumed by signatures and certificates of messages exchanged between vehicles. This avoids the risk of overloading the communication channel, reduces message latency and enhances overall security. We use the cluster concept to build trust between neighbors, which allows vehicles to send unsigned messages. The proposed architecture consists of the following steps:

1) Cluster creation: this architecture enables vehicles to authenticate with each other and verify that their neighbors are authorized to send CAM messages on the road. At the end of this authentication process, all vehicles can have the same list of trusted neighbors, so they will automatically stop sending signed CAMs and switch to unsigned CAMs.

2) Cluster update: when a vehicle moves out of range of its current cluster, it is removed from the latter and can join a new cluster if it enters into communication with other vehicles. Similarly, when a new vehicle enters the scope of an existing cluster, it is added to this cluster and participates in information exchange and collaborative decision-making.

3) Trust management: to maintain the security of the system, we have developed a continuous analysis process to evaluate the behavior of vehicles in the cluster. Every vehicle regularly evaluates data received from its neighbors, such as the speed, direction and position, to predict their trajectories and compare them to the actual data. By constantly monitoring the consistency of this information, the vehicle can detect any inconsistency or malicious activity. This process, integrated into each vehicle, prevents malicious actors from infiltrating the trusted cluster and ensures the security of the information exchange. By constantly checking neighbors' trust, our architecture enables an efficient and secure information exchange.

This solution has been evaluated by simulations on Artery (Riebl et al. 2019), an open-source simulation framework adapted to the Omnet++ network simulator (Varga and Hornig 2010) and connected to the Simulation of Urban Mobility (SUMO) tool to simulate road traffic (Lopez et al. 2018). Artery provides each vehicle with an ITS-G5 instance to choose the routing algorithm and secure traffic warning messages. The system's operation has been verified by simulations with hundreds of vehicles on a 150 km highway. We first applied our cluster architecture with filtering parameters and then compared them with ETSI standard protocols.

The simulation results are considered in four timesteps representing the data accumulated every 300 seconds throughout the simulation. Figure 3.7 shows the difference between the cumulative number of signed and unsigned messages in our cluster-based architecture. The number of unsigned messages represents about 85% of the set of messages exchanged, a fairly good result that enables the bandwidth consumed by security to be efficiently reduced. We also evaluated the size of the messages exchanged during the simulation.

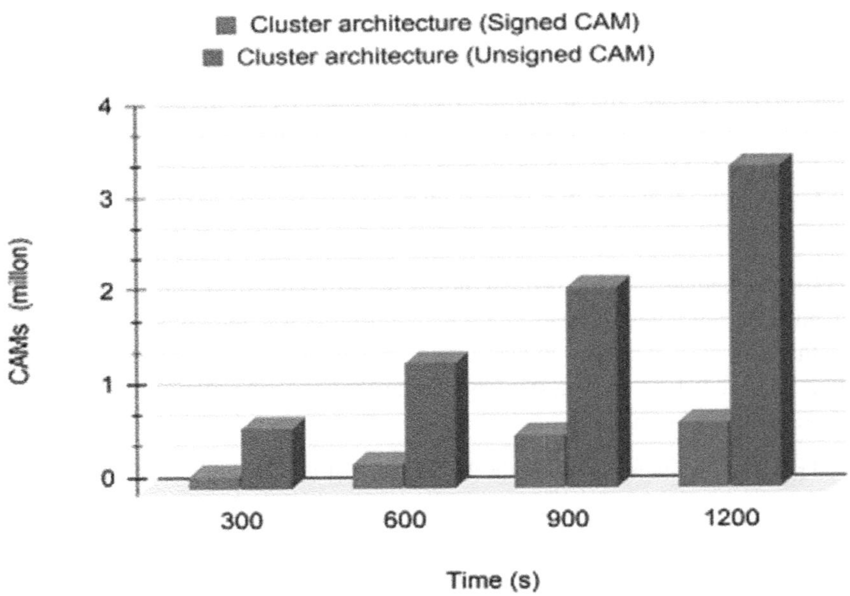

Figure 3.7. *The cumulative number of signed and unsigned CAMs in the proposed cluster architecture*

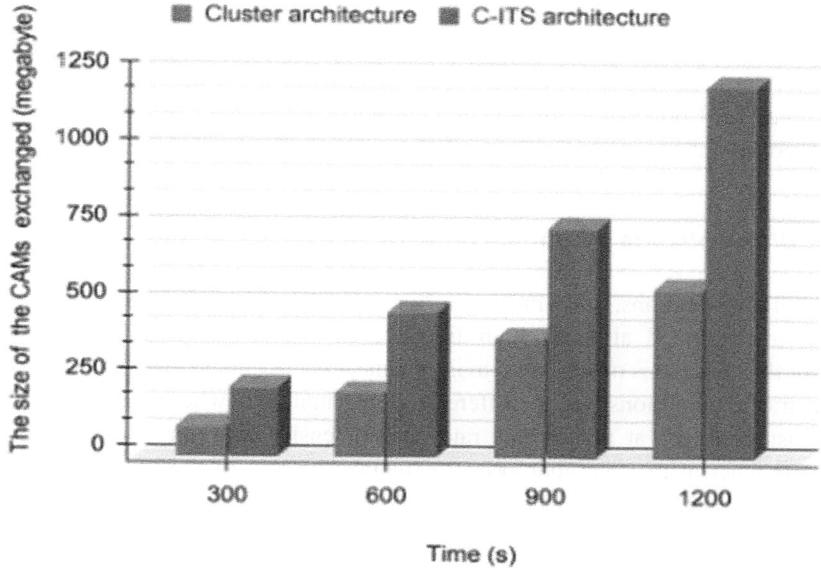

Figure 3.8. *The size of CAMs exchanged in cluster architecture and C-ITS architecture*

In accordance with the C-ITS standard, we considered the size of a signed CAM message to be 300 bytes and that of an unsigned CAM message to be 100 bytes. Figure 3.8 shows a large difference between the two architectures. Indeed, our proposal reduced the size of the messages exchanged by about half, compared to the standard protocol. This is due to its limiting usage of signed messages, including certificates and signatures.

The proposed approach uses ETSI standard protocols to ensure the authenticity and integrity of received messages, while reducing the network overload by using a selective protocol based on trusted vehicle clusters and real-time consistency checking.

The results obtained demonstrate the efficiency of this selective approach in significantly reducing network overload and improving resource utilization. It also reduced the energy consumed by the devices (for signature and for signature verification in particular). This example could reduce the digital footprint of C-ITS messages, while ensuring their security.

3.5. Recommendations and research avenues

The evaluations carried out in this study allowed us to identify certain actions to be planned on Cooperative Intelligent Transport Systems in order to reduce their energy consumption.

3.5.1. *Differentiation by traffic conditions*

In certain situations, valid messages (for example, concerning accidents or road works) are repeated at fairly high frequencies, without necessarily taking into account the conditions of the road environment. It is possible to take greater account of the traffic conditions and to differentiate the moments when it is necessary to broadcast messages at a sustained rate (lower density and high speed); also, the moments when it may be appropriate to reduce the rate of message transmission (absence of neighbors, high density or reduced speed) to optimize the network load.

3.5.2. *Smart broadcasting of constant messages*

Certain messages (MAPEMs in particular) remain valid over long periods of time and undergo few modifications. However, they are broadcast with a high frequency in order to transmit the information to new vehicles arriving in the area. It would seem pertinent to determine whether the vehicle arriving in the area has crossed it before or not, in order to know whether or not this type of message should be transmitted to it. This could limit duplication of information, especially for large messages. This requires the vehicle to keep the MAPEMs in its memory and to have a mechanism to request updates if needed.

Another possibility would be to download MAPEMs from a mapping server using, for example, the MQTT protocol, upon entering a defined area. This solution would avoid overloading the vehicle's memory, while significantly reducing the number of messages sent.

3.5.3. *Smart definition of message broadcast areas*

Message broadcast areas are identical, whatever the situation and whatever the message considered. Although the radius of the information dissemination area has been reduced in recent years from 10 km to 5 km, the area remains large (80 km^2). This dissemination area implies that the messages are sometimes relayed by several vehicles across 80 km^2. In addition to the profusion of messages from such a mechanism, each vehicle mobilizes processing resources to transport this information. However, for certain message types, these dissemination areas need to be smaller or even configurable without jeopardizing the broadcasting of safety messages. They could also depend on the number of Road-Side Units present in a given sector. The finer the mesh, the smaller the dissemination area needs to be. It therefore seems important today to propose new solutions to define message dissemination areas more intelligently, according to the information considered.

3.5.4. *Security-level optimization*

The systematic use of a high security level for messages exchanged between vehicles necessarily implies an overconsumption of energy linked to the implementation of these security mechanisms. As demonstrated in the previous section, reducing the amount of secure data exchanged could potentially significantly increase the efficiency bandwidth utilization. Therefore, it is important today to propose new mechanisms aimed at an optimal security level, depending on the situation considered: traffic conditions, type of information exchanged, etc.

3.6. Conclusion

In this study, we focused on the analysis of data collected during real experiments and simulation results. We observed the behavior of C-ITS devices and the overconsumption of energy caused by certain actions (repetition of packet sending, excessive securing of messages, etc.). The challenge of sobriety, both energy and digital, consists of reducing these parameters without compromising the security and efficiency of the vehicular communication system.

The study enabled us to determine different potential avenues for enabling the system to evolve in this direction. Differentiating the behavior of the system according to the traffic conditions would make it possible to reduce not only the number of messages but also network congestion, particularly in a saturated traffic situation. The dissemination or repetition rules must be able to evolve in order to take into account the long-lasting nature, at least into the medium term, of certain information (for example, MAPEMs). The dissemination areas need to be rethought, in particular, by taking into account the specificities of the use case, as well as the coverage density created by Road-Side Units. Finally, the securing of messages, which is essential to the proper functioning of C-ITS, represents a large proportion of the size of the message. Optimizing the signature by possibly reducing its frequency would allow a considerable gain in terms of the computing power of on-board computers, as well as in terms of network load.

By combining some (or all) of the envisaged solutions, C-ITS could significantly reduce the number and size of sometimes superfluous messages and thus contribute to the general sobriety effort.

3.7. Acknowledgments

This work was carried out as part of the C-Roads and InDiD projects, half funded by CINEA (The European Climate, Infrastructure and Environment Executive Agency).

3.8. References

Aliyu, I., Feliciano, M.C., Van Engelenburg, S., Kim, D.O., Lim, C.G. (2021). A blockchain-based federated forest for SDN-enabled in-vehicle network intrusion detection system. *IEEE Access*, 9, 102593–102608.

Aniss, H. (2016). Overview of an ITS project: SCOOP@ F. In *Communication Technologies for Vehicles: 10th International Workshop, Nets4Cars Nets4Trains Nets4Aircraft 2016*, San Sebastián, June 6–7, 131–135. Springer, Cham.

Bauder, M., Paula, D., Böhm, K., Kubjatko, T., Schweiger, H.G. (2022). Opportunities and challenges of cooperative intelligent transportation systems on accident analysis. In *Proceedings of the 30th Annual Congress*, Strasbourg, 26–28.

Bhattacharyya, K., Laharotte, P.A., Burianne, A., El Faouzi, N.E. (2022). Assessing connected vehicle's response to green light optimal speed advisory from field operational test and scaling up. *IEEE Transactions on Intelligent Transportation Systems*, 24(6), 6725–6736.

Boutahala, R., Ayaida, M., Fouchal, H. (2022). Reducing security overhead in the context of connected vehicles. *GLOBECOM 2022*, 4304–4309.

ETSI (2013). EN 302 663; Intelligent Transport Systems (ITS); ITS-G5 Access layer specification for intelligent transport systems operating in the 5 GHz frequency band. ETSI Standard, Sophia Antipolis.

ETSI (2014). EN 302 637-2; Intelligent Transport Systems (ITS); Vehicular communications; Basic set of applications; Part 2: Specification of cooperative awareness basic service. ETSI Standard, Sophia Antipolis.

ETSI (2019). EN 302 637-3; Intelligent Transport Systems (ITS); Vehicular communications; Basic set of applications; Part 3: Specifications of decentralized environmental notification basic service. ETSI Standard, Sophia Antipolis.

ETSI (2020a). EN 302 636-4-1; Intelligent Transport Systems (ITS); Vehicular communications; GeoNetworking; Part 4: Geographical addressing and forwarding for point-to-point and point-to-multipoint communications; Sub-part 1: Media-independent functionality. ETSI Standard, Sophia Antipolis.

ETSI (2020b). TS 103 301; Intelligent Transport Systems (ITS); Vehicular communications; Basic set of applications; Facilities layer protocols and communication requirements for infrastructure services. ETSI Standard, Sophia Antipolis.

ETSI (2021). TS 136 101 V14.18.0; LTE; Evolved Universal Terrestrial Radio Access (E-UTRA); User Equipment (UE) radio transmission and reception (3GPP TS 36.101 version 14.18.0 Release 14). ESTI Standard, Sophia Antipolis.

Festag, A. (2014). Cooperative intelligent transport systems standards in Europe. *IEEE Communications Magazine*, 52(12), 166–172.

Lokaj, Z., Srotyr, M., Vanis, M., Broz, J. (2020). Technical part of evaluation solution for cooperative vehicles within C-ROADS CZ project. In *2020 Smart City Symposium Prague (SCSP)*, 1–5.

Lopez, P.A., Behrisch, M., Bieker-Walz, L., Erdmann, J., Flötteröd, Y.P., Hilbrich, R., Lücken, L., Rummel, J., Wagner, P., Wießner, E. (2018). Microscopic traffic simulation using sumo. In *2018 21st International Conference on Intelligent Transportation Systems (ITSC)*, 2575–2582.

Maaloul, S., Aniss, H., Mendiboure, L., Berbineau, M. (2022). Performance analysis of existing ITS technologies: Evaluation and coexistence. *Sensors*, 22(24), 9570.

Mannoni, V., Berg, V., Sesia, S., Perraud, E. (2019). A comparison of the V2X communication systems: ITS-G5 and C-V2X. In *2019 IEEE 89th Vehicular Technology Conference (VTC2019-Spring)*, 1–5.

Naik, G., Choudhury, B., Park, J.M. (2019). IEEE 802.11 bd & 5G NR V2X: Evolution of radio access technologies for V2X communications. *IEEE Access*, 7, 70169–70184.

Riebl, R., Obermaier, C., Günther, H.J. (2019). Artery: Large scale simulation environment for ITS applications. In *Recent Advances in Network Simulation: The OMNeT++ Environment and its Ecosystem*, Virdis, A. and Kirsche, M. (eds). Springer, Cham, 365–406.

Sun, L., Li, Y., Gao, J. (2016). Architecture and application research of cooperative intelligent transport systems. *Procedia Engineering*, 137, 747–753.

Taysi, Z.C. and Yavuz, A.G. (2012). Routing protocols for geonet: A survey. *IEEE Transactions on Intelligent Transportation Systems*, 13(2), 939–954.

Varga, A. and Hornig, R. (2010). An overview of the OMNeT++ simulation environment. In *1st International ICST Conference on Simulation Tools and Techniques for Communications, Networks and Systems*, Marseille.

4

Efficient Hybridization of C-ITS Communication Technologies

Badreddine Yacine YACHEUR, Toufik AHMED and Mohamed MOSBAH

LaBRI, Bordeaux INP, France

4.1. Introduction

The primary objective of vehicle-to-vehicle communication is to create a cooperative driving awareness system to improve road safety and efficiency. This awareness system is the first phase of the launch of Cooperative Intelligent Transport Systems (C-ITS). Moreover, V2X applications are divided into three categories. The first category is V2X Day-1 applications, where the driver receives safety warnings from their environment. Day-1 application requirements can be fully met by current V2X radio access technologies (RAT), such as ITS-G5 and LTE-V2X. Day-2 applications focus on sensory driving and more autonomous reactions. Day-3 applications offer cooperative driving and fully autonomous driving to achieve accident-free driving. One of the factors that determine the deployment of Day-2 and Day-3 V2X applications is the availability of more powerful communication technologies or a robust communication architecture that can meet the requirements of these applications. Next-generation standards like IEEE 802.11bd, 5G NR V2X and, within a decade, 6G technology can solve this performance problem. However, as V2X services evolve towards autonomous driving, the use of an architecture that supports only one communication technology may not be sufficient to achieve seamless connectivity and high data rates with 99%

Cooperative Intelligent Transport Systems,
coordinated by Léo MENDIBOURE. © ISTE Ltd 2024.

reliability (Jacob et al. 2018). Indeed, in vehicle networks, the observed performance of a technology is not stable and changes over time due to the high mobility of nodes. ITS-G5 and LTE-V2X PC5 technologies enable short-range, uncoordinated vehicle-to-vehicle communications. As a result, these technologies have similar functionalities, but each has its weaknesses and is better than the other, under certain conditions. For example, a study in Cecchini et al. (2017) shows that LTE-V2X PC5 technology is more reliable over very large communication distances, while ITS-G5 technology is better over short communication distances. They therefore concluded that no optimal technology could cover all the requirements of V2X applications (Lonc and Cincilla 2016)). Studies in Vukadinovic et al. (2018) show that the improvement gain of LTE-V2X PC5 over 802.11p is minor in the case of a short platoon (e.g. up to five cars). However, in the case of long platoons, LTE-V2X PC5 performs better due to a better link budget, leading them to conclude that LTE-V2X PC5 is more suitable than IEEE 802.11p for the platooning use case. Therefore, the choice of communication technology is linked to the requirements of the application and the vehicle's environment.

To overcome the limitations and weaknesses of each technology, we opt for a hybrid vehicular communication architecture. However, the implementation of such a hybrid wireless network presents many research challenges regarding coexistence, radio resource management, RAT selection, overload problems, seamless connectivity and congestion problems. So far, we have studied two approaches to combine IEEE 802.11p-based technologies and V2X cellular technologies. One uses these technologies simultaneously, where both technologies must share the 5.9 GHz spectrum and manage the potentially negative impact of operations on adjacent channels. The other solution is to switch between these technologies according to the quality of service requirements and channel load. This alternative requires a reliable and dynamic RAT selection strategy (Mir et al. 2020)) and a good vertical handover (VHO) to ensure the full availability of V2X services.

In this chapter, we discuss the proposition of a hybrid vehicular network architecture that emphasizes the simultaneous use of ITS-G5 and LTE-V2X PC5, as well as a communication mode selection strategy. We also present two hybrid modes of communication in addition to the single communication mode where only one access technology is used; there are the redundant communication mode and the split communication mode. To select the optimal mode of communication and control the resource usage of each technology used, we propose a real-time selection strategy using deep reinforcement learning (DRL). RL approaches are generally classified as real-time learning algorithms, where agents must undertake actions within an environment to maximize the notion of reward (i.e. feedback). However, we are faced with communication of a very mobile nature in VANETs where the success of the broadcast is difficult to estimate and the link quality is highly variable. Therefore, the DRL reward is difficult to define, based solely on local observations. The distinctive contributions of this chapter are as follows:

– DRL-based communication mode selection to use hybrid modes of communication efficiently;

– the evaluation of the redundant hybrid mode of communication to demonstrate the performance of a hybrid vehicular communication network.

The remainder of this chapter is structured as follows: in section 4.2, we present several related works on hybrid networks of vehicles and the use of RL in selection problems. Section 4.3 deals with the proposed hybrid vehicular architecture and the protocol stack of the ITS station. In section 4.4, we provide several details of our control approach. Later, in section 4.5, we present the simulation environment and our implementation of hybrid architecture for vehicles. Lastly, we present the performance results between traditional vehicular communication architectures and our hybrid architecture in terms of PRR and hybrid communication efficiency.

4.2. Related works

In Zheng et al. (2015), the authors used the term "hybrid architecture" to refer to inter-road communications, combining both V2V and V2I communications. However, in our work, we will use the term "hybrid" to describe communication architectures with multiple access networks.

The access layer of the ETSI C-ITS reference architecture supports numerous radio access technologies. Based on this, a number of research works have attempted to create an abstraction between the application layer and the access layer to activate hybrid communication. This abstraction can be implemented according to two approaches. The first consists of adding a new layer (hybrid communication abstraction layer). The second approach modifies the management layer by adding modules that manage hybrid communication. These modules manage RAT selection and resource allocation. In Jacob et al. (2018), the authors clarified the concept of V2X hybrid communications and studied its potential, the technical challenges and the essential aspects of the design of a hybrid architecture. They studied the selection of a single technology and the selection of several technologies to increase the reliability (using a redundant strategy) or throughput (using an aggregation strategy) of the communication network. These works also introduced a protocol stack with a new layer, which was the hybrid communications management (HCM) layer. However, no concrete evaluation of this architecture has been carried out. In Mir et al. (2020), another approach to hybrid vehicular network architecture is proposed. It combines DSRC (Dedicated Short-Range Communication) technology and LTE technology. The ITS station in this architecture is called CellCar. This architecture defines functionalities such as dynamic RAT selection, a VHO mechanism and a dynamic communication management mechanism. The evaluation of this architecture shows that hybridization of communication technologies offers better reliability and lower communication latencies. However, this work does not include PC5 LTE-V2X, which allows direct communication without going via a base station. CellCar also does not take into account the simultaneous selection of several technologies, as discussed in the previous work of Jacob et al. (2018). Both Zheng et al. (2015) and Zekri and Jia (2018) show comprehensive studies on hybrid vehicular networks that combine DSRC technology and cellular networks. Zheng et al. (2015) studied the requirements of certain safety and non-safety V2X applications in order to propose a hybrid vehicular network architecture. This work is similar to that presented in Jacob et al. (2018); with a new layer, namely the heterogeneous link layer (HLL), introduced. This work concluded that a hybrid vehicular network with LTE for V2I communications and DSRC for V2V communications is one of the best solutions to support new vehicular services. In Abboud et al. (2016), the authors studied an interoperability solution between DSRC technologies and cellular networks, based on a flat or hierarchical hybrid architecture. In a flat hybrid architecture, the choice of technology is decentralized to each vehicle and each vehicle is equipped with both technologies. In contrast, in the hierarchical hybrid architecture, only nodes that belong to a specific hierarchical level are equipped with two interfaces (DSRC/Cellular). Another prospect for hybrid communication architecture consists of using satellite technologies in V2X communications and terrestrial technologies to meet the requirements of delay-tolerant V2X applications, as proposed in Yastrebova et al. (2021).

Decision-making problems such as the choice of RAT can be solved using reinforcement learning (RL) algorithms. In this context, the authors of the article Du et al. (2020) used a RL algorithm to select the appropriate technology in a centralized, heterogeneous network architecture composed of LTE, WLAN and VLC (Visible Light Communication). Zhao et al. (2020) used a RL algorithm based on slow-fade parameters and statistical information to perform V2V communication mode selection and power adaptation in 5G communication networks. In our work, we evaluate the redundant mode of communication and compare our DRL-based selection algorithm with a simple TOPSIS-based MCDM selection and a vehicular network that uses a single technology.

4.3. Definition of a heterogeneous network architecture and design of a protocol stack

The efficiency of a hybrid vehicular communication network depends on the penetration rate of each RAT into an ITS station (i.e. a vehicle or an RSU). If this penetration is not equal to 100%, we would need to use a hierarchical hybrid architecture (Altintas et al. 2018), which means that only vehicles that belong to the upper level of the hierarchy are hybrid and responsible for bridging messages from one technology to another. In this work, we use two short-range communication technologies, ITS-G5 and LTE-V2X PC5. In addition, we assume that the penetration rate of each RAT is 100% in all vehicles. As shown in Figure 4.1, each vehicle can communicate with all vehicles located within its communication area using both technologies.

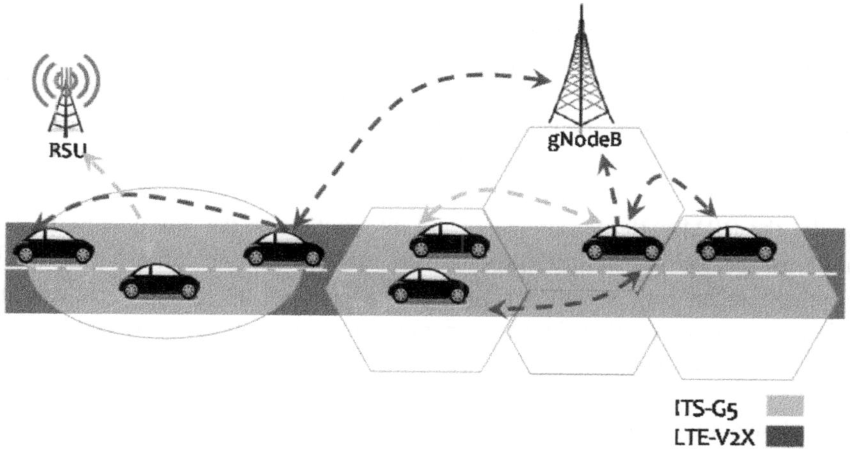

Figure 4.1. *The proposed hybrid vehicular communication architecture. For a color version of this figure, see www.iste.co.uk/mendiboure/transportsystems.zip*

Hybrid communication is achieved by combining several RATs. In this chapter, these combinations are called "modes of communication". We have defined three modes of communication. The first is a singular mode of communication, in which the message is transmitted by a single RAT, as is habitually done in a traditional vehicular communication network. The second mode of communication is the redundant mode; in this mode, the same message is sent to two RATs to improve transmission reliability. This concept of redundancy is called "profile diversity" in Jacob et al. (2018). As an example of a V2X application requiring high reliability, we can cite the emergency braking service in a platoon. With the redundant mode of communication comes a slight complexity on the receiver side; if a copy of the message is successfully received, the second copy must be discarded and deleted. The third mode of communication is a division mode (also called "profile aggregation" (Jacob et al. 2018)); this divides network traffic by two RATs into two independent transmissions. In this mode, complexity is added on both sides. On the one hand, the sender must divide the data according to the performance of each technology. On the other hand, the receiver must reassemble the two parts to maintain the integrity of the message.

The application layer groups all V2X applications, and each application has a set of specific requirements. The facilities layer holds the requirements of each application, stored in the form of an application profile. The network layer uses the position-based transmission strategy (i.e. geoNetworking) for packet routing, and the transport layer uses the basic transport protocol. Short-range vehicular communication technologies can thus be supported. Each technology has a dedicated radio resource management (RRM) entity in the management layer to monitor the channel. The access layer of each communication interface (ITS-G5 and LTE-V2X) is in charge of the channel access procedure of each technology. It also provides the RRM with measurements and information on the link quality, such as frame transmission statistics, SNIR and other channel load indicators which will be used in the technology selection algorithm. In addition, the access layer can also manipulate communication interface parameters, such as the bandwidth, MCS profile and operation channel.

As shown in Figure 4.2, the hybrid communication layer covers all control functions, including selecting the mode of communication, configuring the communication technology and appropriate processing before sending (respectively, after receiving) a message. The communication mode selector must obtain information on the application requirements and channel status from the facilities layer and the access layer, respectively. Based on this information and the availability of the RATs, a mode of communication is selected.

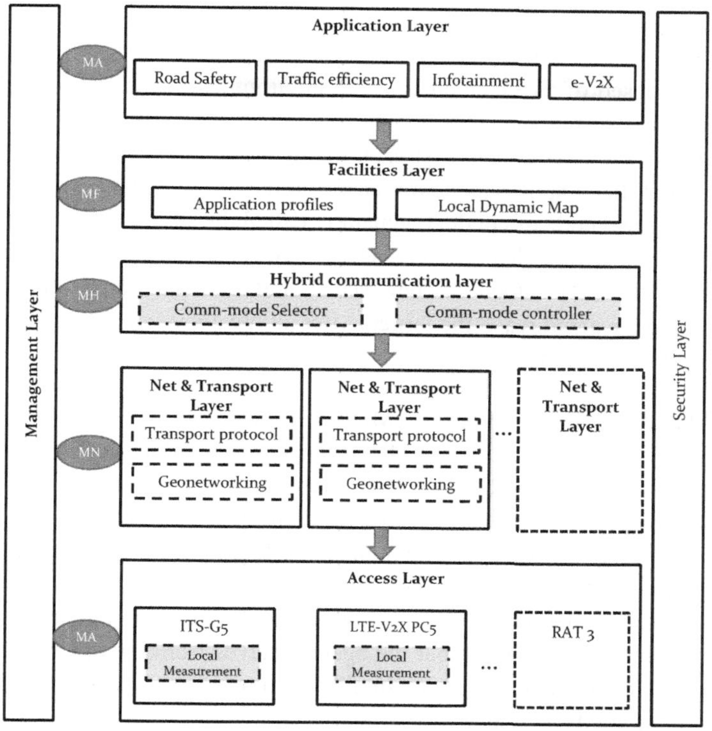

Figure 4.2. *Location of the hybrid communication layer in the protocol stack of an ITS station. For a color version of this figure, see www.iste.co.uk/mendiboure/transportsystems.zip*

4.4. RL for selecting the mode of communication

This section presents the deep reinforcement learning (DRL) framework for our communication mode selection and performance improvement. The simplest solution to benefit from the performance of both technologies is to always transmit the message using both technologies. Indeed, this solution will improve the performance of the vehicular network, but it can lead to a waste of resources and energy. To avoid this, we use a dynamic selection of the mode of communication that is based on RL, real-time measurements and the requirements of the application. RL is a computational approach to real-time learning using continuous evaluation of an environment. It can be used for real-time selection of models to improve the satisfaction of an objective. In our experiments, we evaluated the redundant mode of communication and the objective was to achieve a very high reliability of 99%.

4.4.1. *Deep reinforcement learning*

RL is composed of five elements: agent, environment, state, action and reward. The principle is that the agent learns which action is most beneficial by constantly interacting with the environment. At each moment, the environment will be in one state. The agent then chooses an action based on the observed values in conjunction with their historical code of conduct, generally called the strategy. Once the action is chosen, it triggers certain changes in the environment. These changes generate a new observation of the environment, a "new state", and an evaluation of the action (i.e. reward) to help the agent reach the target. After all the above steps, the agent repeats by choosing a new action.

The Markov decision process (MDP) provides a mathematical basis for representing the decision process of RL agents. An MDP is generally used as a formal means to represent RL. However, we cannot solve our problem using an MDP and a policy evaluation, because the problem of selecting the mode of communication does not have a finite number of states. To this end, we use DRL. In our DRL representation, we use the term s_t to signify the current state of the environment, a_t for the action taken according to the current state at time t, r_t to represent the immediate reward that the agent receives from the environment when the action a_t is completed, and s'_t to signify the new state of the environment after the action is executed a_t.

4.4.2. *Correspondence with key elements of reinforcement learning*

4.4.2.1. *Agent*

An agent represents any vehicle equipped with ITS-G5 and C-V2X technologies. Each V2X application generates messages with a set frequency. To send a message over the vehicular network, the agent must decide to use one of the modes of communication presented above. To make this decision, the agent must interact with their environment and learn from their actions. Therefore, they can gain experience and adjust their strategy in choosing the mode of communication to achieve the optimal policy.

4.4.2.2. *State of the environment*

As mentioned earlier, we have an infinite state space. A local environment state is composed of six elements and is represented as follows: $s_t = [SNIR_{ITS-G5}, SNIR_{LTE}, PRR_{ITS-G5}, PRR_{LTE}, L, R]$; where the values $SNIR_{ITS-G5}$ and $SNIR_{LTE}$ are the measured SNIRs of each RAT network. The values PRR_{ITS-G5} and PRR_{LTE} are the calculated packet reception rate of each RAT.

L and R represent the latency and reliability requirements, respectively, of a V2X service.

These parameters cannot be obtained before sending a message via the sender. The only way to obtain the value of these parameters is to calculate them when a message is received. To this end, we consider three different environment states in order to have accurate state representations. Firstly, s_t is the state at a moment, t, of the environment. Secondly, s'_t represents the environment state after the vehicle has received the first message from another vehicle in the environment before sending a new message. Lastly, we define s_{t+1} as the next state, which is the environment state after the vehicle has received the last message before sending a new one. The agent chooses an action based on s_t, and then measures the link quality each time it receives a message. The agent receives an evaluation of their decision based on the oldest link quality measurement, which is s'_t, and then updates the current state with the value of s_{t+1}. The relationship between each state is shown in Figure 4.3.

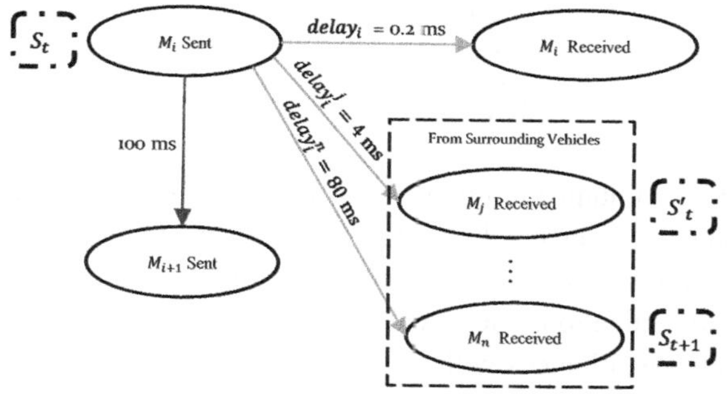

Figure 4.3. *Different representations of the environment*

4.4.2.3. *Action*

The action in our problem is to send a message with a mode of communication. In addition to the hybrid modes of communication, we will consider two instances of a single mode of communication, one for ITS-G5 and the other for LTE-V2X PC5. Therefore, our discrete action space will be given as follows:

$$A = \begin{cases} 0, Single_{ITS-G5} \\ 1, Single_{LTE} \\ 2, Hybrid_{Redundant} \\ 3, Hybrid_{Division} \end{cases} \qquad [4.1]$$

4.4.2.4. *Reward*

The reward in an RL episode is composed of three parts. The first concerns the evaluation of the reception of the message. This verifies whether the message sent using the chosen mode of communication has been successfully received by all surrounding vehicles. It also makes it possible to verify the integrity of a message. The second part is performance satisfaction (PS). This guides the agent in choosing the best mode of communication to meet the requirements of the application. The third part is used to give feedback on the improvement of the link quality (LQ) by comparing the SNIR values of the current state (s_t) with the SNIR values of the next state (s'_t). The reward equation is the following, where α and β are set between 0 and 1 with $\alpha + \beta = 1$:

$$R_{V_i} = \begin{cases} \frac{1}{2}\sum_{j=0}^{n} ReceptionState[V_i][V_j], i \neq j \\ \qquad + \frac{\alpha}{2} PS \\ \qquad + \frac{\beta}{2} LQ \end{cases} \qquad [4.2]$$

In a vehicular communication network, acknowledgment messages are inhibited so as not to congest the channel. As a result, a vehicle cannot know whether its neighbors have received its message correctly or not. However, using ITS-G5 multichannel communication, each vehicle can send an acknowledgment message and avoid creating disturbances with the safety messages. This is how we calculate the first part of our reward function. Every 100 milliseconds, an agent sends a message. It is known that the $delay_i$ (end-to-end latency), shown in Figure 4.3, is always less than 100 milliseconds. Each agent therefore has time to receive the acknowledgment of receipt of a message from surrounding vehicles, before sending another message. These acknowledgments are represented in a vector called the reception evaluation vector. Figure 4.4 shows how the reception of a message M_i is reported with an example of five surrounding vehicles. The reception evaluation vector is updated each time a message is received. It contains reception reports and a successful receptions (SR) counter. If a message is received twice by a vehicle, a value of "2" is reported in the vector; and if it has not been received, a value of "0" is reported. In the opposite case, the message is considered as perfectly received, and a value of "1" is reported in the vector. The SR counter represents the number of perfectly received messages, and it is incremented if all the reception ratios in the vector are equal to "1".

The reception reports are cleared each time a message is sent, and the counter is reset when it reaches "n" perfectly received messages. The value of "n" will be defined in the scenario description.

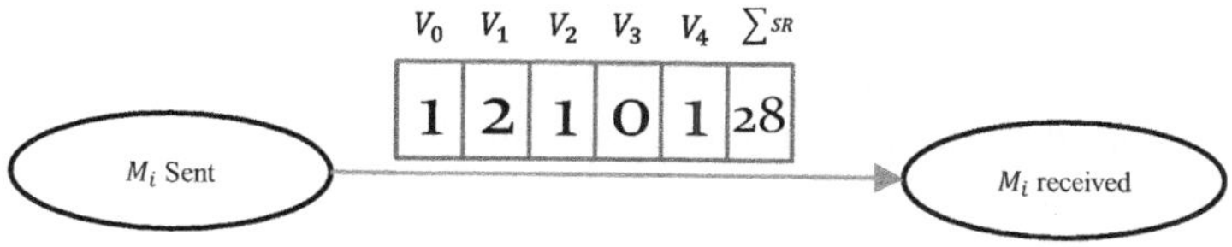

Figure 4.4. *Vector representation of reception evaluation. For a color version of this figure, see www.iste.co.uk/mendiboure/transportsystems.zip*

4.4.2.5. *The double deep Q-learning algorithm*

The DDQL algorithm combines quality learning (Q-Learning) and deep learning. It uses a neural network called the behavior network to make a nonlinear approximation of the quality function $Q(s_t, a_t, \omega)$ using the ω parameters. These parameters are constantly adjusted to correspond to the optimal strategy π by learning from uncorrelated data from past experiences. To do this, DDQL introduces a replay buffer (i.e. experience replay buffer) that collects and stores the tuple s_t, a_t, r_t, s_{t+1} >at each iteration. DDQL also comprises a second neural network (i.e. target network) with the same structure and the same initial ω′ parameters as the approximation neural network. In order to predict the action to be taken, the agent selects the action with the highest quality function (Q) value for each state, according to equation [4.3]; knowing that the values of the quality function (Q) for each state are the behavior network output:

$$\pi(s) = argmax_a Q(s,a) \quad [4.3]$$

During the learning process (as explained in algorithm 3), the DDQL algorithm samples a tuple mini-batch at random, according to a variant of the stochastic gradient descent (SGD) method, called mini-batch SGD. The mini-batch is used to update and minimize the loss function represented by equation [4.4], where M is the size of the sampled mini-batch and r_t is the immediate reward:

$$Loss(\omega) = \left[r\right| \left|t + \gamma \max_{a'} Q(s_t, a', \omega') - Q(s_t, a_t, \omega)\right]^2/M \quad [4.4]$$

Equation [4.4] is the Bellman optimality equation (Sutton and Barto 1998). It is used to calculate the prediction of the quality function, $Q(s_t, a_t, \omega')$, using the output of the target network. This leads to finding the optimal policy π by updating the behavior network:

$$Q_{\pi}\ (s_t, a_t, \omega) = E\big[r_t + \gamma max_{a'} Q_{\pi}\ (s'_t, a'_t, \omega)\big|s_t, a_t \qquad [4.5]$$

These updates are performed to minimize the loss function between the predicted values of the quality function (Q) and the approximation of the target neural network. In RL, agents aim to select the action that maximizes the sum of cumulative rewards through equation [4.3], while focusing on long-term returns. For the agent to consider a long-term reward, we introduce an update factor, γ, into the calculation of the state-action value function $Q_{\pi}(s_t, a_t)$.

RL uses the two principles of exploration and exploitation. Exploration consists of choosing the action at random to see how the environment will react and discover new policies. This operation helps the model to avoid a local maximum. In exploitation, the agent uses equation [4.3] to choose the action, which means that it uses the result of the neural network to choose the action. Exploration and exploitation are implemented in the learning algorithm using the ϵ parameter, which is decremented at each step to reduce the exploration rate.

Input
Initialized parameters of the behavior network and the target network (ω, ω'
Initialized replay buffer of size *L*, and a batch size of 64.
$\epsilon = 1$, and $\epsilon_{decrement} = 10^{-5}$
$\gamma = 0.99$, target = *n*.
Output
None
Start
for episode = 1 to 5,000 do {
  // Initialize the environment and obtain the observations of the initial state
  $o_t = [SNIR_{ITS-G5}, SNIR_{LTE}, PRR_{ITS-G5}, PRR_{LTE}, L, R]$
  while $(reception_{vector}[SR] < target)$ do {
    // As a function of ϵ choose the action and decrement ϵ
    $a_t = \begin{Bmatrix} randomaction \\ a_t = argmax_{a_t} Q(s_t, a, \omega) \end{Bmatrix}$
    $\epsilon = \epsilon - \epsilon_{decrement}$
    // Perform the action a_t, and obtain the immediate result. s'_t, r_t
    If (all values indicated in $reception_{vector} = 1$)
      reception_vector [SR]++
    clean(reception_vector)
    // Obtain the observation of the next state s_{t+1}
    $O_t = O_{t+1}$

store_transition (s_t, a_t, r_t, s_{t+1} >)

// Train the model

If (the buffer size is > 64){

sample a random mini-batch of $\{s_t, a_t, r_t, s_{t+1}\}$

// Calculate the prediction using the mini-batch

$$Q_{target} \rightarrow \begin{Bmatrix} r_t \\ r_t + \gamma \max_{a'} Q(s_t, a', \omega') \end{Bmatrix}$$

$Q_{behavior} \rightarrow Q(s_t, a_t, \omega)$

Perform a mini-batch SGD using the loss function.

Updating of the behavior network parameter

Update the target network parameters at every *T* cycle

}**end if**

} **end while**

$reception_{vector}[SR] = 0$

} **end for**

End

Algorithm 4.1. *DDQL algorithm for selecting the mode of communication in a hybrid network of vehicle-to-vehicle communication*

4.5. Performance evaluation

The evaluation of this contribution is performed by simulation. For this, we use OMNeT++ Frameworks, as well as Artery (Riebl et al. 2015) and SimuLTE (Virdis et al. 2014) to simulate ITS-G5 and LTE-V2X PC5 technologies, respectively. SUMO is used to design our simulation scenario and generate the corresponding traffic flow. In this section, we evaluate the performance of our hybrid vehicular communication architecture. Within the context of this evaluation, we compare our DRL-based selection approach to a simple traditional approach (i.e. a single-RAT vehicle network, ITS-G5 or LTE-V2X).

4.5.1. *Simulation framework and scenario*

In our architecture, we use ITS-G5 and LTE-V2X PC5. We therefore need to implement a new hybrid architecture that supports these short-range technologies in

OMNeT++. Unlike VeinsLTE, our architecture must support the simultaneous use of ITS-G5 and LTE-V2X Sidelink. We implemented the hybrid communication layer using simple OMNeT modules. Each of the ITS-G5 and LTE-V2X protocol stacks is linked to the hybrid communication layer using a sub-application module. When the main application generates a message, this is multiplexed to the appropriate sub-application, according to the selected mode of communication.

We use the platooning service (Kavathekar and Chen 2011) to evaluate our DRL Framework in our simulations. Each vehicle in the platoon is equipped with both the short-range technologies and is considered as a DRL agent. The maximum size of the platoon is set at six vehicles. This means that each vehicle in the platoon is within the coverage of a transmitting vehicle. Messages are exchanged with an average size of 128 bytes and a frequency of 10 Hz. The distance between each vehicle in the platoon is 10 meters, and each member of the platoon follows the CACC (Cooperative Adaptive Cruise Control) car monitoring model defined in the SUMO simulator. The maximum speed for a vehicle in the platoon is set at 10 m/s, and 20 m/s for the other vehicles.

Figure 4.5 describes the platooning scenario on a 2 km highway. We define a two-way highway with two lanes in each direction. As the infrastructure, only two base stations are deployed to assist the LTE-V2X PC5 standard (mode 3). There are no obstacles present near the highway. To evaluate the results of hybrid communication, we use a low traffic flow that does not exceed 60 vehicles. These parameters are summarized in Table 4.1.

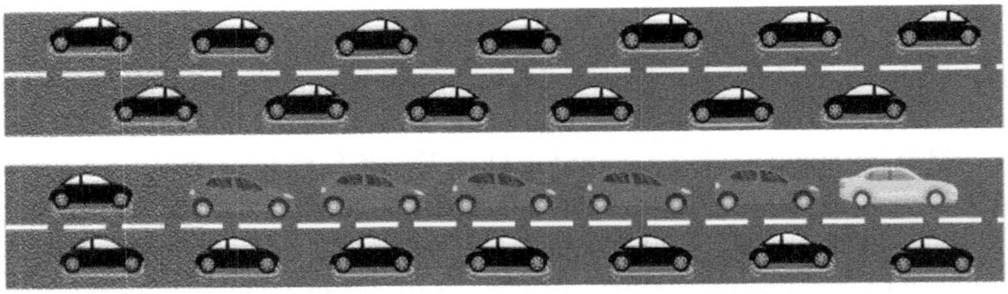

Figure 4.5. *Platoon scenario on a highway. For a color version of this figure, see www.iste.co.uk/mendiboure/transportsystems.zip*

For ITS-G5 technology, the transmitter power is set at 23 dBm, the receiver sensitivity and the energy detection are set at -85 dBm, and the SNIR threshold is set at 4 db. For LTE-V2X, the transmitter power is set at 15 dBm, and the same value as for ITS-G5 is set for the receiver sensitivity and the energy detection. The background noise is set at -90 dBm for ITS-G5 and -110 for LTE-V2X. We use the

Geometry-based Efficient propagation Model (GEMV²) as a radio bearer path loss model for Vehicle-to-Vehicle (V2V) and Vehicle-to-Infrastructure (V2I) communications. As the propagation model, we used the Jakes propagation model in LTE-V2X and the constant speed propagation model for ITS-G5. These parameters are summarized in Table 4.2.

	Value
Number of lanes per direction	2
Number of vehicles	Max = 60
Maximum speed	10.20 m/s
Max. gap between vehicles	10 m
Maximum platoon size	6

Table 4.1. *Road traffic parameters*

	ITS-G5	LTE-V2X
Access mechanism	EDCA	SPS (mode 3)
Access channel	5,875–5,885 MHz	5,895–5,905 MHz
Transmitter power	23 dBm	23 dBm
Receiver sensitivity	-85 dBm	-85 dBm
Energy detection	-85 dBm	-85 dBm
Background noise	-90 dBm	-110 dBm
Propagation model	Constant velocity propagation	Jakes

Table 4.2. *Configuration of standards*

4.5.2. *DDQL algorithm parameters*

We implemented the DDQL algorithm using the C++ libtorch library to be in phase with the OMNeT++ development language. Libtorch is the C++ interpretation of the Python Pytorch Learning Framework. Each vehicle has its own local DRL model. In the latter, the behavior and target networks are composed of four layers, an input layer of six neurons, two hidden layers of 256 neurons and an output layer of three neurons. The activation function in each layer is a "ReLU" function. The ADAM optimizer and the MSE loss function are used to update the neural network parameters. The learning rate of our DDQL algorithm is set at 0.0005, γ is set at 0.99

and the epsilon decay is set at 10^{-5}. The maximum replay buffer size is set at 10^6, and the mini-batch size is set at 64.

A game ends when the target is reached by an agent, and a step in a game is equivalent to the sending of a V2V message. The target is set by a number of messages received with a perfect configuration. This number is set at 100 messages received by all surrounding vehicles. The target network is updated every 100 learning steps. Lastly, we evaluate the reliability of the communication and calculate the PRR using equation [4.6] at the end of each game. Where $N_{sentmessages}$ is the number of messages sent to reach 100 messages received with the optimal configuration:

$$PRR_{game} = \frac{100}{N_{sentmessages}} \quad [4.6]$$

4.5.3. *Simulation results*

To evaluate the performance of our hybrid vehicular communication network and the DRL-based communication mode selection algorithm, we use the following performance measurements: the average PRR during each game play and the number of duplicate messages, when redundant mode is used. The latter allows us to evaluate the efficiency of the use of the redundant hybrid communication mode. In addition, we expose the expected average reward per game play in order to evaluate the convergence of the proposed DDQL algorithm. First, we study the results of our DRL-based selection strategy. Next, we compare our hybrid communication network with two basic networks (i.e. a communication network only with ITS-G5 and a communication network only with LTE-V2X). We also compare our communication mode selection strategy with a static RAT selection, which selects a mode of communication based solely on the application's requirements.

In our scenario, we consider that we have a medium level of congestion, and we note that there are agents that are less resistant to interference and congestion than other agents. This phenomenon is due to the stochastic nature of vehicular communication networks and the mini-batch SGD method. Agents impacted by congestion have long games and slow convergence, as shown in Figure 4.6(a). We can also note certain fluctuations due to the fading of the channel induced by mobility in the V2V communication network. This has an impact on the reliability of the communication, as shown in Figure 4.6(b). These agents need more time and games to achieve the optimal policy. However, despite these fluctuations and the slowness of convergence, we can achieve 80% reliability after a large number of training games, which is almost 2,000 games.

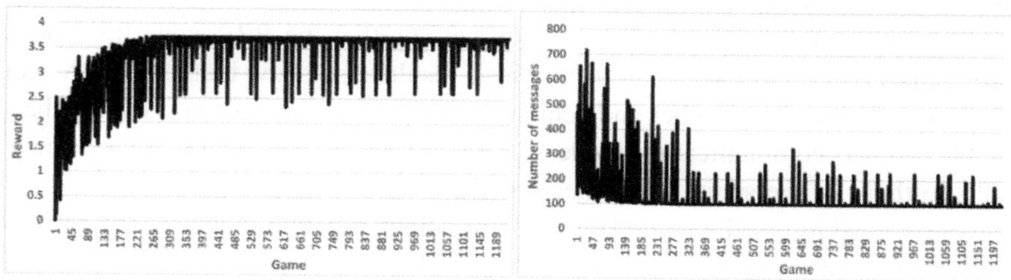

a) b)

Figure 4.6. *Results for agents impacted by congestion*

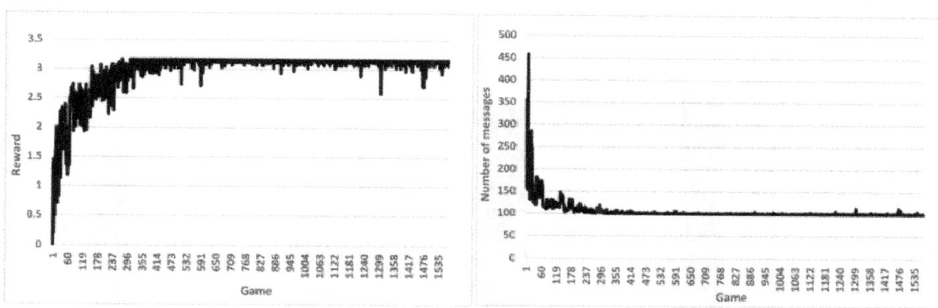

a) b)

Figure 4.7. *Results for agents not impacted by congestion*

If, on the contrary, we focus our study on agents who are not impacted by congestion, we note that the reward converges after 600 game plays. This convergence is followed by an excellent improvement in reliability, as shown in Figure 4.7(b). We note that in the first 400 game plays, transmission reliability rises from 59% to almost 80%. It then reaches 95% at the end of the simulation.

Before implementing the DRL-based selection strategy, we sought to solve the reliability problem with a strategy that uses the redundant hybrid communication mode when the security service requires greater than 90% reliability. Figure 4.8 presents a comparison between a vehicular communication network with a single technology, a hybrid network with a static selection strategy and our DRL-based selection strategy. We can note that hybrid communication with static RAT selection improves PRR in both traffic flow densities, compared to the basic performance with ITS-G5 and LTE-V2X. The greater performance that hybrid communication adds is noted when the network is congested. We can see that the PRR goes from 66% (using only ITS-G5) to more than 90%. These results also show that LTE-V2X and ITS-G5 complement each other and achieve better performance when used together. Our DRL-based communication mode selection strategy, meanwhile, gave similar results to those of the static RAT selection algorithm. We also note that we do not necessarily need to use the redundant mode of communication to satisfy the service requirements and guarantee that the message will be received successfully. This is because if the transmission of the message is assured by both technologies, the message will be received twice. With the hybrid static strategy, we signal up to 7,119 out of 10,718 messages that were received twice (for 250 s of simulation time), meaning that 66% of redundant transmissions were inefficient and are considered a waste of resources.

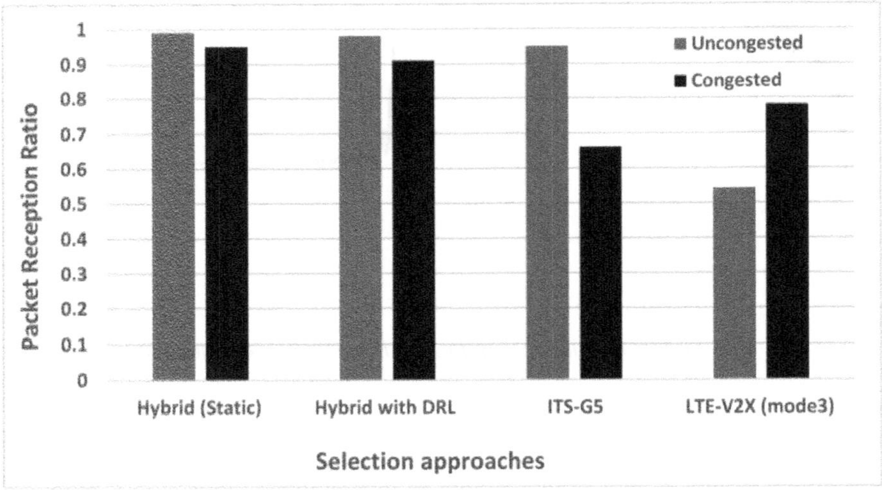

Figure 4.8. *Comparison of RAT selection approaches*

According to the statistics for duplicate messages in Figure 4.9, our strategy reduces the duplicate message rate from 60% to close to 7%. This shows that our DDQL algorithm takes optimal measurements and satisfies application requirements without wasting network resources.

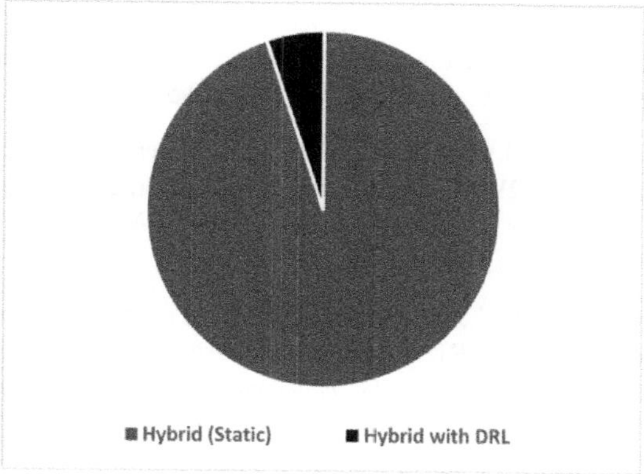

Figure 4.9. *Percentage of messages duplicated at reception between the two hybrid approaches*

4.6. Conclusion

In this paper, we have defined and implemented a hybrid vehicular communication architecture. The purpose of this architecture is, firstly, to satisfy the performance needs of emerging V2X applications; and, secondly, to benefit from the performance and the complementary aspect of vehicular communication technologies, ITS-G5 and C-V2X, in particular, thus overcoming the limitations and weaknesses of each technology. To coordinate these technologies and use them in the best possible way, we proposed a communication mode selection algorithm based on DRL. We compare the performance of our selection algorithm with the performance of a communication with a technology (ITS-G5 or LTE-V2X) in a conventional architecture and with a static selection algorithm. The results show that not only is the DRL selection strategy better than the use of a technology, but also that it makes good use of the redundant mode of communication with fewer redundant receptions. In future works, we will seek to improve other aspects of vehicular communication and to study the coexistence between these short-range communication technologies in the 5.9 GHz frequency band.

4.7. References

Abboud, K., Omar, H.A., Zhuang, W. (2016). Interworking of DSRC and cellular network technologies for V2X communications: A survey. *IEEE Transactions on Vehicular Technology*, 65(12), 9457–9470. doi: 10.1109/TVT.2016.2591558.

Altintas, O., Banerjee, S., Dressler, F., Heijenk, G., Sjoberg, K., Reports, D. (2018). Inter-vehicular communication towards cooperative driving (Dagstuhl Seminar 18202). *Dagstuhl Reports*, 8(5), 31–59. doi: 10.4230/DagRep.8.5.31.

Cecchini, G., Bazzi, A., Masini, B.M., Zanella, A. (2017). Performance comparison between IEEE 802.11p and LTE-V2V in-coverage and out-of-coverage for cooperative awareness. In *2017 IEEE Vehicular Networking Conference (VNC)*, 109–114. doi: 10.1109/VNC.2017.8275637.

Du, Z., Jiang, B., Wu, Q., Xu, Y., Xu, K. (eds) (2020). Learning the optimal network with context awareness: Transfer RL based network selection. In *Towards User-Centric Intelligent Network Selection in 5G Heterogeneous Wireless Networks*. Springer, Singapore, 65–78. doi: 10.1007/978-981-15-1120-2_5.

Jacob, R., Franchi, N., Fettweis, G. (2018). Hybrid V2X communications: Multi-RAT as enabler for connected autonomous driving. In *2018 IEEE 29th Annual International Symposium on Personal, Indoor and Mobile Radio Communications (PIMRC)*, 1370–1376. doi: 10.1109/PIMRC.2018.8580953.

Kavathekar, P., and Chen, Y. (2011). Vehicle platooning: A brief survey and categorization. In *2011 ASME/IEEE International Conference on Mechatronic and Embedded Systems and Applications, Parts A and B*, 829–845. doi: 10.1115/DETC2011-47861.

Lonc, B. and Cincilla, P. (2016). Cooperative ITS security framework: Standards and implementations progress in Europe. In *2016 IEEE 17th International Symposium on A World of Wireless, Mobile and Multimedia Networks (WoWMoM)*, 1–6. doi: 10.1109/WoWMoM.2016.7523576.

Mir, Z.H., Toutouh, J., Filali, F., Ko, Y.-B. (2020). Enabling DSRC and C-V2X integrated hybrid vehicular networks: Architecture and protocol. *IEEE Access*, 8, 180909–180927. doi: 10.1109/ACCESS.2020.3027074.

Riebl, R., Gunther, H.-J., Facchi, C., Wolf, L. (2015). Artery: Extending veins for VANET applications. In *2015 International Conference on Models and Technologies for Intelligent Transportation Systems (MT-ITS)*, 450–456. doi: 10.1109/MTITS.2015.7223293.

Sutton, R.S. and Barto, A.G. (1998). Reinforcement learning: An introduction. *IEEE Transactions on Neural Networks*, 9(5), 1054.

Virdis, A., Stea, G., Nardini, G. (2014). SimuLTE – A modular system-level simulator for LTE/LTE-A networks based on OMNeT++. In *Proceedings of the 4th International Conference on Simulation and Modeling Methodologies, Technologies and Applications*, 59–70. doi: 10.5220/0005040000590070.

Vukadinovic, V., Bakowski, K., Marsch, P., Garcia, I.D., Xu, H., Sybis, M., Sroka, P., Wesolowski, K., Lister, D., Thibault, I. (2018). 3GPP C-V2X and IEEE 802.11p for vehicle-to-vehicle communications in highway platooning scenarios. *Ad Hoc Networks*, 74, 17–29. doi: 10.1016/j.adhoc.2018.03.004.

Yan, M., Feng, G., Zhou, J., Qin, S. (2018). Smart multi-RAT access based on multiagent reinforcement learning. *IEEE Transactions on Vehicular Technology*, 67(5), 4539–4551. doi: 10.1109/TVT.2018.2793186.

Yastrebova, A., Ojanpera, T., Makela, J., Hoyhtya, M. (2021). Hybrid connectivity for autonomous vehicles: Conceptual view & initial results. In *2021 IEEE 93rd Vehicular Technology Conference (VTC2021-Spring)*, 1–6. doi: 10.1109/VTC2021-Spring51267.2021.9448866.

Zekri, A. and Jia, W. (2018). Heterogeneous vehicular communications: A comprehensive study. *Ad Hoc Networks*, 75–76, 52–79. doi: 10.1016/j.adhoc.2018.03.010.

Zhao, D., Qin, H., Song, B., Zhang, Y., Du, X., Guizani, M. (2020). A reinforcement learning method for joint mode selection and power adaptation in the V2V communication network in 5G. *IEEE Transactions on Cognitive Communications and Networking*, 6(2), 452–463. doi: 10.1109/TCCN.2020.2983170.

Zheng, K., Zheng, Q., Chatzimisios, P., Xiang, W., Zhou, Y. (2015). Heterogeneous vehicular networking: A survey on architecture, challenges, and solutions. *IEEE Communications Surveys and Tutorials*, 17(4), 2377–2396. doi: 10.1109/COMST.2015.2440103.

5

Using SDN Technology to Control C-ITS: Towards Decentralized Approaches

Romain DULOUT[1], Lylia ALOUACHE[2], Tidiane SYLLA[3], Léo MENDIBOURE[1], Hasnaâ ANISS[1], Virginie DENIAU[4] and Yannis POUSSET[5]

[1] *ERENA, COSYS, Gustave Eiffel University, Marne-la-Vallée, France*

[2] *ETIS, CY Cergy Paris Université, France*

[3] *LaMIA, Université des Sciences des Techniques et des Technologies de Bamako (USTT-B), Mali*

[4] *COSYS-LEOST, Gustave Eiffel University, Villeneuve-d'Ascq, France*

[5] *XLIM, Université de Poitiers, Chasseneuil-du-Poitou, France*

5.1. Introduction

The applications targeted by Cooperative Intelligent Transport Systems (C-ITS) need to ensure road user safety and the fluidity of road traffic (Sjoberg et al. 2017). They are based on the regular exchange of information between road infrastructure users (vehicles, pedestrians, cyclists, etc.) and roadside infrastructure. These can be messages relating to the detection of an obstacle on the roadway, to a lane change or to the detection of a slowdown. The relevance of the information transmitted to the other entities concerned is directly impacted by the performance level of the underlying communication network: latency, reliability, throughput, security, etc. Yet, the challenges related to the vehicular environment are numerous: high mobility, frequent disconnections, etc. Data transmission therefore constitutes a major challenge for future connected and autonomous vehicles (Camacho et al. 2018).

This is why new solutions are currently being studied to optimize the underlying communication network and guarantee a higher quality of service (QoS) level. A highly promoted approach would consist of integrating software-defined network (SDN) technology into the vehicular communication architecture (Benzekki et al. 2016). We thus speak of software-defined vehicular networks (SDVN) (He et al. 2016). SDN, unlike traditional network architectures, disassociates the control plane and the data plane from the network (see Figure 5.1). Routers, with SDN, are solely responsible for data transfer. A new entity, named the SDN controller, is integrated into the network to select, in real time, the most suitable communication pathways and to indicate to routers the procedure to follow. This offers two main advantages over traditional solutions: 1) optimization of network utilization thanks to the global view of the infrastructure that the SDN control plane has, and 2) significant flexibility and the possibility of having the network topology evolve in real time.

Nevertheless, the characteristics of the vehicular environment (significant number of nodes, high mobility of terminals, frequent breaking of communication links, QoS requests) constitute a significant challenge, including for these SDVN architectures (Zhao et al. 2020b). The issues identified concerns, in particular: 1) the scaling-up of software-defined architectures in the vehicular environment (capability of a single SDVN controller to manage a scalable number of terminals); 2) the level of performance of these architectures (additional cost in terms of computation and latency associated with the control plane) likely to impact C-ITS applications and 3) the level of security of these architectures (integration of software components) likely to disrupt data transfer.

Also, many researchers are today interested in adapting SDN technology to the vehicular environment, in order to allow the advent of SDVN networks. To do this, one promising approach would consist of associating another technology with SDVN architecture: Blockchain (Zheng et al. 2018). Indeed, Blockchain aims to ensure the consistency and security of distributed environments, as well as a significant scale-up. It could therefore possibly address the three issues identified for SDVN architectures (Peng et al. 2020): scaling-up through the coordinated distribution of requests between a set of SDVN controllers, the performance level by reducing the load associated with each SDVN controller (latency reduction) and the security level thanks to the cryptographic primitives integrated into this Blockchain technology.

In this chapter, we propose to study the existing solutions coupling SDN and Blockchain approaches to the service of high-performance vehicular networks, in terms of both QoS and security. Indeed, no state-of-the-art article to date seems to have jointly studied the application of SDN and Blockchain technologies in the vehicular environment, although many studies exist on each of these subjects (Bhatia et al. 2019; Al-Heety et al. 2020; Mendiboure et al. 2020b; Mollah et al. 2020).

We first introduce the main characteristics of SDN-controlled C-ITS architectures (SDVN) and Blockchain. We then discuss the main applications considered so far for Blockchain in SDVN environments, both from the point of view of both the objective and the type of application. We also analyze the different evolutions of Blockchain technology that have been proposed in the literature to meet the characteristics of SDVN architectures (both architectural and software evolutions). Lastly, we identify potential research avenues for future research works that would seek to couple SDVN architectures and Blockchain technology.

5.2. Context

This section aims to introduce contextual elements that may be useful for understanding the chapter as a whole. We thus describe the main characteristics of software-defined vehicular architectures, called SDVNs (see section 5.2.1), as well as certain details relating to the benefit and operating principle of Blockchain technology (see section 5.2.2).

5.2.1. *SDN-controlled C-ITS architectures (SDVN)*

In traditional network architectures, the control of data transmission (in particular selection of the communication path) relies on the use of distributed routing protocols (e.g. Border Gateway Protocol, Open Shortest Path First) combined with decision-making at the router itself. These approaches are often called into question over their lack of flexibility: the rules are defined at the moment the elements are deployed (pre-deployment) and modifying them is then complex. This seems particularly problematic in a highly dynamic vehicular environment.

This is why SDN architectures have proposed the integration of a new element in the communication architecture: the SDN controller (Nunes et al. 2014). In this architecture, data transfer and the associated decision-making (control plane) are thus disassociated. The SDN controller, having a global view of the network, decides on the optimal communication paths and transmits this information to the routers, which will then transmit the packets based on these rules (stored at the routers in a table called the flow table) defined by the SDN controller.

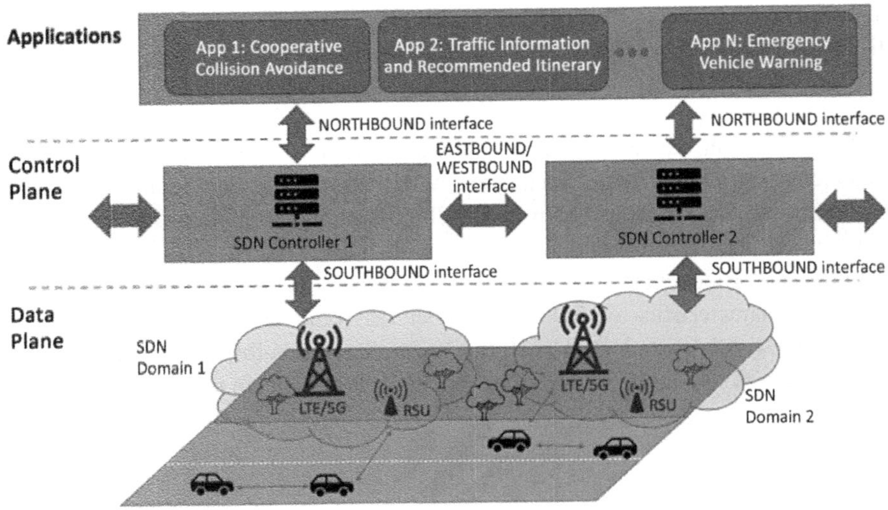

Figure 5.1. *Simplified view of SDVN architecture. For a color version of this figure, see www.iste.co.uk/mendiboure/transportsystems.zip*

Such approaches are today applied to the vehicular environment to enable the optimal management of vehicle mobility, so we speak of SDVN networks (He et al. 2016). The SDVN architecture, based on conventional SDN architecture, is composed of three main planes (see Figure 5.1):

– *The application plane*: this plane/layer integrates the set of applications associated with the SDVN network. These may include the C-ITS services necessary for the smooth operation of connected and autonomous vehicles: emergency management, traffic data transmission, etc. These services are capable of asking the SDN control plane to reserve a specific bandwidth for the application concerned; they are also capable of configuring ports or firewalls, for example. We can add the fact that a concept of priority exists, making it possible to ensure the proper functioning of priority services.

– *The control plane*: this plane generally corresponds to the SDVN controller in charge of managing the data plane. However, in order to ensure a greater scalability, the idea of distributing this control plane both physically and logically is increasingly being considered. Therefore, the SDVN architecture could be based on a control plane integrating not one but a set of SDVN controllers. This control plane transforms the requests of the services of the application layer into rules understandable by the elements composing the SDVN data plane. In parallel, to ensure optimal network management, this control plane also provides information on the status of network devices and the associated communication links.

– *The data plane*: this plane integrates the set of existing network devices/terminals in a SDVN environment: routers, switches, base stations, vehicles, vulnerable users, etc. Certain works break this plane down into two sub-layers: an upper sub-layer corresponding to wired devices, and a lower layer corresponding to wireless devices. Such a distinction aims to optimize the overall functioning of the network by separating the management of these two types of devices.

Beyond these three planes, another element is important in the SDVN architecture: the interfacing between these different planes. This is enabled by two main interfaces: the southbound interface between the SDVN control plane and the SDVN data plane and the northbound interface between the control plane and the application plane. It can also be added that in SDVN architectures based on a distributed control plane (integrating different controllers), a third interface is added to these two main interfaces: the eastbound/westbound interface between the different SDVN controllers.

5.2.2. *Blockchain technology*

Blockchain, and the different implementations associated with this technology (e.g. Bitcoin, Ethereum and Hyperledger Fabric), is the first technological solution to have enabled data storage and transfer without requiring the use of a central control body. This solution, defined for the first time in 2008, therefore represented an important step forward. It enables a set of independent nodes without prior trust to exchange information directly and securely through distributed decision-making. This solution was first applied in vehicular networks to secure inter-vehicle communications. Indeed, in the absence of communication with infrastructure, the establishment of distributed and secure networks (Blockchain) between local nodes appears to represent a promising solution. This solution can more broadly guarantee the integrity of exchanges between different nodes, regardless of their role in the network architecture. The advantages associated with this technology are multiple: the strengthening of security by excluding the intervention of a trusted third party likely to be attacked, the scaling-up enabled by this distributed technology and the establishment of trust in new contexts (absence of Internet connectivity for vehicles, for example).

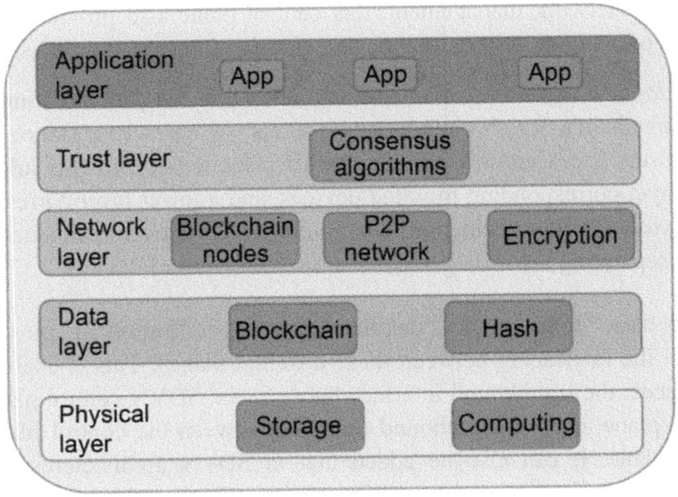

Figure 5.2. *Simplified view of Blockchain architecture. For a color version of this figure, see www.iste.co.uk/mendiboure/transportsystems.zip*

Blockchain technology is generally associated with an architecture composed of five main layers (see Figure 5.2):

– *Physical layer*: the physical device on which the Blockchain node involved in one or more Blockchain networks is deployed. The operation of this Blockchain node involves using the physical resources (storage, computation and communication) of this device.

– *Data layer*: this is the data structure used in Blockchain technology. This solution is based on adding and distributing blocks of data that will be stored by the set of Blockchain nodes. Each block therefore contains a set of information (certificates, transactions, flow rules, etc.) necessary for the proper functioning of the network and secured by Blockchain technology. It may be added that the securing of this chain of blocks is based on the use of hash functions: in each block, the hash of the previous block is stored. This enables a dependency to be created between the set of blocks, thus ensuring that a block cannot be modified without involving the modification of the set of blocks previously transmitted.

– *Network layer*: the layer enabling data exchange between the different Blockchain nodes of a given network. The network layer corresponds to a peer-to-peer network (direct communication between the set of nodes), enabling each entity to receive the set of data exchanged in the Blockchain network. It may also be added that encryption protocols are used in this peer-to-peer network to secure communications.

– *Trust layer*: the main layer for the securing of a Blockchain network. The trust layer integrates the consensus mechanisms that are at the heart of Blockchain technology. These decentralized mechanisms are intended to enable the set of nodes to validate the data that will subsequently be added to the data layer of the associated Blockchain network. This consensus thus ensures that the majority of nodes in the network validate the addition of a new block to the chain of blocks. Different consensus algorithms have already proven themselves, such as proof of work, proof of authority and proof of stake.

– *Application layer*: the set of decentralized applications deployed within a given Blockchain network. These applications aim to automate the processing of data within the Blockchain architecture and to enable interactions with this data. These applications, generally referred to as "smart contracts", correspond to distributed computer programs, designed for Blockchain, and directly deployed in the Blockchain.

5.3. Application of Blockchain to SDVN architectures

Blockchain represents a powerful and promising solution for securing communications in distributed and open environments (see section 5.3). As a result, numerous studies have considered the use of this technology for securing SDVN networks that involve the following: 1) many types of mobile terminals (vehicles, smartphones, surveillance cameras, etc.), with frequent disconnections, and potentially responsible for malicious acts (notably the erroneous transfer of information); 2) a control plane that is potentially distributed and therefore composed of different SDN controllers hosted, or not, in the same physical location; and 3) third-party applications, not designed by the operators of the SDVN network, that may present limitations in terms of security. In this context, the use of Blockchain was considered.

At numerous levels

– *Data layer*: the exchange of information between vehicles is a process that needs to be secure. This mitigates the risks of accidents that may be related to the transmission of erroneous messages: obstacles, lane changes, etc. Within this context, Blockchain thus appears like a potential solution to secure these communications without involving the use of a remote Public Key Infrastructure that is only accessible in the presence of Internet connectivity. Many works, such as Xie et al. (2019), Mershad (2020b) and Vishwakarma et al. (2022), have therefore proposed the application of Blockchain to inter-vehicle exchanges and more broadly between the set of nodes that may be involved in vehicular networks. In this scenario, Blockchain nodes are therefore deployed at the level of the vehicles themselves. The consensus reached between these vehicles aims to ensure that the information transmitted by one of them is real. Two main ideas were considered for this purpose: 1) the real-time crossing of different information sources to ensure the veracity of information before transmitting it to the set of network nodes; and 2) continuous analysis of vehicle behavior to detect malicious actors and prohibit them from transmitting information to other members of the Blockchain network.

– *Control layer*: when a distributed control plane is considered, securing exchanges between SDVN controllers (eastbound/westbound interface) becomes necessary. Blockchain could therefore be used for this purpose, by deploying nodes at the SDVN controller level. This is notably what was proposed by the authors of Mendiboure et al. (2020). Going further, the authors of Zhang et al. (2019) introduced the idea of using Blockchain to ensure the consistency of this distributed control plane. The consensus algorithm aims here to guarantee that the flow rules proposed by each of the SDVN controllers are not likely to be in opposition and to disturb the operation of the network devices. The Blockchain therefore becomes responsible not only for the security but also for the reliability of this control plane. Beyond this first potential application to the control layer, like the solution described in Choudhary et al. (2022), Blockchain could also be used to secure exchanges between SDN controllers and network devices (southbound interface). The authors of this article consider the possibility of storing flow rules emitted by SDN controllers in a Blockchain so that a malicious entity cannot intercept and modify this rule before it is transmitted to the network device. Similarly, the information reported by the network devices, relating to their level of performance/statistics, could also be transmitted via this communication channel.

– *Application layer*: The applications deployed within the application layer of SDVN architecture are likely to be third parties and as such are neither guaranteed nor secured by the SDVN network operator. It therefore appears essential to control the behavior of these applications to ensure that they do not attempt to compromise the Blockchain network. Within this context, the authors of Mendiboure et al. (2018) proposed to define a tool intended for the continuous analysis of the actions of SDVN applications in operation. This tool integrates three main components: 1) authentication and access control for SDVN applications; 2) relaying of information provided by the SDVN controllers, relating to the requests/actual usage of resources of these applications; and 3) adaptation of SDVN application rights based on analysis of this data. The use of the Blockchain in this context aims to enable the decentralization and security of this tool and its provision to the set of controllers likely to be deployed at the SDVN control plane level.

With multiple objectives

– *Security*: securing communications is the primary objective associated with the use of Blockchain technology in SDVN networks. This securing can be applied both to internal communications at each layer (Data and Control in particular) and to cross-layer communications (interfaces of any kind). In this context, Blockchain aims to authenticate the different entities involved in the network as well as to ensure the confidentiality, availability and integrity of the data shared between the different nodes. It therefore replaces traditional solutions by offering the advantage that it can be deployed in both the wired part and the wireless part.

– *Data reliability*: enhancing the reliability of communications between SDVN entities constitutes a second objective related to the application of Blockchain technology (Li et al. 2020). This idea has mainly been applied to inter-vehicle/inter-terminal communications and to the control of SDVN applications. Two main strategies have been put in place to ensure this data reliability: 1) real-time controlling of information/requests emitted both by applications and data-layer entities (including cross-checking with other information sources); and 2) controlling the behavior of these entities over time by recording and evaluating their past actions (definition of a trust index for each entity). In this context, Blockchain offers significant traceability of data, which represents a definite advantage for their analysis.

– *Consistency*: consistency is an idea that has mainly been applied to the control plane of SDVN architecture. The objective here is to ensure that decisions taken by the various controllers, even when they are not malicious, do not conflict with other decisions taken previously or in parallel by other SDVN controllers. The absence of consistency could have a variety of consequences: multiple rules defined for the same data transfer, allocation of the same resources to two applications, etc. Within this context, the Blockchain and the associated consensus algorithms ensure that such scenarios cannot be encountered: the set previously transmitted rules are systematically analyzed so that a conflicting rule cannot be integrated into the Blockchain. Similarly, two contradictory rules are systematically rejected. Blockchain therefore constitutes an interesting approach to guaranteeing consistency in a distributed environment.

– *Scaling-up*: as noted by the authors of Mendiboure et al. (2020a), the scaling-up offered by this technology is considered by many authors as an additional advantage justifying the application of Blockchain. Therefore, their idea is that ultimately SDVN networks will be deployed on a large scale and will involve the management of a significant number of nodes, which traditional approaches will not be able to achieve in real time. The Blockchain based on a distributed architecture, having demonstrated, according to them, its ability to scale-up, therefore represents a long-term solution for ensuring the security, reliability and consistency of SDVN networks.

Also different strategies

– *Integration into SDN architecture*: this is the approach that is most commonly considered in the literature. We can mention, in particular, the solutions defined by the authors Yahiatene et al. (2019) and Zhang et al. (2019). In this context, SDVN nodes and Blockchain nodes are directly associated. Blockchain nodes are thus deployed at the mobile terminal level, a network device or an SDVN controller and their operations are interdependent. Therefore, the Blockchain network aims here to directly optimize the operation of the associated network nodes and offers the possibility of deployment at various levels (in both the wired part and the wireless part).

– *Deployment in parallel with SDN architecture*: in this type of approach, Blockchain architectures are considered as a third-party application that the SDVN network will rely on to perform certain functions, particularly in relation to data securing and identity control. Without a basis for comparison, the suitability of such approaches appears complex to evaluate in relation to solutions that consider the integration of Blockchain into the SDVN architecture. Today, nevertheless, this idea of a distributed Public Key Infrastructure, based on Blockchain, is put forward in many works, including outside of the vehicular environment (Orman 2018; Yakubov et al. 2018) owing to its ability to strengthen the architecture's overall security. Although limited by its applications (wired part only), such an approach would also appear to be capable of responding to certain use cases and is worth exploring further.

5.4. Optimization of Blockchain technology for SDVN architectures

Due to the numerous works linking Blockchain technology and SDVN architectures, some researchers have focused on adapting Blockchain technology for this highly mobile environment that vehicular networks represent. In this section, we highlight the main improvements proposed in the literature both in terms of architecture (see section 5.4.1) and mechanisms (see section 5.4.2).

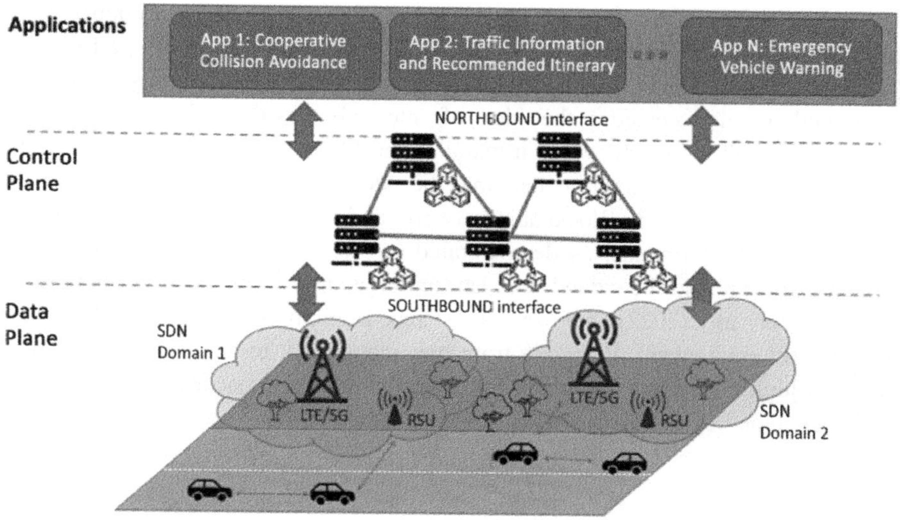

Figure 5.3. *Example of advanced Blockchain architecture intended for SDVN networks. For a color version of this figure, see www.iste.co.uk/mendiboure/transportsystems.zip*

5.4.1. *New architectures*

The main idea of the works focusing on the adaptation of Blockchain architecture to SDVN networks is to consider that SDVN networks will ultimately be deployed on a large scale and will require a large scalability capacity (management of several million SDVN nodes). The architectural solutions proposed by these authors therefore aim to promote the scaling-up of the underlying Blockchain architecture, considered as the basis of SDVN networks. These authors position themselves within a context in which Blockchain nodes are integrated into SDVN equipment and focus on the network layer of the Blockchain (see Figure 5.2).

The approaches described in articles focusing on this subject (Lin et al. 2020; Mendiboure et al. 2020a; Mershad et al. 2020a; Alouache et al. 2022) are very similar: they propose to consider not a single global Blockchain network but a set of geographically distributed Blockchain subnetworks (see Figure 5.3). The main idea is to guarantee low latency between neighboring Blockchain nodes and therefore requiring regular exchanges. This can be applied both to SDVN controllers and at the data layer level.

Each subnetwork manages a limited number of SDVN devices (or controllers or applications), which limits the associated load and thus ensures high performance (latency and throughput). To manage exchanges between the different subnetworks, these authors also consider the idea of inter-subnetwork nodes (edge nodes), involved in different subnetworks in parallel and likely to verify the reliability and consistency of any information that may be added. These nodes thus ensure the overall consistency of the Blockchain network, and being limited in number offer a high level of performance, as demonstrated by the authors of Mendiboure et al. (2020a). It can also be added that an initial solution has been introduced in Alouache et al. (2022) for real-time reconfiguration of these Blockchain subnetworks (movement of a node from one subnetwork to another, for example). This has highlighted the importance of such dynamic approaches and the need to further develop them in the future.

5.4.2. *New mechanisms*

To adapt the Blockchain to SDVN architectures, beyond the evolution of the underlying architecture (see section 5.4.1), another solution proposed by many authors in the literature consists of optimizing other Blockchain layers, notably the data layer and the trust layer (see Figure 5.2). The idea is therefore for these authors to propose new mechanisms intended to optimize the functioning of these layers.

At the data layer level, the main proposal defined in the literature was introduced by the authors of Busygin et al. (2018). Under normal operation, every node must have the same block initiating its Blockchain (Genesis Block). This implies long recovery times for the set of blocks when a new node wishes to join an existing Blockchain network. Yet, this appears unlikely in a highly mobile vehicular context that can at any moment require the integration of new vehicles arriving in a given area. This is what led the authors to introduce the idea of a floating genesis block, in order to enable a new node to instantly join an existing Blockchain network at any moment, while guaranteeing the security of this network. This new mechanism therefore guarantees a much higher level of flexibility for Blockchain solutions and application to a new, wider use case in the SDVN environment.

At the trust layer level, works such as Qiu et al. (2018), Zhang et al. (2019), Zhao et al. (2020a), Ren et al. (2021) and Gao et al. (2022) can be mentioned in particular. The main idea of these works, applied to different contexts (SDVN control layer and SDVN data layer), is to optimize the consensus algorithms used by the Blockchain network. This objective is achieved through the application of different methods, including 1) the selection of a limited number of Blockchain nodes in charge of data validation and 2) the selection of an optimal number of nodes involved in the block validation process, by relying on artificial intelligence tools and the measuring of the performance level of these nodes (computing capacity, for example). These works have demonstrated the potential benefit in defining solutions at the trust layer level, notably in terms of throughput and latency.

The definition of solutions, combining both architectural and protocol could also be considered an interesting prospect in this context.

5.5. Future research avenues

This section introduces research avenues that could be considered within the framework of works aiming to develop Blockchain-based SDVN networks. The optimal positioning of Blockchain nodes (see section 5.5.1), reducing energy consumption (see section 5.5.2), the integration of artificial intelligence tools into the Blockchain (see section 5.5.3), a more complete integration between SDN and Blockchain (see section 5.5.4) could in particular offer interesting prospects.

5.5.1. *Optimal positioning of Blockchain nodes*

In the works presented in section 5.3, different deployments of Blockchain nodes are considered: at the terminal device level, at network infrastructure level and at SDVN controller level. However, the proposed architectures apply to specific use cases (data layer security, controller consistency, etc.) and none of the architectures presented in the literature aims to deal jointly with these different applications. As a result, it appears necessary today to conduct new studies that will lead to the global definition of a Blockchain architecture for SDVN networks. This study should notably help us to identify the different potential use cases of this technology as well as the implications in terms of the architecture of these solutions. It can also be added that such implications will involve the establishment of an emulation environment for Blockchain technology in vehicular networks, an emulator that does not exist for the time being and could be of definite interest to many current works.

5.5.2. *Energy consumption reduction*

Energy consumption is a problem frequently associated with Blockchain technology (Zhang et al. 2020). This idea is generally associated with the consensus algorithm commonly used in these Blockchain networks (Proof of Work, first algorithm with proven efficiency) involving continuous computation at the network equipment level. Nevertheless, other algorithms have also been proposed for this technology that guarantee much lower levels of energy consumption. It therefore seems important today to design new algorithms, adapted to the vehicular environment, and guaranteeing limited energy consumption. This has been achieved to a certain extent but applied to specific contexts (vehicle-to-vehicle data exchange, for example). Beyond this problem in relation to consensus algorithms, the storage of the Blockchain itself seems likely to represent a problem in the long term. Indeed, the storage of ever-larger data volumes, at the level of a significant number of nodes, also seems likely to increase the overall consumption of the Blockchain network. As a result, it would seem appropriate today to propose an adaptive Blockchain network (Mendiboure et al. 2021) likely to limit the amount of data stored by each of the Blockchain nodes and optimize exchanges between these nodes. This area therefore presents many research avenues yet to be explored.

5.5.3. *Integration of AI and Blockchain*

The integration of artificial intelligence (AI) tools such as machine learning, deep learning or federated learning into Blockchain architecture would today seem to constitute an important step forward. Indeed, the deployment of AI tools within an environment guaranteeing a high level of security and traceability of stored data would seem likely to increase the level of trust associated with these tools. Many works therefore aim to combine Blockchain and AI, notably by considering federated learning approaches, so the concept seems to be directly applicable to the Blockchain context. In this context, smart contracts, applications of Blockchain, are used to interface the different processes. In the context of SDVN networks, the integration of AI into the Blockchain architecture would seem to open up many prospects whether at the data layer level (reliability of exchanges thanks to AI), the control layer level (consistency of the control plane) or the application layer level (Smart Application Control). It would therefore seem worthwhile at present to reflect on the definition of new solutions in this context, as well as on the application of solutions proposed in other environments.

5.5.4. *A more complete integration between SDN and Blockchain*

The works presented in this chapter focus on the integration of Blockchain into the SDVN architecture and on the optimization of this architecture, particularly in terms of security, thanks to this technology. However, many works in the literature, such as Rahman et al. (2022), seek, conversely, to apply the SDN approach with a view to optimizing the functioning of Blockchain technology. The idea is thus to optimize the information broadcasting process and maximize the performance of Blockchain technology, an element that may appear to be essential, including in the context of SDVNs. It would therefore seem appropriate to reflect more broadly on how to enable the full integration of these two technologies by applying the advantages of Blockchain technology to SDVN architecture and vice versa. This could eventually lead to the introduction of architectures presenting high levels of performance (latency, use of available network resources and flexibility) and security. This has already been demonstrated in various works that have highlighted how environments integrating Blockchain and SDN could, for example, enable the implementation of Intrusion Detection Systems that are more efficient than existing systems (Aliyu et al. 2021) or indeed open the door to the definition of new types of services, such as those relating to the management of future electrical systems (Chaudhary et al. 2019; Jindal et al. 2019).

5.6. Conclusion

Blockchain technology could occupy an important place in software-defined vehicular networks (SDVNs) in future.

This is why in this chapter, after briefly describing SDN and Blockchain technologies, we studied the different existing applications of Blockchain within the context of SDVN. In particular, we highlighted: 1) the different levels to which this technology could be applied (data layer, control layer and application layer); 2) the different associated objectives (communication security, data reliability, consistency of the control plane and scaling-up of architecture); and 3) the different strategies considered (integration into the SDVN architecture, and use of Blockchain as a third-party tool).

We also highlighted some of the optimizations of Blockchain technology that have been proposed in the literature to ensure a better integration of this technology into the SDVN architecture. These optimizations are notably in terms of architecture (introduction of local Blockchain subnetworks) and in terms of mechanisms (optimization of consensus algorithms).

Lastly, we listed several research avenues that could be considered for future works relating to Blockchain and SDVN networks, notably the definition of an optimal positioning for Blockchain nodes, the reduction of Blockchain network energy consumption, the integration of artificial intelligence tools into Blockchain architecture and better integration between Blockchain and SDN.

5.7. References

Al-Heety, O.S., Zakaria, Z., Ismail, M., Shakir, M.M., Alani, S., Alsariera, H. (2020). A comprehensive survey: Benefits, services, recent works, challenges, security, and use cases for sdn-vanet. *IEEE Access*, 8, 91028–91047.

Aliyu, I., Feliciano, M.C., Van Engelenburg, S., Kim, D.O., Lim, C.G. (2021). A blockchain-based federated forest for SDN-enabled in-vehicle network intrusion detection system. *IEEE Access*, 9, 102593–102608.

Alouache, L., Sylla, T., Mendiboure, L., Aniss, H. (2022). A fuzzy approach for load balancing in blockchain-based software defined vehicular networks. In *2022 18th International Conference on Wireless and Mobile Computing, Networking and Communications (WiMob)*, 235–242. IEEE, Thessaloniki.

Benzekki, K., El Fergougui, A., Elbelrhiti Elalaoui, A. (2016). Software-defined networking (SDN): A survey. *Security and Communication Networks*, 9(18), 5803–5833.

Bhatia, J., Modi, Y., Tanwar, S., Bhavsar, M. (2019). Software defined vehicular networks: A comprehensive review. *International Journal of Communication Systems*, 32(12), e4005.

Busygin, A., Konoplev, A., Kalinin, M., Zegzhda, D. (2018). Floating genesis block enhancement for blockchain based routing between connected vehicles and software-defined VANET security services. In *Proceedings of the 11th International Conference on Security of Information and Networks*, 1–2.

Camacho, F., Cárdenas, C., Muñoz, D. (2018). Emerging technologies and research challenges for intelligent transportation systems: 5G, HetNets, and SDN. *International Journal on Interactive Design and Manufacturing (IJIDeM)*, 12(1), 327–335.

Chaudhary, R., Jindal, A., Aujla, G.S., Aggarwal, S., Kumar, N., Choo, K.K.R. (2019). BEST: Blockchain-based secure energy trading in SDN-enabled intelligent transportation system. *Computers & Security*, 85, 288–299.

Choudhary, S. and Dorle, S. (2022). Secured SDN based blockchain: An architecture to improve the security of VANET. *International Journal of Electrical and Computer Engineering Systems*, 13(2), 145–153.

Gao, L., Wu, C., Du, Z., Yoshinaga, T., Zhong, L., Liu, F., Ji, Y. (2022). Toward efficient blockchain for the internet of vehicles with hierarchical blockchain resource scheduling. *Electronics*, 11(5), 832.

He, Z., Cao, J., Liu, X. (2016). SDVN: Enabling rapid network innovation for heterogeneous vehicular communication. *IEEE Network*, 30(4), 10–15.

Jindal, A., Aujla, G.S., Kumar, N. (2019). SURVIVOR: A blockchain based edge-as-a-service framework for secure energy trading in SDN-enabled vehicle-to-grid environment. *Computer Networks*, 153, 36–48.

Li, W., Meng, W., Liu, Z., & Au, M.H. (2020). Towards blockchain-based software-defined networking: Security challenges and solutions. *IEICE Transactions on Information and Systems*, 103(2), 196–203.

Lin, H., Garg, S., Hu, J., Kaddoum, G., Peng, M., Hossain, M.S. (2020). Blockchain and deep reinforcement learning empowered spatial crowdsourcing in software-defined internet of vehicles. *IEEE Transactions on Intelligent Transportation Systems*, 22(6), 3755–3764.

Mendiboure, L., Chalouf, M.A., Krief, F. (2018). Towards a blockchain-based SD-IoV for applications authentication and trust management. In *International Conference on Internet of Vehicles*, 265–277. Springer, Cham.

Mendiboure, L., Chalouf, M.A., Krief, F. (2020a). A scalable blockchain-based approach for authentication and access control in software defined vehicular networks. In *2020 29th International Conference on Computer Communications and Networks (ICCCN)*, 1–11. IEEE, Honolulu.

Mendiboure, L., Chalouf, M.A., Krief, F. (2020b). Survey on blockchain-based applications in internet of vehicles. *Computers & Electrical Engineering*, 84, 106646.

Mendiboure, L., Maaloul, S., Aniss, H. (2021). Towards an adaptive blockchain for internet of vehicles. In *Communication Technologies for Vehicles: 16th International Workshop, Nets4Cars/Nets4Trains/Nets4Aircraft 2021*, 16–17 November, 37–48. Springer International Publishing, Madrid.

Mershad, K. (2020). SURFER: A secure SDN-based routing protocol for internet of vehicles. *IEEE Internet of Things Journal*, 8(9), 7407–7422.

Mershad, K. and Said, B. (2020). A blockchain model for secure communications in Internet of vehicles. In *2020 IEEE/ACS 17th International Conference on Computer Systems and Applications (AICCSA)*, 1–6. IEEE, Antalya.

Mollah, M.B., Zhao, J., Niyato, D., Guan, Y.L., Yuen, C., Sun, S., Lam, K.Y., Koh, L.H. (2020). Blockchain for the internet of vehicles towards intelligent transportation systems: A survey. *IEEE Internet of Things Journal*, 8(6), 4157–4185.

Nunes, B.A.A., Mendonca, M., Nguyen, X.N., Obraczka, K., Turletti, T. (2014). A survey of software-defined networking: Past, present, and future of programmable networks. *IEEE Communications Surveys & Tutorials*, 16(3), 1617–1634.

Orman, H. (2018). Blockchain: The emperors new PKI? *IEEE Internet Computing*, 22(2), 23–28.

Peng, C., Wu, C., Gao, L., Zhang, J., Alvin Yau, K.L., Ji, Y. (2020). Blockchain for vehicular Internet of things: Recent advances and open issues. *Sensors*, 20(18), 5079.

Qiu, C., Yu, F.R., Xu, F., Yao, H., Zhao, C. (2018). Blockchain-based distributed software-defined vehicular networks via deep q-learning. In *Proceedings of the 8th ACM Symposium on Design and Analysis of Intelligent Vehicular Networks and Applications*, 8–14. Association for Computing Machinery, Montreal.

Rahman, A., Montieri, A., Kundu, D., Karim, M.R., Islam, M.J., Umme, S., Nascita, A., Pescapé, A. (2022). On the integration of blockchain and SDN: Overview, applications, and future perspectives. *Journal of Network and Systems Management*, 30(4), 73.

Ren, Y., Chen, X., Guo, S., Guo, S., Xiong, A. (2021). Blockchain-based VEC network trust management: A DRL algorithm for vehicular service offloading and migration. *IEEE Transactions on Vehicular Technology*, 70(8), 8148–8160.

Sjoberg, K., Andres, P., Buburuzan, T., Brakemeier, A. (2017). Cooperative intelligent transport systems in Europe: Current deployment status and outlook. *IEEE Vehicular Technology Magazine*, 12(2), 89–97.

Vishwakarma, L., Nahar, A., Das, D. (2022). LBSV: Lightweight blockchain security protocol for secure storage and communication in SDN-enabled IoV. *IEEE Transactions on Vehicular Technology*, 71(6), 5983–5994.

Xiao, Y., Liu, Y., Li, T. (2020). Edge computing and blockchain for quick fake news detection in IoV. *Sensors*, 20(16), 4360.

Xie, L., Ding, Y., Yang, H., Wang, X. (2019). Blockchain-based secure and trustworthy internet of things in SDN-enabled 5G-VANETs. *IEEE Access*, 7, 56656–56666.

Yahiatene, Y., Rachedi, A., Riahla, M.A., Menacer, D.E., Nait-Abdesselam, F. (2019). A blockchain-based framework to secure vehicular social networks. *Transactions on Emerging Telecommunications Technologies*, 30(8), e3650.

Yakubov, A., Shbair, W., Wallbom, A., Sanda, D. (2018). A blockchain-based PKI management framework. In *The First IEEE/IFIP International Workshop on Managing and Managed by Blockchain (Man2Block) colocated with IEEE/IFIP NOMS 2018*, 23–27 April, 1–6. IEEE, Tapei.

Zhang, R. and Chan, W.K.V. (2020). Evaluation of energy consumption in block-chains with proof of work and proof of stake. *Journal of Physics: Conference Series*, 1584(1), 012023.

Zhang, D., Yu, F.R., Yang, R. (2019). Blockchain-based distributed software-defined vehicular networks: A dueling deep Q learning approach. *IEEE Transactions on Cognitive Communications and Networking*, 5(4), 1086–1100.

Zhao, L., Li, J., Al-Dubai, A., Zomaya, A.Y., Min, G., Hawbani, A. (2020a). Routing schemes in software-defined vehicular networks: Design, open issues and challenges. *IEEE Intelligent Transportation Systems Magazine*, 13(4), 217–226.

Zhao, N., Wu, H., Zhao, X. (2020b). Consortium blockchain-based secure software defined vehicular network. *Mobile Networks and Applications*, 25(1), 314–327.

Zheng, Z., Xie, S., Dai, H.N., Chen, X., Wang, H. (2018). Blockchain challenges and opportunities: A survey. *International Journal of Web and Grid Services*, 14(4), 352–375.

6

Application of Network Slicing in C-ITS Systems

Abdennour RACHEDI, Toufik AHMED and Mohamed MOSBAH
LaBRI, Bordeaux INP, France

6.1. Introduction

Cooperative Intelligent Transport Systems (C-ITS) have generated a lot of interest over the past decade. It involves a smart network whose connected components (vehicles, pedestrians and RSU road infrastructure) share data. A C-ITS aims to improve road safety, optimize traffic, reduce emissions and improve the driving experience.

Vehicle-to-Everything (V2X) communications play an essential role in facilitating data exchange to enable a variety of applications. These applications include cooperative autonomous and remote driving, traffic efficiency and infotainment services, which includes advertisements, as well as software updates and diagnostics. The diversity of vehicular applications calls on several requirements in terms of connectivity. In particular, ultra-low latency (less than 10 milliseconds), ultra-high reliability (almost 100%) and a high data rate (in the Gbps range) present significant challenges (Campolo et al. 2021, pp. 239–257). The high mobility and density of vehicles significantly reduce their realization.

The fifth generation of cellular networks (5G) is expected to support enhanced mobile broadband communication services, as well as a multitude of new use cases in vertical industries. 5G systems integrate considerable technological advances in communication, automation, detection and positioning, which are set to revolutionize the future of transportation and quality of life.

Network slicing is one of the decisive technological factors for 5G networks. The idea behind it is to provide a dedicated, end-to-end virtual network for each use case, with resources such as bandwidth and processing power allocated according to the specific needs of that use case. Network slicing is a significant aspect of 5G networks, as it enables the creation of customized and efficient networks for different use cases and services. It also enables more efficient use of network resources and improved network performance. For example, a Network Slice (NS) for an autonomous-driving car would have low latency and high reliability to ensure smooth and safe operation. Another slice for a virtual-reality gaming service would have high bandwidth and low jitter to provide a seamless gaming experience.

We begin this chapter by presenting several aspects of the literature concerning Vehicle-to-Everything (V2X) communications and Network slicing. Next, we will discuss some recent research on Network slicing for V2X communication technologies based on cellular technologies. Lastly, we present our main contribution, an end-to-end network slicing architecture for vehicular communications based on the European ITS-G5 standard. The proposed approach to slicing aims to improve end-to-end (E2E) latency for highly priority traffic and user groups.

6.2. Vehicle-to-everything (V2X) communications

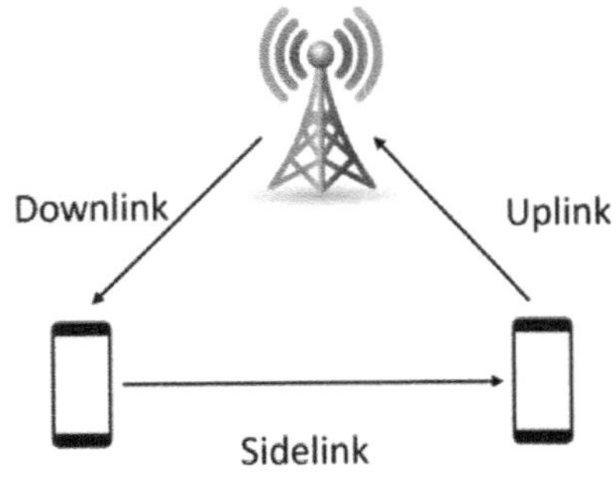

Figure 6.1. *V2X communication types. For a color version of this figure, see www.iste.co.uk/mendiboure/transportsystems.zip*

V2X refers to communication between vehicles and other entities, called ITS stations (ITS-S), such as other vehicles, Road-Side Units (RSU) and devices. Within the V2X context, Sidelink (SL) communication is used for direct communication between ITS stations, while Uplink/Downlink communication is used for communications between an ITS station and a base station of the cellular network. Sidelink communication is often used for sharing short-range, real-time data, such as speed, positioning and braking status. Uplink/Downlink is used for longer-range applications such as Internet access, telephone calls or sending data to the cloud.

Several V2X communication scenarios have been developed and tested, including the following (Hasan et al. 2020, pp. 693–713):

– Vehicle-to-Vehicle (V2V) communication: vehicles share information such as speed, positioning and braking status with nearby vehicles. This can improve road safety by enabling vehicles to alert each other to potential hazards and coordinate their movements.

– Vehicle-to-Pedestrian (V2P) communication: vehicles communicate with pedestrians and vulnerable road users, such as cyclists, to improve road safety and provide them with relevant information.

– Vehicle-to-Infrastructure (V2I) communication: vehicles communicate with traffic lights, road signs, Road-Side Units (RSU) and other infrastructure to exchange real-time information on traffic conditions and optimize their routings.

– Vehicle-to-Network (V2N) communication: vehicles connect to a cellular network to access the Internet and other services. This can enable drivers and passengers to access navigation, entertainment and other services.

– Vehicle-to-Cloud (V2C) communication: enables vehicles to communicate with cloud-based services to access real-time data and perform advanced analytics.

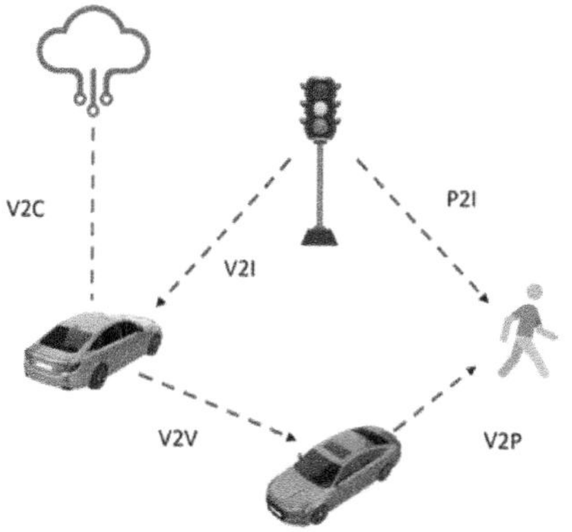

Figure 6.2. *V2X communications. For a color version of this figure, see www.iste.co.uk/mendiboure/transportsystems.zip*

6.3. Presentation of V2X technologies

In 1990, the United States Federal Communications Commission (FCC) allocated 75 MHz of licensed spectrum at 5.9 GHz (5.885–5.925 GHz) for Intelligent Transportation System (ITS) applications.

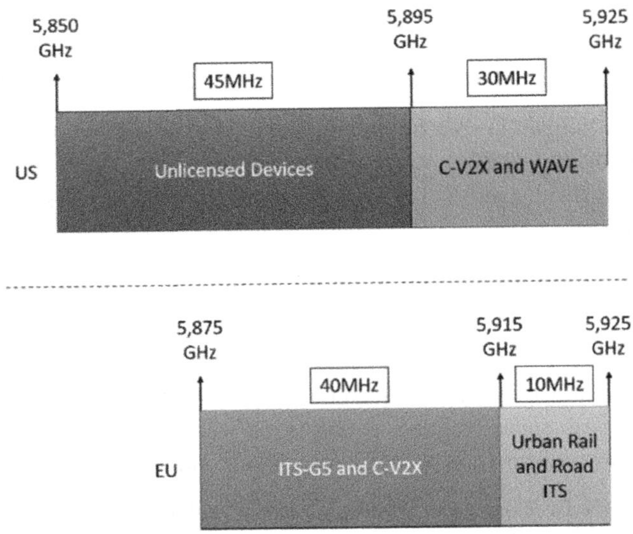

Figure 6.3. *Choice of spectrum in the United States and in the EU (Canis and Gallagher 2019). For a color version of this figure, see www.iste.co.uk/mendiboure/transportsystems.zip*

There are two categories of V2X communication technologies: Wi-Fi technologies and cellular technologies. The first group is based on the IEEE 802.11p standard introduced in 2009, which is an extension of the 802.11a (Wi-Fi) standard. This standard was developed for vehicle-to-vehicle communication and ensures safety applications and better traffic management with the following requirements (Kambiré et al. 2020):

– speeds of up to 200 kph;

– response time of approximately 100 milliseconds;

– a communication range of up to 1,000 m.

In the following section, we will present the European technology based on ITS-G5 Wi-Fi and 4G/5G-V2X cellular technologies.

6.3.1. *ITS-G5*

ITS-G5 is the European baseline technology for V2X communications in vehicular ad hoc networks (VANET). This technology was developed by the European Telecommunications Standards Institute (ETSI), based on the IEEE 802.11p standard, and operates on the 5.9 GHz frequency band (ERTICO 2015). It enables vehicles to share real-time information, such as their speed, positioning and braking status, which can be used to improve traffic safety and efficiency. The ITS-G5 standard is considered a key tool for the deployment of Cooperative Intelligent Transport Systems (C-ITS) and autonomous vehicles.

ETSI has specified a set of message formats for various potential applications, notably Cooperative Awareness Messages (CAM) and Decentralized Event Notification Messages (DENM) (Bellache et al. 2017). CAMs are messages transmitted periodically by ITS stations (ITS-S) to share information about their presence, position and speed with neighboring ITS stations. DENMs are messages exchanged when an event is detected in the vicinity of an ITS-S, such as accidents and road warnings. These messages are transmitted by direct link in broadcast mode.

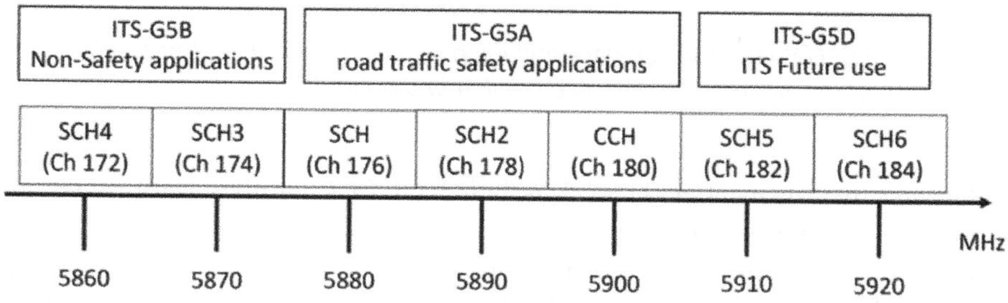

Figure 6.4. *Frequency allocation for ETSI ITS-G5 in the EU*

6.3.1.1. *Decentralized congestion control (DCC)*

In high-density scenarios, vehicles generate massive traffic, particularly CAM messages, leading to congestion problems in the transmission channels (Bellache et al. 2017). In such scenarios, a cross-layer congestion control mechanism is needed to alleviate the channel load. ETSI has developed the decentralized congestion control (DCC) mechanism, a set of algorithms that dynamically adjust and manage traffic parameters to improve the quality of data exchanged on the channel (ETSI 2011). DCC ensures network stability, channel load limitation and equitable resource allocation. Figure 6.5 illustrates the ITS-G5 protocol stack as well as the corresponding DCC functionalities (ETSI 2012).

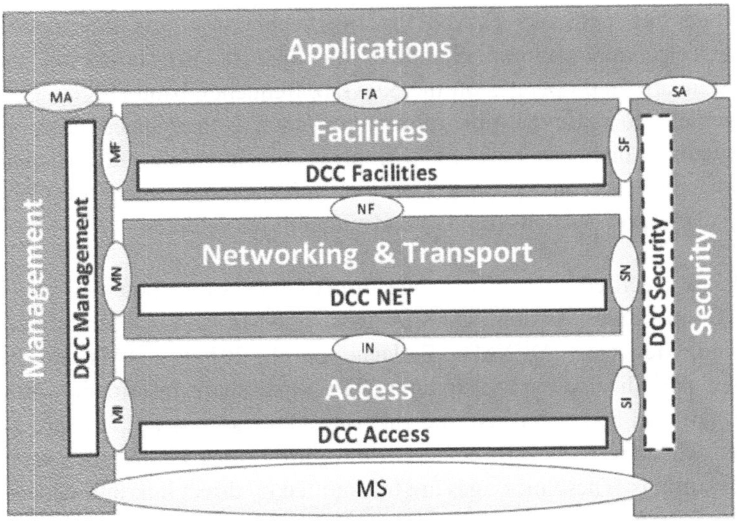

Figure 6.5. *ITS-G5 protocol stack including DCC functionalities. For a color version of this figure, see www.iste.co.uk/mendiboure/transportsystems.zip*

Resource management in ITS-G5 is based on DCC profiles (DP), a set of transmission parameters used to identify and control a traffic flow (such as CAM, DENM, etc.). This constitutes the basis for traffic discrimination at the access, network and transport layers (ETSI 2012). Therefore, depending on the DP of a message and the state of the transmission channel, a DCC algorithm (or several algorithms) can be applied to a traffic flow to reduce channel congestion.

DCC profiles are mapped to enhanced distributed channel access (EDCA) traffic classes (TC), which are used to classify and rank data traffic at the network and transport layer level. The MAC layer can rank the data traffic using four different queues. The highest-priority TC has fewer listening periods (i.e. the Arbitration Inter-Frame Space/AIFS parameters) and contention window (CW) (ETSI 2013), both of which enable faster priority access to the channel. These parameters will later serve as the basis for our slicing mechanism implemented for ITS-G5. Table 6.1 summarizes the AIFS and CW default values for the different access categories (Wu et al. 2006).

AC	CWmin	CWmax	AIFS
Voice (VO)	3	7	58 μ
Video (VI)	7	15	71 μ
Best effort (BE)	15	1,023	110 μ
Background (BK)	15	1,023	149 μ

Table 6.1. *Default values for AIFS and CW at the access layer level*

6.3.1.2. *ITS network architecture*

Figure 6.6 represents a global abstraction of the architecture of the ITS network (ETSI 2014) composed of an ad hoc network, an access network and a core network. The ITS network architecture comprises network elements connecting the ITS stations, such as the vehicle's ITS-S (referred to as the On-Board Unit/OBU), personal ITS-S, Road Side (RSU) ITS-S and the core ITS station. These networks are managed by a road operator, a telecoms operator or a national authority. The ad hoc network uses ITS-G5 technology to provide direct communications between ITS stations. The access network is a dedicated network that collects traffic from different ad hoc networks and provides access to specific ITS services and applications. It aims to interconnect RSUs and to enable vehicle-to-vehicle communications via road infrastructure. The core network is a high-speed network of routers and switches used to redirect and converge network traffic to core ITS infrastructure that hosts the V2X application servers. This network also interconnects with the cloud infrastructure, automotive Original Equipment Manufacturers (OEM), third-party networks and public networks such as the Internet.

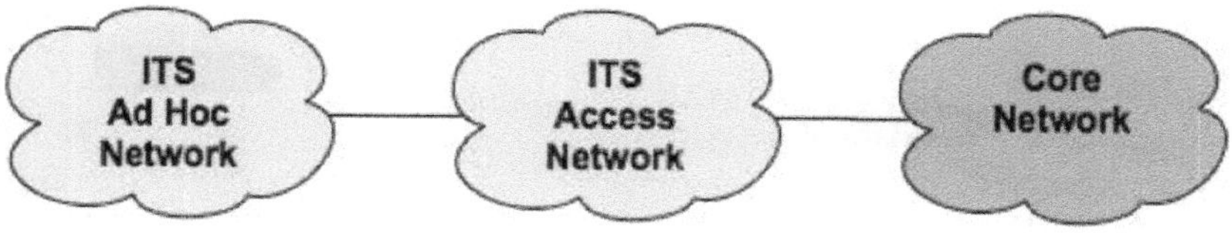

Figure 6.6. *High-level architecture of the ITS network*

6.3.2. *LTE-V2X*

To align with the expansion of vehicular communication and the evolution of use cases, 3GPP defined the V2X standard based on LTE as the underlying technology in 2017 with the specifications of version 14 (Rel-14). It is generally referred to as "cellular V2X" (C-V2X) or LTE-V2X to differentiate it from V2X technology based on the 802.11p standard. Several improvements to V2X communications were subsequently made by 3GPP to support advanced Cooperative, Connected and Automated Mobility (CCAM) scenarios with 5G specifications in its Rel-15 and Rel-16, known by the name "New Radio V2X" (NR-V2X).

6.3.2.1. *C-V2X operating modes*

There are two operating modes for C-V2X communication (Hasan et al. 2020, pp. 693–713), as shown in Figure 6.7:

– V2X communication on the PC5 interface (Sidelink): the PC5 interface directly connects ITS stations and enables the direct exchange of messages between ITS stations.

– V2X communication via the LTE-Uu interface ("Uplink/Downlink"): the LTE-Uu interface connects ITS stations to the eNB (E-UTRAN NodeB), which acts as a base station in LTE networks.

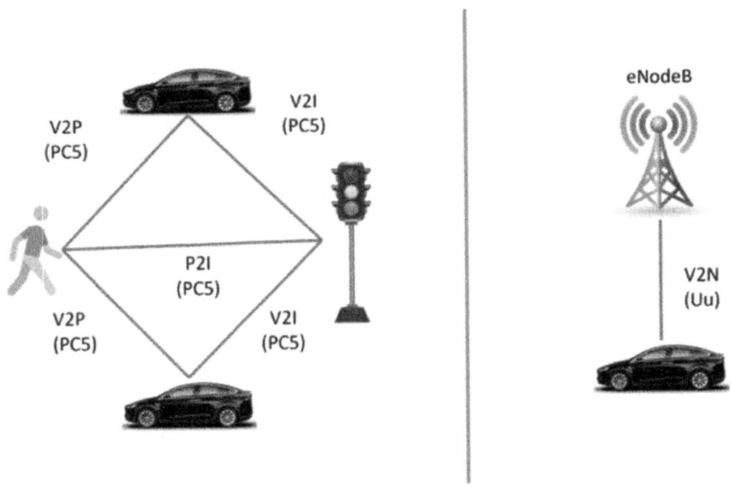

Figure 6.7. *C-V2X operating modes. For a color version of this figure, see www.iste.co.uk/mendiboure/transportsystems.zip*

6.3.2.2. *LTE-V2X modes of communication*

The LTE-V2X PC5 interface uses two modes of communication: mode 3 and mode 4 (Figure 6.8). In mode 3, the cellular network allocates and manages the radio resources used by vehicles for their communications; this mode therefore requires being within the coverage of a base station. In mode 4, the vehicles autonomously select radio resources with (or without) the help of the network for their V2V direct communications. Mode 4 uses congestion-control distributed scheduling. Mode 4 is considered the V2V-based mode, since road safety applications cannot depend on cellular coverage availability.

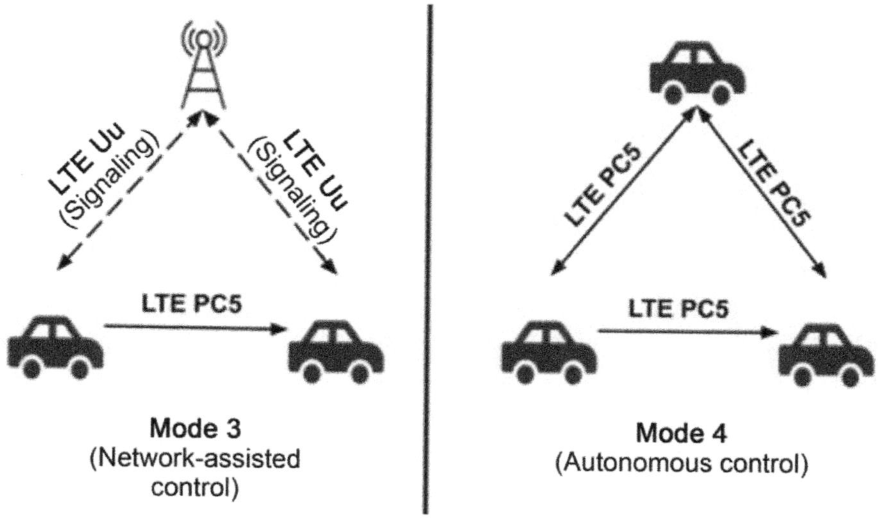

Figure 6.8. *LTE-V2X communication. Mode 3: in coverage and mode 4: out of coverage (Mendiboure 2020)*

6.3.3. *5G-V2X*

This is also called New Radio V2X (NR-V2X). NR-V2X Rel-16 was the first NR Sidelink standard focused on improving V2X reliability, latency, capacity and flexibility. This version includes advanced use cases, such as platooning, extended sensors, advanced driving, remote driving and road safety. Different broadcast types, such as unicast and multicast (groupcast), are also specified for V2X communication in addition to the broadcast. Furthermore, the NR-V2X in version 16 supports the HARQ (Hybrid Automatic Receive reQuest) function to improve the reliability of SL communication (Harounabadi et al. 2021, pp. 12–21).

Two frequency bands are specified as the V2X operating bands on the PC5 interface for SL NR in Rel-16 (ETSI 2020): the 5.9 GHz band (n47) and the 2.5 GHz band (n38). The 2.5 GHz band is used exclusively in certain regions for NR-V2X. The two frequency bands (5.9 and 2.5 GHz) belong to Frequency Range 1 (FR1), which goes from 410 MHz to 7,125 MHz. The baseline design of Rel-16 is based on FR1. Frequency Range 2 (FR2) of Millimeter Waves (MMW), which goes from 24.25 GHz to 52.65 GHz, is not the primary focus of Rel-16 studies.

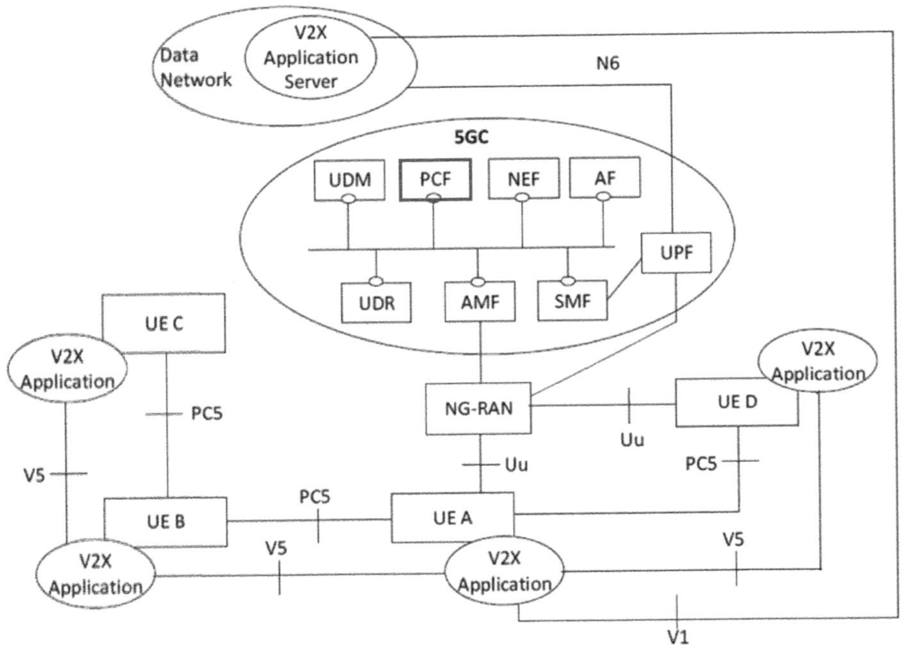

Figure 6.9. *5G system architecture for PC5- and Uu-based V2X communication: Alternative 1. For a color version of this figure, see www.iste.co.uk/mendiboure/transportsystems.zip*

6.3.3.1. *5G-V2X mode of communication*

Rel-16) specifies two additional modes, modes 1 and 2 (Garcia et al. 2021, pp. 1,972–2,026), for the selection of the subchannel in SL NR V2X communications using the PC5 NR V2X interface. These two modes correspond to mode 3 and mode 4 in LTE V2X. NR V2X supports broadcast, groupcast and unicast SL communications, but LTE V2X only supports broadcast SL communications.

6.4. Network slicing for 5G-V2X

Network slicing is one of the technological tools of 5G networks. It consists of creating multiple independent logical networks (called "slices") on the same underlying physical infrastructure, where dedicated resources are allocated and virtual network functions are instantiated for specific services, users or traffic types. Network slices (NS) are built using technologies such as network function virtualization (NFV) and software-defined network (SDN). End-to-end network slicing includes a radio access network (RAN), a core network (CN) and slicing of end devices.

3GPP defines network slicing as a technology that "enables the operator to create customized networks to provide optimized solutions for different market scenarios that demand diverse requirements, for example, in terms of functionality, performance and isolation" (3GPP 2016). Network slicing is based on the following four main principles to provide the necessary services:

– resources: including PNF (Physical Network Function) or VNF (Virtual Network Function) resources;

– virtualization: a key technology that enables NS scalability and flexibility;

– orchestration: controls allocation and scaling of resources, NS lifecycle management, etc;

– isolation: a fundamental property of network slicing that indicates the independence of NS to guarantee the performance and security desired by users.

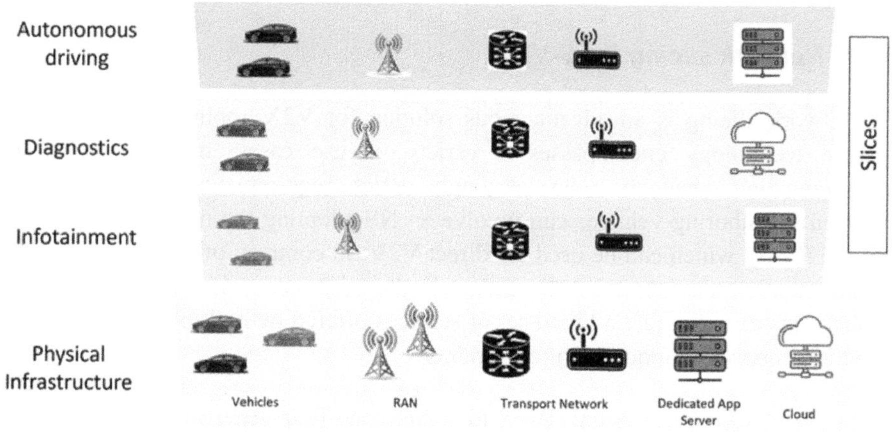

Figure 6.10. *An example of NS V2X. For a color version of this figure, see www.iste.co.uk/mendiboure/transportsystems.zip*

An NS Instance (NSI) is an entity managed in the operator's network with a lifecycle independent of the lifecycles of the other instances of the service (3GPP 2017). The lifecycle of an NSI generally comprises an instantiation, configuration and activation phase, a run-time phase, and a decommissioning phase, as shown in Figure 6.11. To correctly manage an NSI, the following procedures must be taken into account: 1) breakdown management; 2) performance management; 3) configuration management; and 4) policy management.

The service provider is responsible for the management of services, while the network operator controls the management of NS. Such a paradigm requires a business-to-business interface, such as a Service-Level Agreement (SLA). On the basis of this commercial connection, a network operator can offer the service provider different levels of control, such as monitoring only, limited control to compose NSs from a catalog, or extended control, where the service provider instantiates its VNFs and its MANO stack.

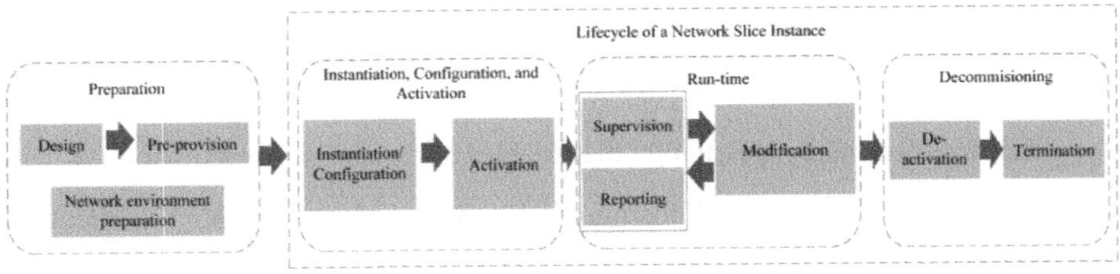

Figure 6.11. *Management of the lifecycle of an NSI. For a color version of this figure, see www.iste.co.uk/mendiboure/transportsystems.zip*

6.4.1. *Network slicing for C-V2X*

Network slicing is an advantageous solution for V2X applications, given that vehicle technology encompasses a variety of use cases, traffic models and communication channels with varyingly strict requirements. Communication between neighboring vehicles can involve an NS adopting the new radio sidelinks, NR-SL (PC5), which can be used for direct V2V. In contrast, other NS can provide communication services for V2N links via the edge network. Furthermore, V2X applications can integrate a broad set of services offered by multiple suppliers, each of which imposes a unique set of constraints.

In 3GPP systems, it is envisaged to support the four generic V2X use cases in Table 6.2 (Campolo 2021).

Application	Payload (bytes)	Latency (ms)	Reliability	Throughput (Mbps)
Vehicle platooning	50–6,500	10–20	90–99.999	0.012–65
Advanced driving	300–12,000	3–100	90–99.999	0.096–53
Extended sensors	1,600	3–100	90–99.999	10–1,000
Remote driving	-	5	99.999	25 (UL) 1 (DL)

Table 6.2. *Connectivity requirements for V2X autonomous driving applications*

Many scientific articles on network slicing for V2X have been published. Although the set of 5G radio, network, applications and services need to be customized to meet V2X needs via network slicing, the existing literature focuses on certain network segments and domains. Most works focus on resource allocation in the RAN, omitting the allocation of network resources in the CN segment and the attributing of computing resources. Research is mainly analytical and/or simulation-based, and only a few articles focus on experimental results.

An example of reference network slicing architecture is presented in Campolo et al. (2018, pp. 400–405) (Figure 6.12); this is based on the three-layer paradigm: the infrastructure layer, which comprises user devices, RAN nodes, edge and remote compute, storage and network resources; the service layer, which includes Virtual Network Functions (VNF) and logical network behaviors that activate an NS when chained together, the business layer, which comprises the mechanisms and tools to describe the behavior of an NS at a high level and grasp the requirements of a given SLA for a vertical segment, whose assurance is controlled by the Operational/Business Support System (OSS/BSS); and the MANagement and Orchestration (MANO) layer, which comprises all the tools to enable an operator to monitor, orchestrate, and scale components of the slice in order to comply with the negotiated SLAs.

Figure 6.12 also illustrates the instantiation of an NS for autonomous driving. It uses V2V as the main connection mode of Radio Access Technology (RAT) and other RAN/CN characteristics, such as network-controlled resource allocation on the PC5 interface (in eNodeB), mobility, authentication, authorization and subscription management. In addition, extremely low latency and reliable V2N communication with an edge-based V2X Application Server (AS) are required for video/data exchange, allowing vehicles to process high-definition maps of the surrounding region and increase their visual perception.

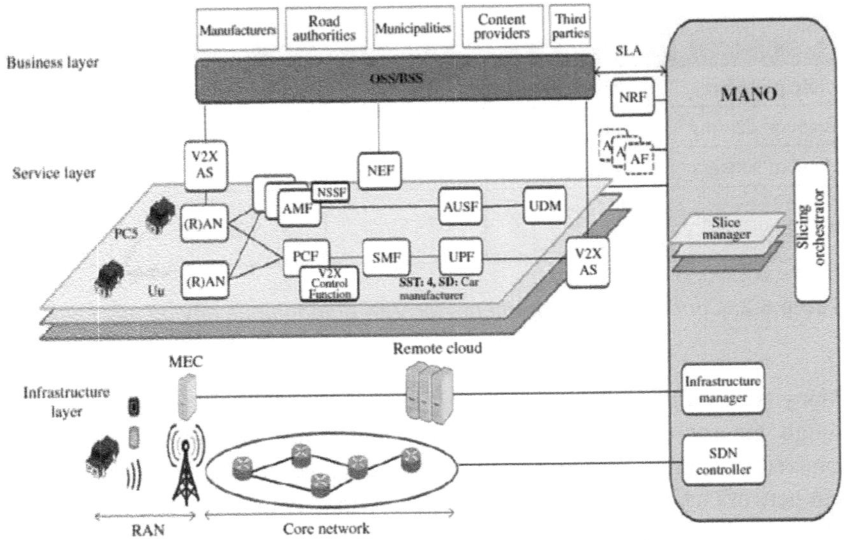

Figure 6.12. *Network slicing architecture and NS instantiation for autonomous driving services. For a color version of this figure, see www.iste.co.uk/mendiboure/transportsystems.zip*

6.4.2. *ITS-G5 network slicing*

Network slicing could also be advantageous when used for ITS-G5 communication technology. Indeed, the messages generated by ITS-G5 in C-ITS scenarios can be exchanged between the vehicles, the communicating infrastructures and the application servers. Such a network architecture includes vehicular ad hoc networks and access/aggregation/core networks (ETSI 2014). Access and aggregation networks can host edge nodes. In addition, a remote ITS data center runs numerous services to support advanced V2X applications (traffic-jam and accident prediction, real-time geolocation, traffic efficiency improvement, etc.), such as Application Servers (AS), Public Key Infrastructure (PKI), the Authentication, Authorization and Accounting (AAA) server, etc. In such scenarios, network slicing enables network operators to provide customized and optimized network services to different users and applications, all whilst sharing the same physical infrastructure.

6.4.2.1. *End-to-end network slicing for ITS-G5*

This section describes our end-to-end network slicing architecture for ITS-G5 by presenting the ITS-G5 slicing system architecture and its management.

In addition to the network architecture shown in Figure 6.6, we consider an aggregation network between the access network and the core network to collect traffic from various road infrastructures. It aggregates the data received from the access networks and transmits them to the ITS-G5 core network. The ITS data center has high compute and storage resources and can host the different V2X application servers (PKI, AAA server, etc).

The proposed slicing approach is based on traffic classification for ITS-G5 to provide improved End-to-End (E2E) latency for high-priority traffic and user groups. Slice resources are orchestrated across the radio access, aggregation, and core networks of ITS-G5, and the ITS data center. The proposed system architecture enables a message sent/received by an ITS station to be transported to/from remote V2X Application Servers (AS) to be processed and used to provide advanced ITS services. Certain task processing needs to be performed closer to the data sources in a specific peripheral node.

A slice covers the ITS ad hoc networks, the access network, the aggregation network and the core network up to the remote ITS data center, as described by the architecture presented in Figure 6.13. Our slicing mechanism enables the creation of as many slices as necessary to cover various scenarios and use cases, each of which can have a specific priority. For example, a higher-priority slice can be provided by a combination of ITS-G5 service parameters (high-priority DCC profile, appropriate generation rate, etc.) and dedicated resources (bandwidth, CPU, storage and memory) to ensure quality of service requirements (latency, throughput, reliability, etc.) and specific security. The end-to-end slice is composed of the following elements:

– radio resources, expressed in the form of ITS-G5 Physical Network Functions (PNF) corresponding to RAN slicing;

– access/aggregation resources and core resources, expressed in the form of end-to-end bandwidth allocated to links connecting road ITS-Ss to V2X application servers.

These links pass through different points of presence, such as edge nodes and data centers, in which compute resources are allocated (CPU, storage and memory) to instantiate Virtual Network Functions (VNF).

To focus on slice management functionalities in the ITS-G5 network, slice Management and Orchestration (MANO) are added to extend the ITS-G5 network architecture and provide control over physical resource allocation and the lifecycle of NS. MANO aims to perform configurations and optimize physical resources in order to guarantee the quality of service requirements for each slice/service. It is composed as follows (Campolo et al. 2018):

– The slice manager is responsible for describing slices by means of well-defined models, instantiating slices and managing their lifecycle.

– The slice orchestrator manages the brokering of physical infrastructure resources among slices of the same operator. It also exchanges information with other road operators to enable seamless platform sharing between operators on a worldwide scale to improve quality of service and user experience (particularly for V2X services, where vehicles can move between different road operators).

– The SDN controller enables network resources to be managed using a logically centralized view. It fulfills several functions such as traffic engineering, monitoring network statistics and controlling infrastructure nodes via a southbound interface (e.g. OpenFlow). The SDN controller directs traffic to the appropriate edge nodes and manages the migration of the User Plane (UP) and Control Plane (CP) functions, resource reconfiguration and transfer procedures between different edge nodes.

– The infrastructure manager supports infrastructure orchestration by monitoring the storage and processing status of point-of-presence nodes, such as edge nodes and data centers.

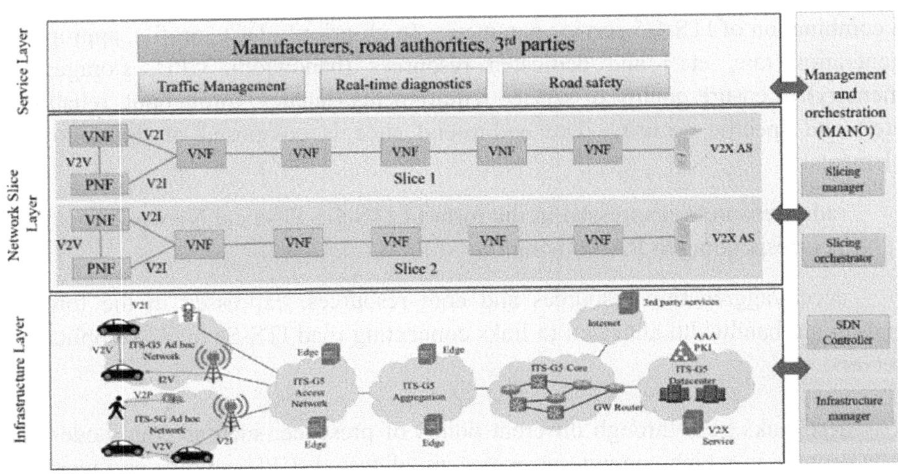

Figure 6.13. *Overview of the global architecture of the ITS-G5 system. For a color version of this figure, see www.iste.co.uk/mendiboure/transportsystems.zip*

The proposed MANO aims to determine the appropriate resource allocation parameters for ITS stations in order to guarantee quality of service to the applications and isolation of the slices. The parameters are sent to all ITS stations (On-Board Units, Road-Side Units, etc.). These parameters include service priorities (DCC profile), transmission rates, transmission powers, etc. The area to which an ITS-G5 service configuration is applied may only be a coverage area of one RSU or a larger geographic area containing multiple RSUs. The decision on the slicing configuration of each area depends on the nature of the environment (urban, highway, etc.), the density of the ITS stations, the type of vehicles (ordinary or priority), etc.

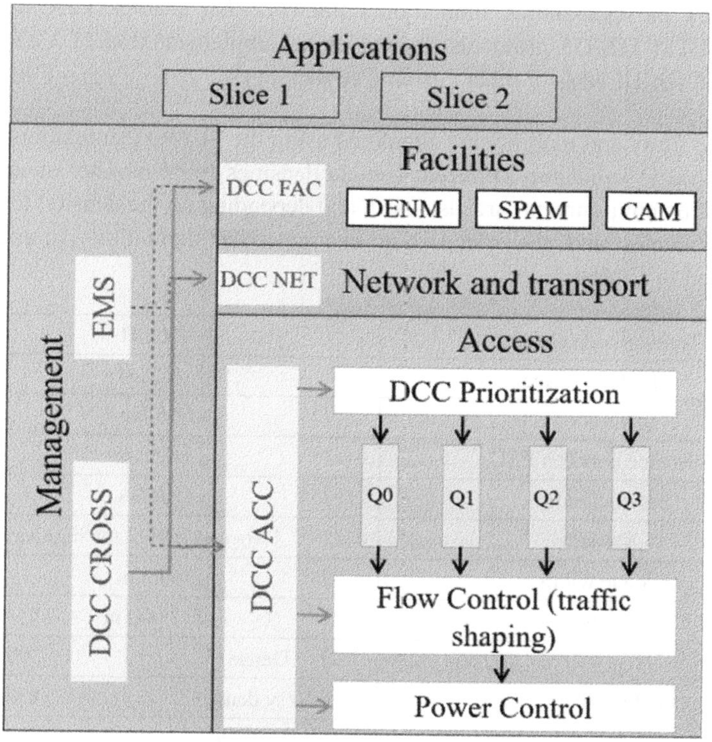

Figure 6.14. *Protocol stack of the ITS station + EMS. For a color version of this figure, see www.iste.co.uk/mendiboure/transportsystems.zip*

In order to achieve these objectives, we introduce an element of the ITS-S Element Management System (EMS) into the management layer, as described in Figure 6.14. This element is another cornerstone of the slicing mechanism, as it receives and applies slicing configuration parameters. It configures services by defining different parameters in each layer: priority in the management layer, traffic control and shaping in the access layer, definition of the Leaky/Token bucket shaping mechanism parameters in order to conform to a specific rate, configuration of the message generation rate in the facilities layer, etc. The slice could be continuously reconfigured based on the state of the transmission channel or upon reception of an update of the quality of service parameters. For example, a vehicle can dynamically switch from slice 1 to slice 2.

6.4.2.2. *End-to-end network slicing for ITS-G5*

The simulation tool used to implement our solution is the Artery simulator (Riebl et al. 2015), an open-source simulation framework that enables V2X simulations based on ETSI ITS-G5 protocols. It enables the implementation of V2X services through CAMs, DENMs or other customized messages.

Vehicle flows and mobility are simulated using the SUMO (Simulation of Urban Mobility) traffic simulator. Different vehicle densities (total number of vehicles in the simulation environment) are also used, and depending on the density, the spacing between vehicles and the speed may change. The simulation parameters are summarized in Table 6.3.

<table>
<tr><td>Transmission channel</td><td colspan="2">CCH</td></tr>
<tr><td>ITS-G5 service</td><td colspan="2">CAM</td></tr>
<tr><td>Transmission rate</td><td colspan="2">5 Hz</td></tr>
<tr><td>Distance between RSUs</td><td colspan="2">750 m</td></tr>
<tr><td>Simulation time</td><td colspan="2">200 s</td></tr>
<tr><td>Scenario</td><td colspan="2">Two-way eight-lane highway</td></tr>
<tr><td>Lane width</td><td colspan="2">3.5 m</td></tr>
<tr><td>Length</td><td colspan="2">2,000 m</td></tr>
<tr><td rowspan="3">Vehicle density</td><td>Dense</td><td>300</td></tr>
<tr><td>Very dense</td><td>450</td></tr>
<tr><td>Extremely dense</td><td>600</td></tr>
<tr><td>Maximum speed</td><td colspan="2">15 m/s (dense) /
13.3 m/s (very and extremely dense)</td></tr>
<tr><td>Vehicle length</td><td colspan="2">3 m</td></tr>
<tr><td>Minimum space</td><td colspan="2">5 m</td></tr>
<tr><td>Flow type</td><td colspan="2">Police (HPS)
Regular vehicles (OPS)</td></tr>
</table>

Table 6.3. *Simulation parameters*

In our work, we implement a slicing approach for ITS-G5 to demonstrate its efficiency in a use case where a high density of vehicles is present. The main objective of this network slicing approach is to improve the latency of the end-to-end CAM flow for high-priority vehicles.

We define two slices for two groups of users with different priorities: the high-priority slice (HPS), which sends high-priority CAMs (HP CAM), and the ordinary-priority slice (OPS), which sends ordinary-priority CAMs (OP CAM). The number of slices can be increased to support more services and applications (in our case, we only use two slices). This mapping aims to ensure the transmission of high-priority messages as quickly as possible by allocating them less waiting time and more possibilities for accessing the transmission channel. We allocated the same VNF resources to focus on slicing the RAN at access-network and core-network levels.

The slice configuration sets the HPS priority to DP0 (the highest priority) for the priority flow. Therefore, the priority flow will be mapped to the highest-priority queue, Q0, which contains the VO traffic class. For OPS, the default priority is used (corresponding to DP2, which is mapped to Q2, and contains the BE traffic class).

For this simulation, we used a 2 km-long, two-way, eight-lane highway with four lanes for each direction. We allocated the vehicle flows to a specific slice as follows: police car flows are allocated to the HPS slice, and ordinary vehicle flows to the OPS slice. The HPS flows vary from 5% to 30% of the total number of flows, and the rest of the OPS flows varies from 95% to 70%.

In addition, to simulate a real congestion scenario, we added two services for all environment vehicles: the first is a DENM service with an access priority set at DP1 (mapped to queue Q1). The second is a Background service (BK), whose access priority is set at DP3 (mapped to queue Q3). The transmission speed of the two services is set at 1 Hz.

The targeted measurement analyzed in these simulations is end-to-end latency. Since we assume the same resources for access/aggregation and the core network, end-to-end latency is closely linked to the latency of the ITS-G5 ad hoc network. It is measured as the time elapsed between the sending of a CAM message by the CAM services of the application layer of the vehicle and reception of this message by the RSU. Latency includes the message propagation delay on the transmission channel and the queuing delay (the waiting time in the system queues) needed for transmission.

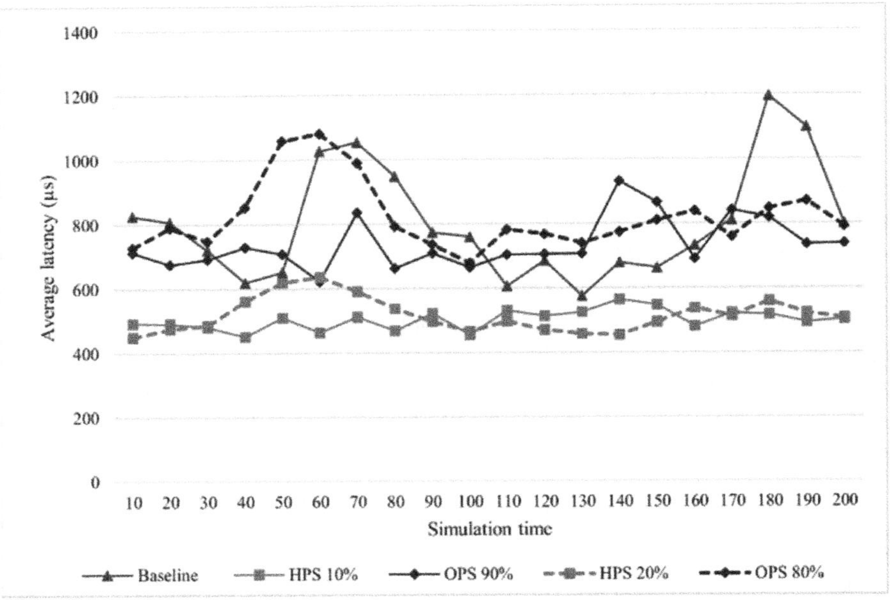

Figure 6.15. *Average RAN latency (density: 450). For a color version of this figure, see www.iste.co.uk/mendiboure/transportsystems.zip*

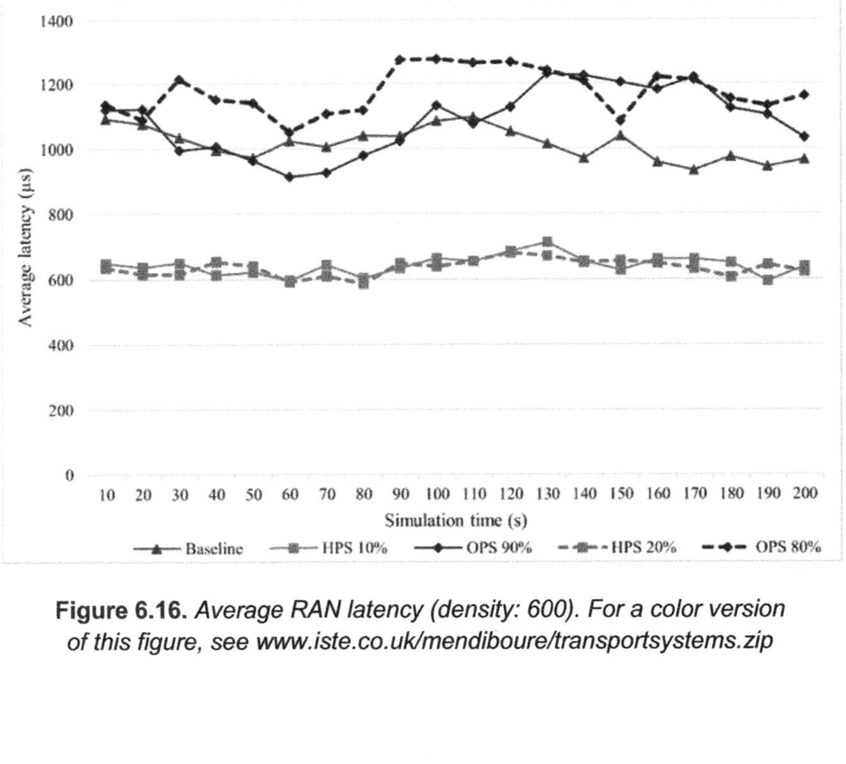

Figure 6.16. *Average RAN latency (density: 600). For a color version of this figure, see www.iste.co.uk/mendiboure/transportsystems.zip*

To better demonstrate the contribution of our work, we simulated a scenario using default CAM priorities as a baseline comparison. In this case, the CAM message priority is set at DP2. A sample of the average latency obtained for the different simulations is summarized in the graphs of Figures 6.15 and 6.16.

In Figure 6.15, the results obtained for a density of 450 vehicles show an improvement in the latency of CAM messages sent on the HPS, compared to the OPS. The latency of OPS flows varies between 700 μs and 1,000 μs on average. In contrast, the latency of the HPS slice is almost stable, between 500 μs and 600 μs for the 10% and 20% HPS scenarios, respectively. Compared to the baseline scenario, the OPS performance is not very different, since latency varies between 600 μs and 1,000 μs on average.

When the density is increased to 600 vehicles, the improvement offered by our proposed slicing solution becomes more apparent. In Figure 6.16, for the 10% and 20% HPS scenarios, the latency remains stable between 600 μs and 700 μs throughout the duration of the simulation; whereas, for the OPS scenario and the basic scenario, latencies vary between 1,000 μs and 1,200 μs on average.

The improved latency achieved with our solution supports advanced V2X use cases, such as autonomous driving and remote driving, where latency is a major concern as it must not exceed several milliseconds, for example, 5 ms for autonomous driving (Campolo et al. 2020).

In the following, we study the isolation property of the proposed slicing mechanism. We increased the CAM generation frequency for OPS flows to 10 Hz compared to the HPS, which remains set at 5 Hz. Therefore, we chose to overload the OPS and monitor its impact on HPS latency.

We used the worst-case scenario where vehicle density is set at 600, and OPS flows represent 95% of the total traffic (5% HPS flows). The simulation results are presented in Figure 6.17, in which HPS_O and OPS_O represent scenarios where OPS is overloaded.

The results obtained showed that the average latency after OPS overload had no impact on HPS performance. As a result, the isolation characteristic is assured by our solution.

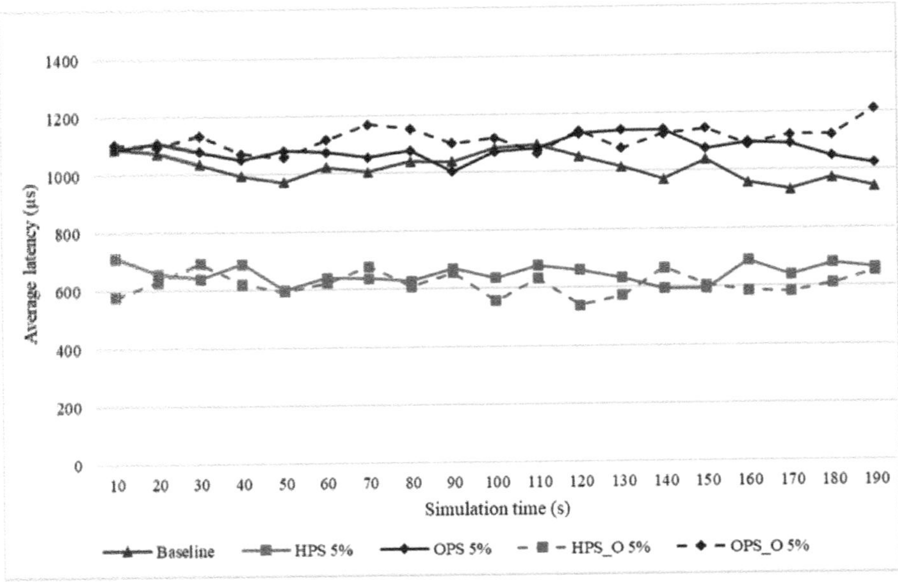

Figure 6.17. *Average RAN latency (density: 600; HPS: 5%). For a color version of this figure, see www.iste.co.uk/mendiboure/transportsystems.zip*

6.4.2.3. *End-to-end slicing*

In this section, we deploy in a testbed with two slices: HPS and OPS, from the slicing of the RAN ITS-G5 (simulated part) to the network core (real part). We used open-source MANO (OSM) version 6 as a management and orchestration stack aligned with the evolution of the ETSI-NFV framework. This tool enables instantiation of NSs/network services and their lifecycle management, including NS monitoring. Each slice is composed of three VNFs and a V2X application server. All of these components are deployed in Docker containers. An overview of the deployed architecture is shown in Figure 6.18.

We deployed a client/server socket on both endpoints to measure end-to-end slicing performance. This socket is used to transport the CAM transmissions coming from the ITS-G5 network (simulated part). We are considering the scenario where vehicle density is set at 600, and HPS flows represent 10% of the total traffic. We configured the link parameters using the TC (Linux Traffic Control) tool with the parameters shown in Table 6.4 to highlight slice priority, while emulating the WAN property.

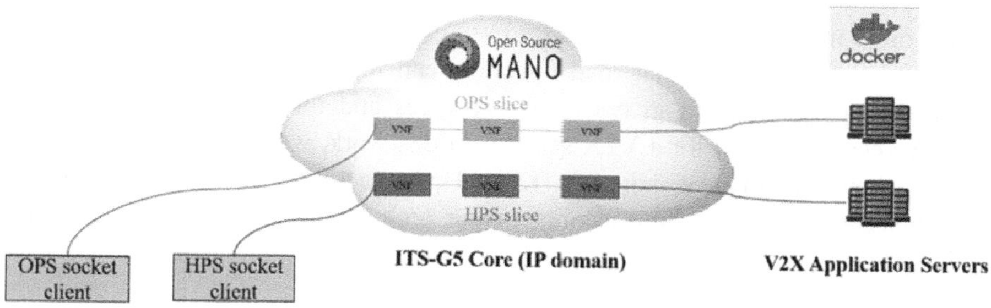

Figure 6.18. *Overview of the deployed E2E architecture. For a color version of this figure, see www.iste.co.uk/mendiboure/transportsystems.zip*

Slice	Maximum link latency in the CN	Link reliability	% flow by slices
HPS	50 ms	99%	10%
OPS	200 ms	95%	90%

Table 6.4. *Configuration of link parameters*

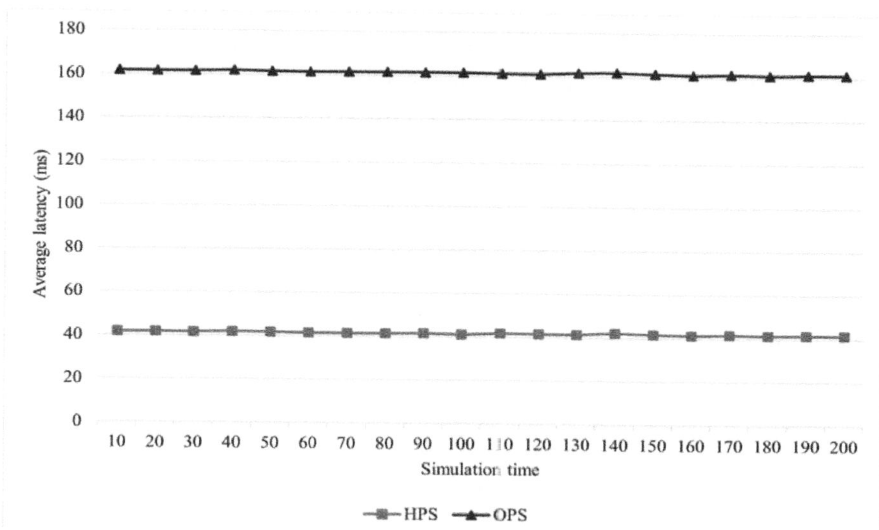

Figure 6.19. *Average E2E latency (density: 600; HPS: 10%). For a color version of this figure, see www.iste.co.uk/mendiboure/transportsystems.zip*

The results obtained from the simulation are shown in Figure 6.19. An average latency of 40 ms for HPS and 160 ms for OPS can be noted. These results represent an example of parameters that road operators can implement to provide different priority groups transported in two different slices and thus implement a Service Level Agreement (SLA) for a V2X service. In this work, we focused on latency as the main key performance indicator (KPI). However, other NS key performance indicators could be taken into account in our architecture to meet the quality of service (QoS) requirements of V2X services.

6.5. Conclusion

In this chapter, we presented certain aspects of V2X, the description of the European ITS-G5 standard and the 4G-V2X and 5G-V2X standards, and we also presented network slicing and its significance for V2X. We then presented the advances in network slicing research for V2X communications in cellular networks. Lastly, we detailed the architecture that we propose to support end-to-end slicing for V2X applications in vehicular communications based on ITS-G5.

The proposed slicing mechanism improves end-to-end (E2E) latency for specific traffic types and user groups. To this end, we demonstrated the efficiency of the slicing mechanism on two different priority groups using two slices, namely HPS and OPS. The result is that HPS has lower latency than OPS. Consequently, the quality of service of the slices can be differentiated according to their priorities. This improvement could be beneficial for many advanced V2X applications, where the latency envelope must be as low as possible, as in the case of autonomous driving or platooning scenarios.

6.6. References

3GPP (2016). TR 23.799; Study on architecture for next generation system. 3GPP Technical Report, Release 14.

3GPP (2017). TR 28.801; Study on management and orchestration of network slicing for next generation network. 3GPP Technical Report, Release 15.

3GPP (2019). TR 23.786 (V2.0.0); Study on architecture enhancements for the Evolved Packet System (EPS) and the 5G System (5GS) to support advanced V2X services. 3GPP Technical Report.

Bellache, T., Oyunchimeg, S., Samir, T. (2017). DCC-enabled contention based forwarding scheme for VANETs. In *2017 IEEE 13th International Conference on Wireless and Mobile Computing, Networking and Communications (WiMob)*. doi: 10.1109/WiMOB.2017.8115767.

Campolo, C., Molinaro, A., Iera, A., Menichella, F. (2017). 5G network slicing for vehicle-to-everything services. *IEEE Wireless Communications*, 24(6), 38–45.

Campolo, C., dos Reis Fontes, R., Molinaro, A., Esteve Rothenberg, C., Iera, A. (2018a). Slicing on the road: Enabling the automotive vertical through 5G network softwarization. *Sensors*, 18(12), 4435.

Campolo, C., Molinaro, A., Iera, A., Fontes, R.R., Rothenberg, C.E. (2018b). Towards 5G network slicing for the V2X ecosystem. In *2018 4th IEEE Conference on Network Softwarization and Workshops (NetSoft)*, 400–405.

Campolo, C., Molinaro, A., Sciancalepore, V. (2021). 5G network slicing for v2x communications: Technologies and enablers. In *Radio Access Network Slicing and Virtualization for 5G Vertical Industries*, Zhang, L., Farhang, A., Feng, G., Onireti, O. (eds). John Wiley & Sons, Hoboken, 239–257.

Canis, B., and Gallagher, J.C. (2019). Smart cars and trucks: Spectrum use for vehicle safety. *Congressional Research Service*, No. IF11260.

ERTICO – ITS Europe (2015). Version 1.1: Report about technologies for future C-ITS service scenarios.

ETSI (2011). TS 102 687 V1.1.1; Intelligent Transport Systems (ITS); Decentralized congestion control mechanisms for intelligent transport systems operating in the 5 GHz range; Access layer part. ETSI Standard.

ETSI (2012). TS 102 724 V1.1.1; Intelligent Transport Systems (ITS); Harmonized channel specifications for intelligent transport systems operating in the 5 GHz frequency band. ETSI Standard.

ETSI (2013). TS 102 636-4-2 V1.1.1; Part 4: Geographical addressing and forwarding for point-to-point and point-to-multipoint communications; Sub-part 2: Media-dependent functionalities for ITS-G5. ETSI Standard.

ETSI (2014). EN 302 636-3 V1.2.1; Intelligent Transport Systems (ITS); Vehicular communications; GeoNetworking; Part 3: Network Architecture. ETSI Standard.

ETSI (2020). TS 138 101-1 V16.5.0; 5G; NR; User Equipment (UE) radio transmission and reception; Part 1: Range 1 Standalone (3GPP TS 38.101-1 version 16.5.0 Release 16). ETSI Standard.

Garcia, M.H.C., Molina-Galan, A., Boban, M., Gozalvez, J., Coll-Perales, B., Şahin, T., Kousaridas, A. (2021). A tutorial on 5G NR V2X communications. *IEEE Communications Surveys & Tutorials*, 23(3), 1972–2026.

Harounabadi, M., Soleymani, D.M., Bhadauria, S., Leyh, M., Roth-Mandutz, E. (2021). V2X in 3GPP standardization: NR sidelink in release-16 and beyond. *IEEE Communications Standards Magazine*, 5(1), 12–21.

Hasan, M., Mohan, S., Shimizu, T., Lu, H. (2020). Securing vehicle-to-everything (V2X) communication platforms. *IEEE Transactions on Intelligent Vehicles*, 5(4), 693–713.

Kambiré, S.J., Aniss, H., Krief, F., Maaloul, S., Berbineau, M. (2020). Selection of radio access technologies in vehicular networks: State-of-the-art and perspectives for autonomous connected vehicles. In *Communication Technologies for Vehicles: 15th International Workshop, Nets4Cars/Nets4Trains/Nets4Aircraft 2020*, Bordeaux, 99–112. Springer, Cham.

Mendiboure, L. (2020). Distribution géographique de données dans l'Internet des Véhicules : une approche logicielle et sécurisée utilisant les réseaux cellulaires. Thesis. Bordeaux.

Riebl, R., Günther, H.J., Facchi, C., Wolf, L. (2015). Artery: Extending veins for VANET applications. In *2015 International Conference on Models and Technologies for Intelligent Transportation Systems (MT-ITS)*.

Wu, H., Wang, X., Zhang, Q., Shen, X. (2006). IEEE 802.11e enhanced distributed channel access (EDCA) throughput analysis. In *2006 IEEE International Conference on Communications*, 1.

PART 3

New Approaches to Data Processing in Cooperative Intelligent Transport Systems

7

A Novel Cloud Approach for Connected Vehicles

Geoffrey WILHEM[1], Marwane AYAIDA[2] and Hacène FOUCHAL[1]

[1] *CReSTIC, Université de Reims, Champagne-Ardenne, France*

[2] *IEMN, Université Polytechnique Hauts-de-France, Valenciennes, France*

Across the world, Cooperative Intelligent Transport Systems (C-ITS) are being deployed. Soon, not just vehicles, but also bicycles, pedestrians, buses and all mobile equipment will be compatible with C-ITSs. These systems are connected in a wireless LAN network, based on the IEEE 802.11p Wi-Fi standard. The growing number of these systems and the services associated with them will inexorably lead to data overload on the network. We propose a new approach, called Geographical Vehicular Central Data Networking (GeoVCDN), aimed at both reducing bandwidth occupancy and increasing network freshness; it builds on the paradigm of information-centric networks (ICN) in a cloud environment. To do this, we exploit standardized application beacons and geographic routing natively used by C-ITSs.

REndez-vous NEtwork (RENE) and Named Data Networking (NDN) are two well-known ICN-based methods. Each of these is the leading figure of a set of proposals based on similar principles. We compared them to the performance of our approach. To do this, we first proposed a probabilistic model that evaluates the network freshness and network load of each method in a given environment. Then, we compared them in a simulator. It appeared that our proposal made it possible to improve these two performance indicators.

7.1. Introduction

The deployment of Cooperative Intelligent Transport Systems (C-ITS) began several years ago. In the long term, they should improve road safety and comfort. From this perspective, it is important to maintain a certain quality of service, namely

Cooperative Intelligent Transport Systems,
coordinated by Léo MENDIBOURE.

low latency, little packet loss, low bandwidth consumption, etc. To achieve this, one of the strategies proposed is to change the communication paradigm in the network; from a host-oriented network to a content-oriented network (ICN2020: Horizon 2020 Project 2020). Vehicular Cloud Networking (Musaddiq et al. 2020) combines the notions of Vehicular Cloud and ICN and helps manage mobility through mechanisms such as network cache or sessionless connections. In any case, we note that in a secure environment, an additional load is created on the network because of the security mechanisms. As a reminder, the bandwidth at the MAC layer of IEEE 802.11p is 5.6 Mb per second (700 KB/s). In this chapter, we propose a VCN approach, named Geographical Vehicular Central Data Networking (GeoVCDN), which aims, compared to other approaches, to improve network performance in terms of freshness and load. To do this, we take advantage of the application beacons sent by all C-ITSs (vehicles, Road-Side Units (RSUs), pedestrians (Hamdani et al. 2020), etc.), between 1 and 10 times per second. Rather than sending independent packets, ICN exchanges are incorporated into the beacons. This mechanism helps reduce bandwidth consumption. However, to improve network freshness, we use multi-hop (vehicle-to-vehicle) broadcasting, which enables data to be broadcast more efficiently and more deeply in the network.

We measure the efficiency of our proposal and compare it with the REndez-vous NEtwork (RENE) and Named Data Networking (NDN) approaches through simulations and an analytical evaluation, based on a probabilistic model. The results show that GeoVCDN improves the chosen performance indicators. Finally, and to illustrate the functional efficiency of our approach, we propose extending the comparison to a scenario: fastest pathfinding in a smart, connected city.

The chapter is organized as follows. Section 1.2 describes the works relating to the ICN, C-ITS, vehicular cloud computing (VCC) and C-ITS standards. Section 1.3 details the mechanisms and architecture of our approach. Section 1.4 defines the probabilistic model that we propose and presents the results for each approach concerning the network load and freshness. Finally, section 1.5 describes our simulator and presents the results obtained for each approach.

7.2. State of the art

The field of study on vehicular networks is vast. Research problems are multiplying and taking different forms. We can nevertheless highlight some of the major areas. Among these, we find cloud computing, applied to vehicular networks. Vehicular cloud computing reuses already-known forms or introduces new ones. On the contrary, vehicles will be both producers and consumers of large amounts of data. It is therefore necessary to find communication methods that support both the restrictions imposed by physical technologies, such as ITS-G5, and the data stream specific to Big Data. Information-centric networking (ICN) is one of those methods that should be explored. Finally, ITS-G5 serves as the basis for ETSI (European Telecommunications Standards Institute) standards, which are standards adopted for the deployment of connected vehicles in Europe. For these three topics, we provide a presentation of what is established, a link with our proposal and possibly a critical discussion.

7.2.1. *ETSI standards for C-ITSs*

Figure 7.1 is a representation of the communication stack used by Intelligent Transport Systems (Festag 2014) and standardized by ETSI; it presents the physical (ETSI 2014) and MAC layers (ETSI 2015) (IEEE 802.11p), the network and transport layers (GeoNetworking (GN) and Basic Transport Protocol (BTP), respectively, the data layer specifying the messages exchanged on the channel (CAM (Cooperative Awareness Message) (ETSI 2019)), DENM (Decentralized Environmental Notification Message) (ETSI 2019), SPAT (Signal Phase and Time) and MAP (Map Data) (ISO 2019), and finally, we find the application layer. The security and management layers are transversal to all of these layers.

G5 technology is based on the IEEE 802.11p standard, a type of Wi-Fi (IEEE 802.11) dedicated to short-range communications and between potentially mobile nodes. In this study, we will focus on three types of communication: V2V (Vehicle-to-Vehicle), V2I (Vehicle-to-Infrastructure) and I2V (Infrastructure-to-Vehicle).

7.2.1.1. *Routing*

The GeoNetworking protocol, associated with the BTP, is used to route and transport data. It is a geographic routing protocol; packets are intended for a geographic area and not for a specific node. ETSI defines three routing algorithms: Simple, CBF (Contention-Based Forwarding) and Advanced.

The three algorithms use the same process for selecting the next router; they choose the node closest to the destination in their neighborhood. When the packet is received in the destination area, it is broadcast to everybody present. In this study, we will only consider Simple GeoNetworking, since this is what is currently used by ITS.

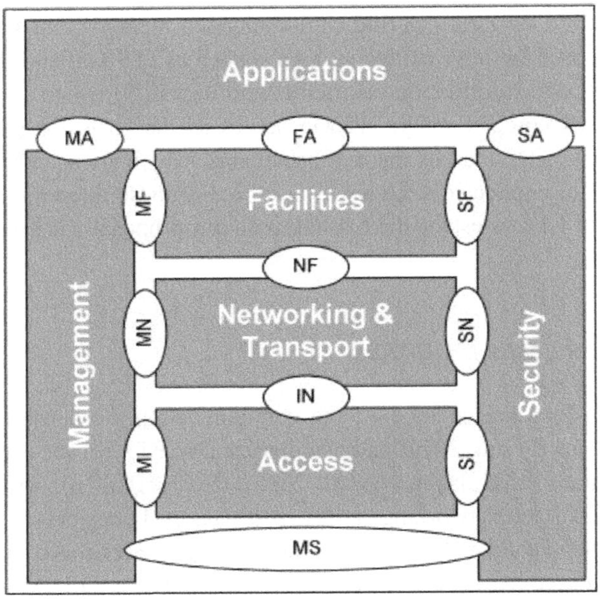

Figure 7.1. *Architecture of an ITS station*

7.2.1.2. *Security and beacons*

Security data, used to verify the integrity and authenticity of messages, is included in the GeoNetworking header. In addition, we will mainly focus on CAM messages, which are the most common messages. These are application beacons, transmitted by all stations. In the case of vehicles, these messages give information on speed, position, orientation, etc. They are transmitted in a single hop at a frequency of 1 Hz to 10 Hz (10 Hz most of the time).

Its structure is defined as being extensible, which allows add proprietary content to be added without changing the nature of the message. It is this characteristic that we exploit to include our own data.

7.2.2. *Vehicular cloud computing*

The smart city is one of the main fields of application of the Internet of Things. The world's population is increasingly centered in urban areas (Bitam and Mellouk 2012, Petrolo et al. 2015), increasing the number of connected objects in these areas. The Internet of Things is exploding; over four years, the amount of stored data has multiplied by 7 (Zaslavsky et al. 2013). At the same time, the amount of data circulating on the network has been multiplied by three. It is therefore necessary to design a network adapted to manage such traffic (Ijemaru et al. 2022). Rao et al. (2012), Suciu et al. (2013) and Whaiduzzaman et al. (2014) are unanimous that cloud computing is one of the effective solutions.

Rao et al. (2012) and Morello et al. (2017) introduced the sensor-cloud paradigm with two resulting services: Sensing as a Service (SnaaS) and Sensor Event as a Service (SEaaS). SnaaS is a request/response service, while SEaaS is a subscriber/publisher service. Using either will have an impact on bandwidth consumption. The objectives of sensor clouds are to compute complex data, support mobility and ensure real-time data processing. In Petrolo et al. (2015), the Internet of Things is divided into five layers: equipment, network, middleware, application and business. It is often mentioned that security is a challenge for sensor networks and they see the ICN as an architecture that provides concrete solutions, both for security and for interoperability as middleware.

As the number of connected objects grows, their computing power and storage capacity does the same. This enables a portion of the IaaS to be relocated locally, reducing the traffic caused by data destined for the remote cloud (Feng et al. 2022); this is possible thanks to vehicular cloud computing (VCC). With VCC, the network is evolving towards data (Gerla 2012) and edge computing (Khan et al. 2019). ICN covers the first point fully and the second partially, keeping the data as close as possible to the relevance area.

VCC can take different forms: cloud computing for vehicles (infrastructure cloud and remote cloud) (Ghafoor et al. 2013) and cloud computing by vehicles (dynamic cloud and context cloud) (Qiu et al. 2018). The first aims to render a service to mobile equipment and the second to transform vehicles into computing nodes to provide an infrastructure service (Whaiduzzaman et al. 2014).

7.2.3. *Information-centric networking*

In 1999, a new paradigm emerged: information-centric networking (ICN); this rethought the architecture of conventional IP networks, which is focused on the host of a content, in order to focus on the content itself (Cheriton and Gritter 2000). This new paradigm is adapted to both the Internet of Things and mobility, two characteristics of vehicular networks.

Several articles have already introduced the concept of ICN for vehicular networks, such as Bo Yu (2010) and Thomas et al. (2020). However, none take into account the nature of the communication link between vehicles or the ETSI standards currently used in C-ITS deployment in Europe.

Ahlgren et al. (2012) presented the main principles of ICN. The first principle is named content (Named Data Objects: NDO). Naming is the foundation of ICN exchanges; it makes it possible to identify data (as a server is identified with its IP), to route the "requests" and possibly to verify data integrity. Each approach differs in the naming method. However, two methods can be highlighted: the hierarchical method (as for URLs) and the flat method (not readable for humans, but may contain security data to verify the integrity of the message and authenticate the sender). The exchanges are made according to a request/response model, denoted as interest/data in ICN. The client/consumer submits an interest in an NDO in the network and the network responds to them with the data, when found. There are two routing phases, one consisting of routing the interest and one that returns the data to the client. For interest routing, we can either rely on a name resolution, in the same way that we currently do on the Internet with DNS servers or the routing is made based on the name of the object. The data then takes the same return path to reach the client or a better path if possible. The fourth characteristic is caching. Caching takes place during the data return process. Devices that are capable of it store the data, which makes it possible to provide the latter more quickly afterwards, rather than routing the request back to the source again.

All these principles are generic; there are several ways to implement them as we can see in these projects. Furthermore, we believe that the nature of the communication link and the nature of the exchanged data cannot be ignored in designing an efficient ICN method (Yi et al. 2021).

Ahlgren et al. (2012) present the advantages of the ICN approach. Initially, caching reduces the amount of information exchanged on the network (Alhowaidi et al. 2021; Amadeo et al. 2021) and better manages node mobility. Naming makes it possible to store the data on several devices without considering the copies as a single data item. In addition, for security aspects, it is not necessary to trust the data provider, since the data itself transports the security data.

Today's networks are designed for communications between two static nodes. The advent of wireless and mobile networks is leading to a change in the nature of connections; they are becoming ephemeral and opportunistic. In addition, thanks to network caching and the absence of a session, ICN is designed to manage mobility. Since the Internet was not natively designed to counter malicious users, it has been necessary to design mechanisms to counter them. Often, these security mechanisms are deployed on servers and do not penetrate deep into the network. Therefore, there are fraudulent messages continuously circulating on it. By applying the security part to the content itself, as is the case with the ICN, each node of the network is able to identify a malicious packet and reject it as soon as it enters the network, not at the end.

Amadeo et al. (2016) and Liu et al. (2017) proposed a state of the art of the different ICN mechanisms. The routing methods are classified into three categories: proactive routing, reactive routing and opportunistic routing. Proactive routing is the process of a node regularly transmitting the route to reach it; nodes in the network can then keep their routing table updated. Reactive routing consists of discovering the network when needed. Opportunistic routing consists of using received data to update the routing table (for example, beacons received by a vehicle). Concerning data naming, according to Amadeo et al. (2016), the hierarchical method makes consensus, but the details remain to be determined. Finally, the authors discussed the way to keep the network updated (at the data level): either by flooding the network with interests (for example, NDN) or by the initiative of the producer of the data (RENE). They believed that the first method was better suited to VANET, but additional mechanisms were needed to avoid overloading the bandwidth. This is a problem that we address with our approach.

Amadeo et al. (2016) and Liu et al. (2017) discussed the contribution of the ICN for the Internet of Things. In these studies, the ICN is not seen as an alternative method to the existing one, but as a supplement for the emerging field of connected objects. The ICN improves data access latencies and bandwidth consumption. Amadeo et al. (2016) mentioned further reasons; these include scalability, quality of service, energy efficiency and node heterogeneity. All these items are very important for C-ITSs.

Finally, it is important to determine how to optimize the use of each type of physical communication link. Xylomenos et al. (2014) discussed the possibility of using ICN on different layers. This idea is also mentioned in Amadeo et al. (2016) and is also incorporated into our proposal.

Liu et al. (2017) and Grewe et al. (2018) highlighted that research on naming, routing and the security part are the first steps, and that future research areas concern the physical communication link. Our approach explores these four topics.

ICN has been seen by some researchers as a substitute for IP architecture; for others, as a complement to the latter for specific fields, such as the Internet of Things. Concerning the Internet of Vehicles, none of the proposals take into account the choices made in the deployment of cooperative intelligent transport systems in Europe or the nature of the communication link. Bouk et al. (2017) concluded that it is necessary to bring modifications to the C-ITS if we wish to use ICN. However, with our approach, we managed to include ICN exchanges without modifying the standards currently used.

7.3. The GeoVCDN approach

Our approach is named GeoVCDN, for Geographical Vehicular Central Data Networking. Geographical for the geographical routing protocol used and central data networking for the process applied to the data and the delivery method. We detail this in the present section.

7.3.1. *A centralized context-cloud architecture*

In the field of intelligent transport systems (ITS), the relevance of the data is determined by a position and is limited in time. Considering this, the context-cloud paradigm gives us the possibility to filter the data by relevance area (district, city, harbor, ski resort, etc.). Furthermore, the context cloud is hierarchical (e.g. a building belongs to a district that belongs to a city). For ICN, the hierarchical method of data naming seems the most efficient (Dannewitz 2009). In our architecture, the cloud is composed of the static part (RSU and central ITS) and a dynamic part (the vehicles).

Figure 7.2 describes the message exchanges between each actor of the architecture.

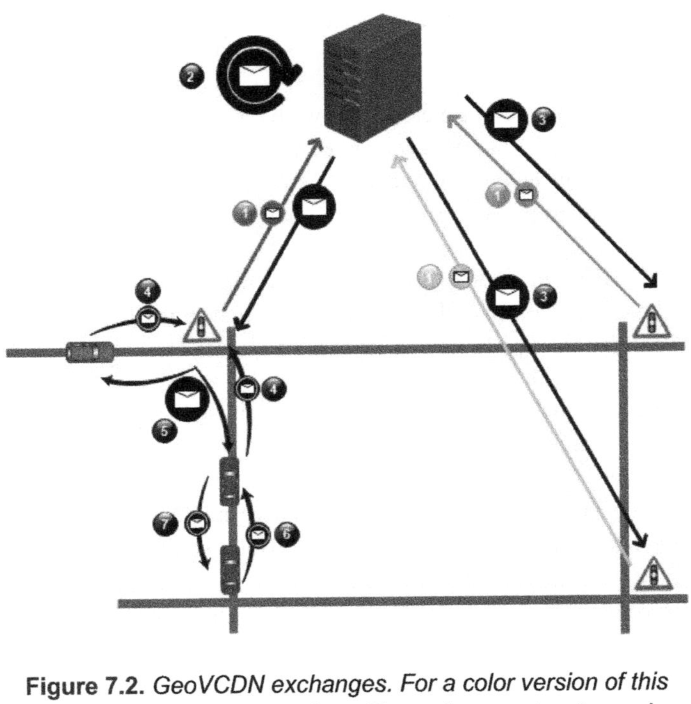

Figure 7.2. *GeoVCDN exchanges. For a color version of this figure, see www.iste.co.uk/mendiboure/transportsystems.zip*

Step 1: Sensors send their data to the central C-ITS station.

Step 2: The central C-ITS station generates metadata.

Step 3: The metadata is distributed over the RSUs.

Step 4/6: Vehicles broadcast their interests over the CAMs.

Step 5: RSUs broadcast the metadata after receiving interests from vehicles (step 4).

Step 7: Vehicles broadcast the metadata after receiving interests from other vehicles (step 6).

7.3.1.1. *Static part*

The static part of the context cloud is composed of the RSUs and the central C-ITS station. RSUs are the gateways between the parts of the cloud. They communicate using G5 with the mobile nodes and are wired to the central station. The central station is responsible for computing the metadata and distributing it to the RSUs. We consider that the connectivity between the RSUs, the central station and the compute resources suffer from no constraints: the central station can be distributed according to the needs and size of the context. This is a topic related to edge computing that we do not examine in this chapter; however, this would address all the issues related to latency, bottleneck, packet loss, etc.

7.3.1.2. *Dynamic part*

The mobile part of the cloud is composed of vehicles. A vehicle is considered to belong to the cloud if it has access, via its G5 channel, to the static part. This communication channel and the ephemeral nature of the connections raise issues: firstly, the bandwidth is limited to 750 KB/s and is not duplicable (wireless network); secondly, because a vehicle is not permanently connected to the cloud; and finally, for data to be useful to a vehicle, it must be received rapidly.

7.3.1.3. *Centralization*

Data centralization is the key process in an environment with many sensors. It aggregates data when relevant and minimizes exchanges in the network to keep static nodes synchronized and updated. The central C-ITS station therefore facilitates the creation of metadata (which groups together a set of smaller data) and the application of a version number and a hop limit; three mechanisms that we propose in our approach and that aim to reduce bandwidth consumption. To determine whether a mechanism (metadata or version number) should be used for a given service, in a given context, we give two expressions, based on the parameters presented in Table 7.1.

Designation	Description	Value
C	Number of sensors concerned	
Pi	Data ID size (bytes)	1
Pv	Data version number size (bytes)	1
P_d	Data size (bytes)	4
$FCAM$	CAM frequency (Hz)	10
F_d	Data update frequency (Hz)	1
Nv	Number of vehicles in the range of an RSU	

Table 7.1. *Mechanism parameters*

7.3.1.4. *Metadata*

With our approach, we propose grouping a set of sensor data into one data item, known as metadata. Each time a sensor records new data, the information is sent to the central station. New metadata is generated with the same identifier and a new version number. This new metadata is distributed to the routers (here, the RSUs) of the static part of the context cloud. Grouping together multiple data makes it possible to have only one name and one version number. For example, in a city context, if we wish to access the information provided by approximately 100 traffic lights, only one ICN interest will be needed with our approach, compared to 100 with the other approaches. In contrast, the data produced contains information from all traffic lights, even those that did not change. Therefore, it is necessary to ensure that the use of metadata for a given service will not be detrimental to the bandwidth; it depends on the context, the number of sensors and the data update frequency. The following in equation $C * Fcam * (Pi + Pv) * Nv > C * (Pi + Pv + Pd) * Fd$ can be used to determine whether the metadata mechanism is beneficial or not. The left-hand side of the inequation represents the weight of the interests generated every second by all vehicles in range of an RSU without the metadata mechanism. The right-hand side gives the weight of the metadata sent on the G5 channel by an RSU at the frequency *Fd*.

For example, let us take $NV = 100$ and $C = 100$; therefore:

$$\begin{cases} C * Fcam * (Pi + Pv) * Nv = 200{,}000 \ bytes \\ C * (Pi + Pv + Pd) * Fd = 600 \ bytes \end{cases}$$

In this system, it is preferable to build metadata. If Pd = 100, Fd = 10 and Nv = 10, then:

$$\begin{cases} C * Fcam * (Pi + Pv) * Nv = 20{,}000\ bytes \\ C * (Pi + Pv + Pd) * Fd = 102{,}000\ bytes \end{cases}$$

This time, it is better not to build any metadata.

7.3.1.5. *Version number*

The version number avoids sending data if the latter has not changed. Is this additional information that causes an additional load on the bandwidth but can be compensated because it reduces the number of sendings? When $Pv * Nv * Fcam < (Pd + Pi) * C * Fcam$, the version number mechanism reduces the network load caused by ICN exchanges on the G5 channel.

Below is another example. With $Pv = 4$, $Nv = 20$ and $C = 10$:

$$\begin{cases} Pv * Nv * Fcam = 800\ bytes \\ (Pd + Pi) * C * Fcam = 500\ bytes \end{cases}$$

In this case, the version number will not reduce bandwidth consumption.

Now, if $C = 50$, then $(Pd + Pi) * C * Fd = 2{,}500\ octets$; in this case, the version number reduces the network load.

7.3.1.6. *Hop limit*

The hop limit is an essential mechanism for exchanges between vehicles. It avoids propagating data too far from its relevance area and too long after its relevance period. Indeed, a vehicle that is far from an RSU will not be aware of the validity status of the data that it receives. Natively, the CAM is a bounce-free message. Its routing header has no information linked to this mechanism, so it needs to be added to the ICN message body.

7.3.2. *Geographic routing ICN protocol*

The static part manages the data preprocessing. To deliver this to the dynamic part of the cloud, we designed an ICN approach that we describe below.

7.3.2.1. *Packet structure*

CAMs are sent on the G5 channel every 100 ms and are encapsulated with security data.

The CAM structure is extensible, which enables us to introduce proprietary data. We chose to exploit this mechanism to include our ICN packets (Figure 7.3). Therefore, they benefit from the CAM security data without any additional bandwidth load.

Figure 7.4 shows the structure of the ICN message body with our approach. Compared to that of an NDN packet (ICN2020: Horizon 2020 Project 2020), there is no security layer (managed by the CAM) and we add some additional information: the version number and the hop limit.

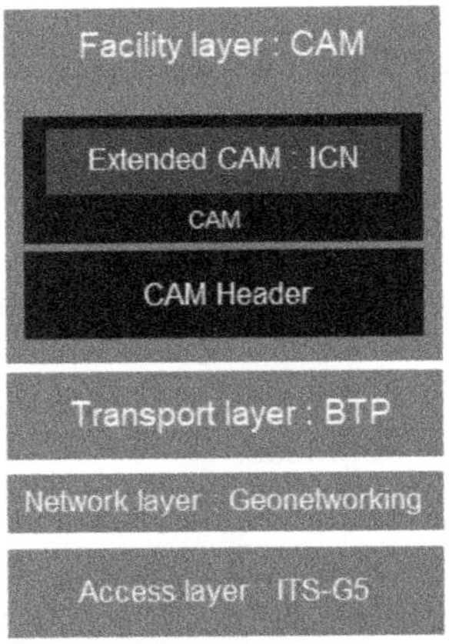

Figure 7.3. *ICN data in the CAM. For a color version of this figure, see www.iste.co.uk/mendiboure/transportsystems.zip*

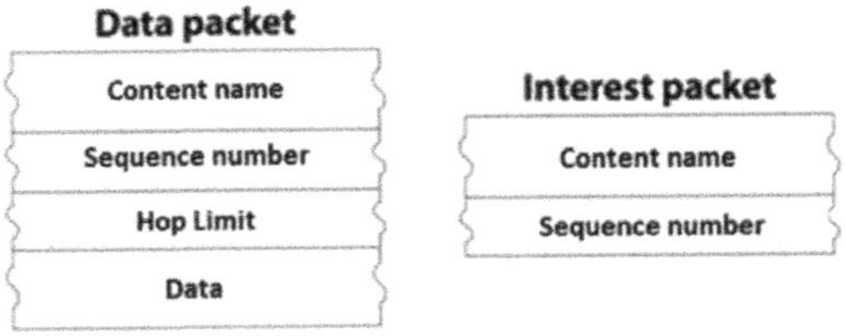

Figure 7.4. *ICN packets*

7.3.2.2. *Packet naming*

In the conventional architecture of networks (including the Internet), finding data requires first finding a host. The naming of this host is understandable for the network (for example, IP address). In contrast, in ICN, naming is totally opaque to the network; each application is therefore free to choose any naming scheme for its data.

Despite this, there is a tendency to use hierarchical naming. Hierarchical naming helps scalability, a very significant consideration in the world of connected objects. In addition, it is perfectly suited to the architecture of context clouds, which are hierarchical in nature. Therefore, the name only needs to be unique in its context but not necessarily worldwide.

7.3.2.3. *Content store (CS)*

The content store contains the data, the version number and the number of hops remaining for this packet. Before storing data, a node will check that the number of hops remaining is greater than 1. Before updating the content store with the received data, the node will check that the version number of the data that it possesses is lower than the version number of the received data. If these two conditions are met, the data received is stored with its new version number and the number of remaining hops decremented by 1 (Figure 7.7). For an RSU, data received from the central station are automatically stored and data from vehicles are automatically rejected.

7.3.2.4. *Pending data table (PDT)*

The pending data table references which content store data has to be included in the next CAM. When a node receives data with the same version number as the one the latter possesses, the data is removed from the PDT (Figure 7.7). When a node receives an interest for data with a version number lower than the one it possesses, the data is added to the PDT (Figure 7.5).

7.3.2.5. *Pending interest table (PIT)*

The pending interest table is filled in by the applications. When an application needs data, it identifies the latter's name and fills in the PIT with the name and version number of the last data received, where the latter has already been received in the past. Interests are transferred at regular intervals, from vehicle to vehicle, and from vehicle to RSU (Figures 7.5 and 7.6).

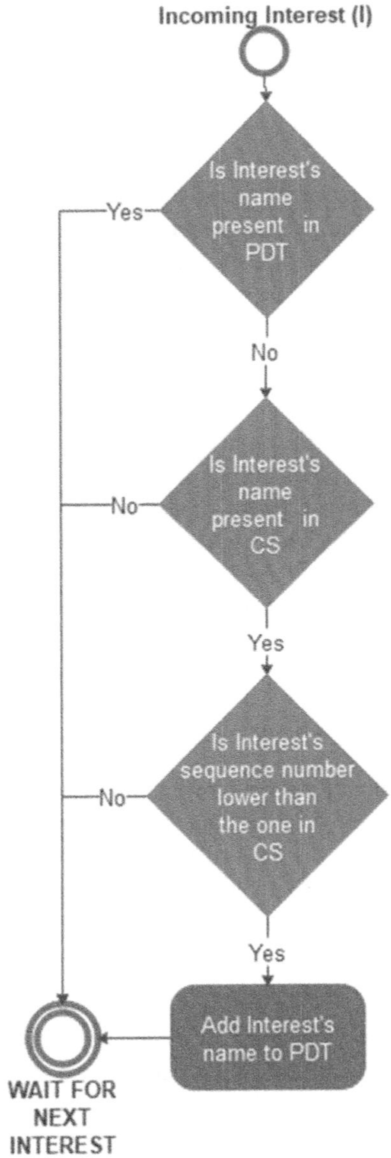

Figure 7.5. *Reception of an interest. For a color version of this figure, see www.iste.co.uk/mendiboure/transportsystems.zip*

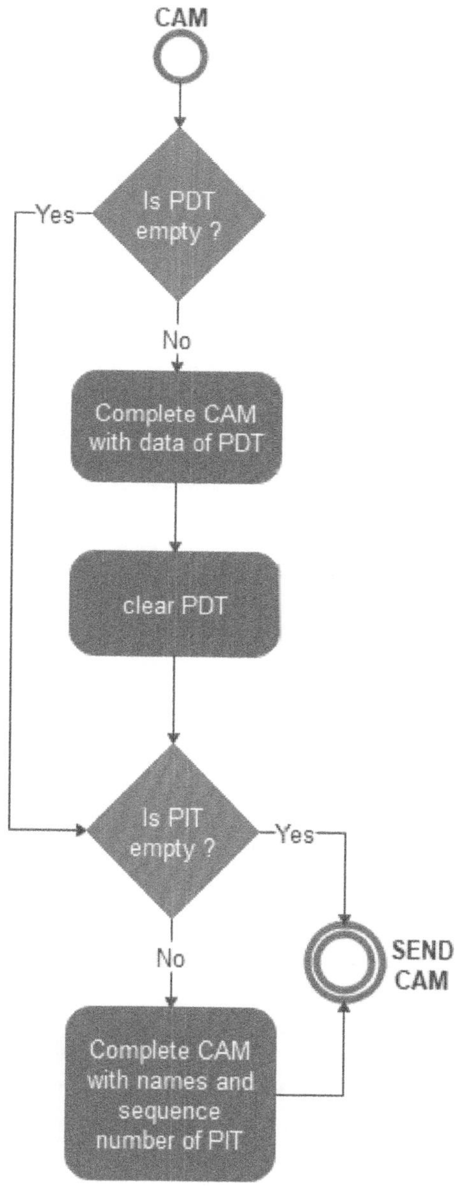

Figure 7.6. *Vehicle sending a CAM. For a color version of this figure, see www.iste.co.uk/mendiboure/transportsystems.zip*

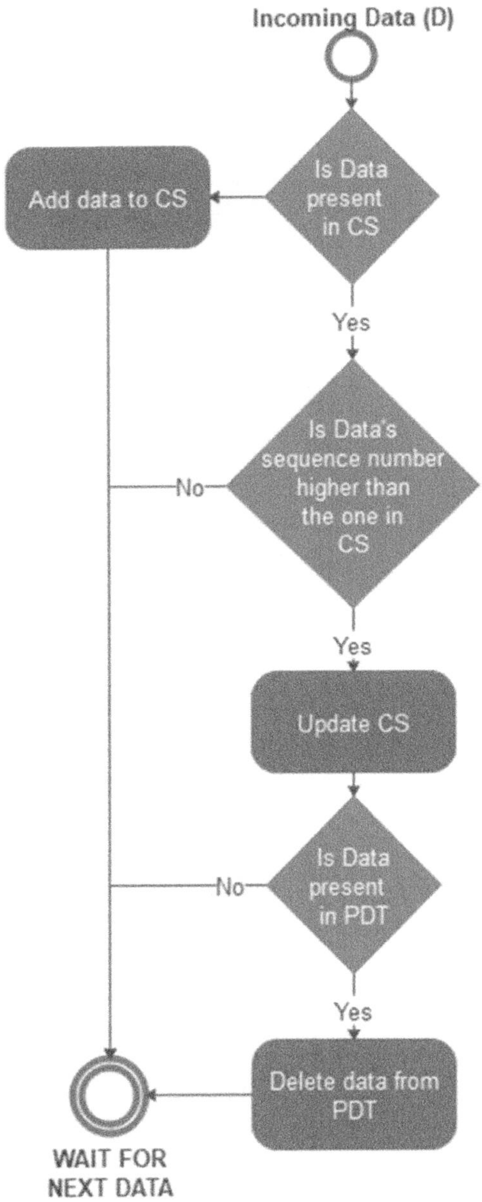

Figure 7.7. *Reception of data. For a color version of this figure, see www.iste.co.uk/mendiboure/transportsystems.zip*

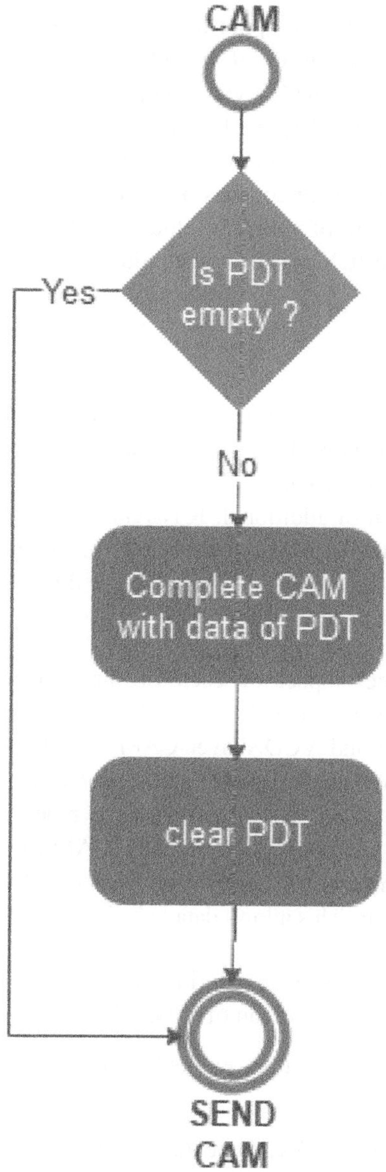

Figure 7.8. *RSU sending a CAM. For a color version of this figure, see www.iste.co.uk/mendiboure/transportsystems.zip*

7.3.3. *Discussion*

To summarize, VCC offers the opportunity to manipulate the data provided by a large amount of sensors: creation of metadata, adding a version number and computing a hop limit. This is with a view to optimizing bandwidth occupation and the data-reception node rate. The next section of this chapter consists partly of expressing and measuring the effects of these three mechanisms. In addition, we present our ICN approach, which consists of taking advantage of the CAM security data. Our approach is a form of vehicular cloud networking (Musaddiq et al. 2020) and enables the size of our ICN packets to be reduced.

7.4. Analytical model

We wish to compare our proposal with other ICN approaches on two performance indicators: network load and freshness. By freshness, we mean the rate of vehicles having obtained a piece of information before it became obsolete. By load, we mean the amount of additional, ICN-related information exchanged on the ITS-G5 channel. To do this, we model the network, the movement of the vehicles and the dissemination of the messages. Then, we use this model to give an analytical expression of each approach studied.

In order to study the impact of each of the mechanisms of our approach (GeoVCDN), we also have two, more basic, versions of it, namely vehicular centralized data networking (VCDN) and VCDN over CAM (VCDNoCAM). VCDN functions like NDN but replaces the unique data of each sensor with a versioned metadata that groups together the data of all sensors of the same type. With VCDNoCAM, ICN exchanges are integrated into application beacons (CAM), rather than transiting as an independent packet. Finally, GeoVCDN is an extended VCDNoCAM approach in which vehicles act as routers. This allows data to be broadcast using CAMs.

Different working hypotheses are defined to facilitate the modeling, but these will have no impact on the comparison of the models:

– to facilitate the representation without distorting it, a Manhattan network topology is used;

– for our approach (GeoVCDN), we express the worst case for each performance indicator (namely that for network load we give the upper limit and for freshness we give the lower limit);

– the vehicles are evenly distributed throughout the network;

– the speed of the vehicles is constant.

7.4.1. *Description of the model*

The main purpose of the model is to provide a tool for expressing the behavior of each approach from a network point of view in a smart city. First, we model the network itself: size, number of RSUs, coverage, etc. Then, we model the message exchanges in this network.

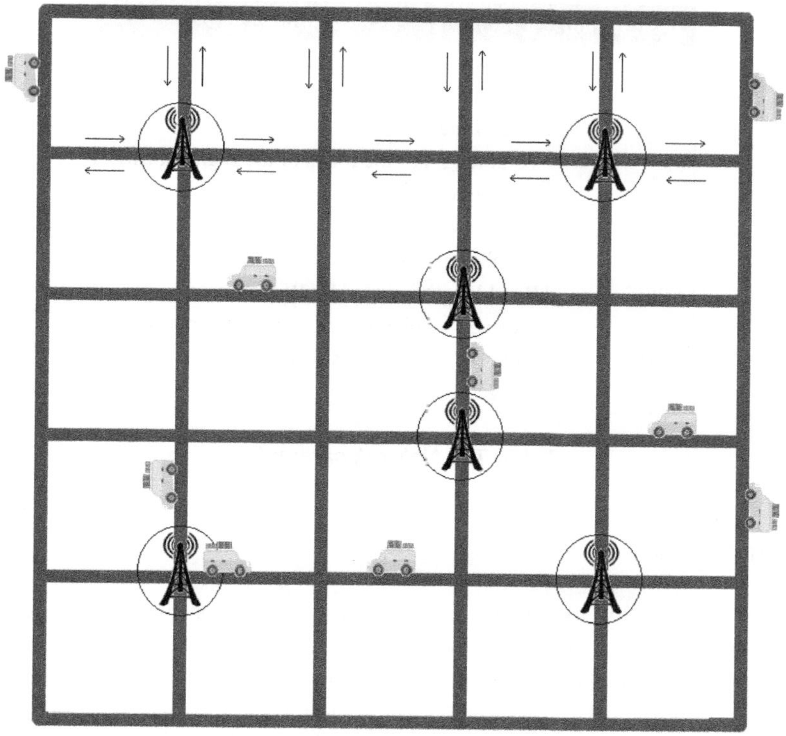

Figure 7.9. *Diagram of the environment. For a color version of this figure, see www.iste.co.uk/mendiboure/transportsystems.zip*

7.4.2. *Network modeling*

DEFINITION of R_I.–

R_I is the distance between two adjacent intersections, divided by two.

R_I = L/2

DEFINITION of ϵ.–

In a Manhattan model, the number of intersections, ϵ, is given by the square of dimension D. Then:

$$\epsilon = D^2$$

Designation	Description
R_I	Radius of an intersection (m)
L	Length of a segment (m)
D	Size of the grid
L_R	Total length of the roads of the segment (m)
N_S	Number of road segments
ϵ	Number of junctions
β	Number of RSUs
LRSU	Road length covered by RSUs (m)
φ	Average number of segments per junction
δRSU	Area covered by an RSU (%)
PRSU	Range of an RSU (m)

Table 7.2. *Network parameters*

DEFINITION OF L_R.–

Cx is temporarily expressed as being an intersection with x entries. In a Manhattan topology, x can be 2, 3 or 4. $C2$ = 4, as there are four corners. $C3$ is the set of intersections on the edge of the city that are not corners.

Also, $C4$ is all the remaining intersections. The road length, RL, is obtained by multiplying the sum of the (Cx x) values with the radius of an intersection, R_I.

$$C2 = 4 \text{ x } 2$$

$$C_3 = 4(D - 2) \text{ x } 3$$

$$C4 = (D - 2)2 \text{ x } 4$$

$$RL = (C2 + C3 + C4) \text{ x } R_I$$

DEFINITION OF Φ.–

We compute the percentage of each type of intersection with the following expressions:

– $\varphi_4 = \frac{(D-2)^2}{D^2} * 100$, the percentage of intersections linked to four segments;

– $\varphi_3 = \frac{4(D-2)}{D^2} * 100$, the percentage of intersections linked to three segments;

– $\varphi_2 = \frac{4}{D^2} * 100$, the percentage of intersections linked to two segments.

The average number, φ, of segments linked to an intersection is expressed as follows:

$$\varphi = \frac{(4 * 2) + (D - 2) * 4 * 3 + (D - 2)^2 * 4}{\varepsilon}$$

DEFINITION OF L_{RSU}.–

The road length covered by the RSUs is expressed as a function of φ and the range of the RSUs (P_{RSU}), which we multiply by the number of RSUs:

$$L_{RSU} = \varphi * P_{RSU} * \beta$$

DEFINITION OF δ_{RSU}.–

The coverage area is the percentage of road covered by an RSU:

$$\delta = L_{RSU}/L_r * 100$$

Designation	Description
P_V	Vehicle communication range (m)
V	Vehicle speed (m/s)
α	Number of vehicles
σ	Data generation frequency (Hz)
ω	Number of sensors
γ	Inter-vehicle distance (m)
ρ	Vehicle density (vehicle/m)
δ_V	Percentage of the network where vehicles are connected to an RSU
λ	Vehicle arrival rate (m/s)
F_V	Frequency of arrival of new vehicles in the range of an RSU (Hz)
F_λ	Number of new vehicles within range (Hz)
$\delta(A, B)$	Function that returns A if $A < B$, B otherwise

Table 7.3. *Communication parameters*

7.4.3. Communication environment modeling

DEFINITION OF γ.–

The vehicles are evenly distributed throughout the network; we define the inter-vehicle distance as the quotient of the road length by the number of vehicles on it:

$$\gamma = L_R/\alpha$$

DEFINITION OF ρ.–

We have the following expression that gives the density, ρ, of vehicles, with the velocity, V ∈ [a; b]:

$$\rho = \lambda \int_a^b \left(\frac{F(V)}{V}\right) dV$$

DEFINITION OF λ.–

Finally, we annotate with "λ" the arrival rate of vehicles on a road. Since *V* is constant.

$$\rho = \frac{\lambda}{V} \text{ and } \lambda = \rho * V$$

DEFINITION OF δ_V.–

The range of a vehicle is usually smaller than that of an RSU. Therefore, if a vehicle is connected to an RSU, it is not necessarily capable of detecting it. We express this number, δ_V, as the surface of the network where the RSU is able to receive messages from vehicles:

$$\Delta_V = (\varphi \times P_V \times \beta)/L_R \times 100$$

DEFINITION OF F_λ.–

The number of vehicles entering (and also leaving) the coverage area each second is obtained by the following expression:

$$F_\lambda = (\lambda \times \varphi \times \beta)$$

DEFINITION OF F_λ.–

By dividing F_λ by the number of RSUs, we obtain the number of vehicles within the range of an RSU every second:

$$F_V = (\lambda \times \varphi)$$

DEFINITION OF ϑ.–

ϑ is a function that takes two parameters and returns the smaller of the two.

1) Based on the following observations:

$$\forall v < 1, \frac{[v]}{v} = 0, \left\lceil \frac{[v]}{v} \right\rceil = 0$$

$$\text{Also, } \forall v \geq 1,\ \ [v] \leq v,\ \ 0 < \frac{[v]}{v} \leq 1,\ \ \left\lceil \frac{[v]}{v} \right\rceil = 1$$

We can say that if v is less than 0, its integer part is 0, and therefore, no matter what it is divided by, the result will be 0; and, when rounded, its value will remain 0. Otherwise, we will obtain 1.

2) We can generalize this with the following function that returns A if $A < B$ and B otherwise:

$$\vartheta(A, B) = B - (B - A) \left\lceil \frac{[B/A]}{B/A} \right\rceil$$

7.4.4. *Message dissemination modeling*

We model the dissemination of the message in the network to consider vehicle-to-vehicle communications. Every second, vehicles send data to other vehicles. Therefore, we need to estimate how many vehicles will send data every second. To do this, the distribution of RSUs is significant.

Figures 7.10(a) and (b) shows the difference between two RSU distributions. Each color represents a dissemination step (i.e. the list of new segments where data is disseminated). For example, step 1 is marked in red, step 2 in blue, etc. In the second case (Figure 7.10(b)), more exchanges are needed (seven steps) to cover the entire network, while in the first case (Figure 7.10(a)), it is only four steps. Therefore, the number of segments reached in step 2 is different: 24 (Figure 7.10(a)) versus 6 (Figure 7.10(b)).

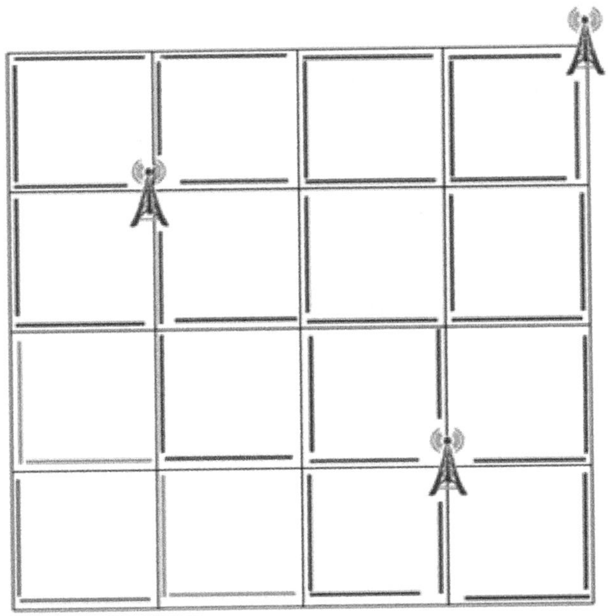

a) Best dissemination scenario case

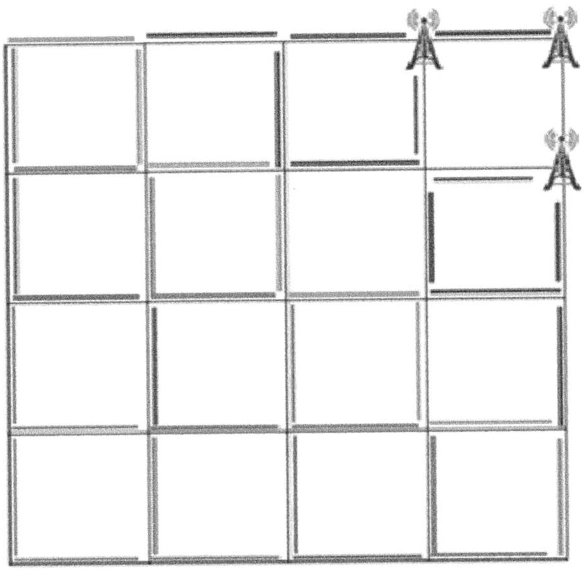

b) Worst dissemination scenario case

Figure 7.10. *Dissemination scenario. For a color version of this figure, see www.iste.co.uk/mendiboure/transportsystems.zip*

Designation	Description
L	Length of a segment (m)
X_I	Indicates the presence (1) or absence (0) of an RSU at an intersection, I
S_X	Probability for a segment to have X RSUs
M_X	Minimum number of V2V communications needed to cover a segment of a configuration x
H_X	Maximum number of V2V communications needed to cover a segment of a configuration x
HM_X	Average number of V2V communications needed to cover a segment of a configuration x
O_X	Number of segments linked to X other segments
$N0$	Number of segments with one RSU
$NRSU$	Number of segments with two RSUs

Table 7.4. *Dissemination parameters*

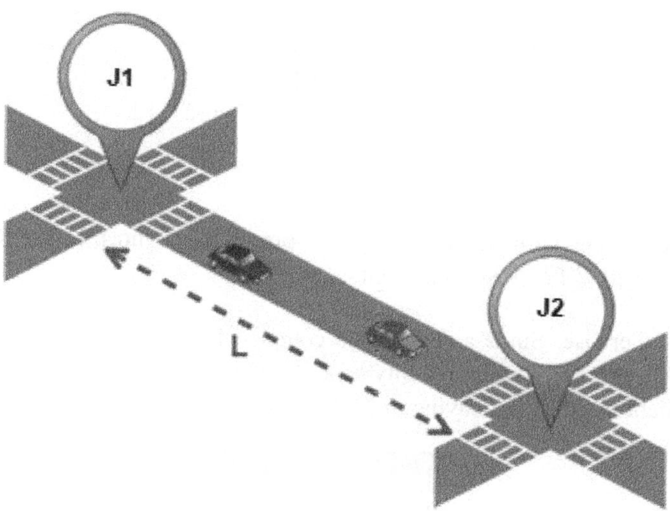

Figure 7.11. *Segment. For a color version of this figure, see www.iste.co.uk/mendiboure/transportsystems.zip*

DEFINITION OF L.–

Figure 7.11 schematically shows the generic configuration of a segment with $L = 2 * R_I$.

7.4.4.1. *Distribution of RSUs*

In order to model dissemination, we first need to determine how many RSUs are placed in the network. Figure 7.12 presents the different possible configurations for a segment.

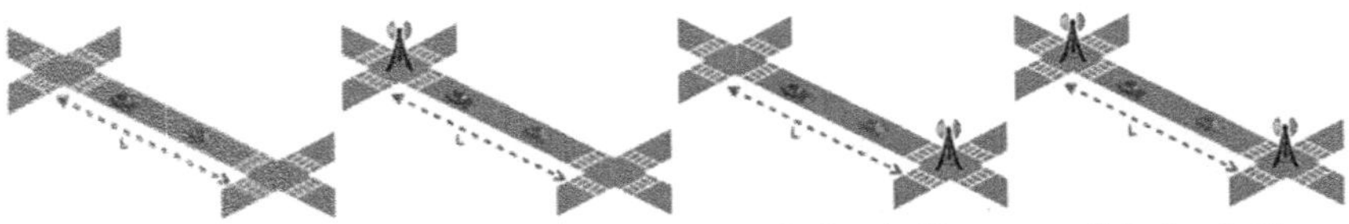

Figure 7.12. *Possible segment configurations. For a color version of this figure, see www.iste.co.uk/mendiboure/transportsystems.zip*

Figures 7.12(a) and (c) represent the segments that have only one RSU. Figure 7.12(a) shows a segment with no RSU, and Figure 7.12(d) shows a segment with two RSUs. Note that Figure 7.12(b) is the same as Figure 7.12(c).

DEFINITION OF *X*.–

We note X_I the variable indicating the presence (or not) of an RSU at an intersection. $X_I = 1$ if an RSU is present,

$X_I = 0$ otherwise. Since a segment is composed of two intersections, we will work with X_{I1} and X_{I2}. The probability of X_I having an RSU is given by $P(X_I = 1)$, and the probability of X_I not having one by $P(X_I = 0)$.

	X_{I1}	X_{I2}	$P(X_{I1})$	$P(X_{I2})$
Config. 1	0	0	$\frac{\varepsilon-\beta}{\varepsilon}$	$\frac{\varepsilon-1-\beta}{\varepsilon-1}$
Config. 2.1	1	0	$\frac{\beta}{\varepsilon}$	$\frac{\varepsilon-\beta}{\varepsilon-1}$
Config. 2.2	0	1	$\frac{\varepsilon-\beta}{\varepsilon}$	$\frac{\beta}{\varepsilon-1}$
Config. 3	1	1	$\frac{\beta}{\varepsilon}$	$\frac{\beta-1}{\varepsilon-1}$

Table 7.5. *Probability of RSU*

DEFINITION OF S_X.–

S_0 is the probability of a segment not having an RSU, S_1 is the probability of having one, and S_2 is the probability of having two. These probabilities are defined as follows:

$$S_0 = \left(\frac{\varepsilon - \beta}{\varepsilon}\right)\left(\frac{\varepsilon - 1 - \beta}{\varepsilon - 1}\right)$$

$$S_1 = 2\left(\frac{\varepsilon - \beta}{\varepsilon}\right)\left(\frac{\beta}{\varepsilon - 1}\right)$$

$$S_2 = \left(\frac{\beta}{\varepsilon}\right)\left(\frac{\beta - 1}{\varepsilon - 1}\right)$$

7.4.4.2. *Vehicles within the transmission range of an RSU*

Let P_{RX} be the probability of a vehicle being within the range of an RSU for a configuration X, and P_{ORX} be the probability that a vehicle is outside the range of an RSU for a configuration X.

	P_R	P_{OR}
Config. 1	0	1
Config. 2	$\frac{P_{RSU}}{L}$	$\frac{L - P_{RSU}}{L}$
Config. 3	$\frac{2P_{RSU}}{L}$	$\frac{L - 2P_{RSU}}{L}$

Table 7.6. *Probability of segments*

7.4.4.3. *Segments*

As shown in Figure 7.13, a segment can connect three (area 1), four (area 2), five (area 3) or six (area 4) segments.

All possible configurations are shown in Figure 7.14.

DEFINITION OF O_X.–

O_X, is the number of segments connected to X segments.

$$O_3 = 8.$$

$O_4 = (\mathrm{D} - 3) * 4.$

$O_5 = (\mathrm{D} - 2) * 4.$

$O_6 = 2 * (\mathrm{D} - 3) * (\mathrm{D} - 2).$

DEFINITION OF N_0.–

The number of segments with a single RSU is defined by $N_0 = N_s * S_1$.

DEFINITION OF N_{RSU}.–

The number of segments connected with 2 RSUs is defined by $N_{RSU} = N_s * S_2$.

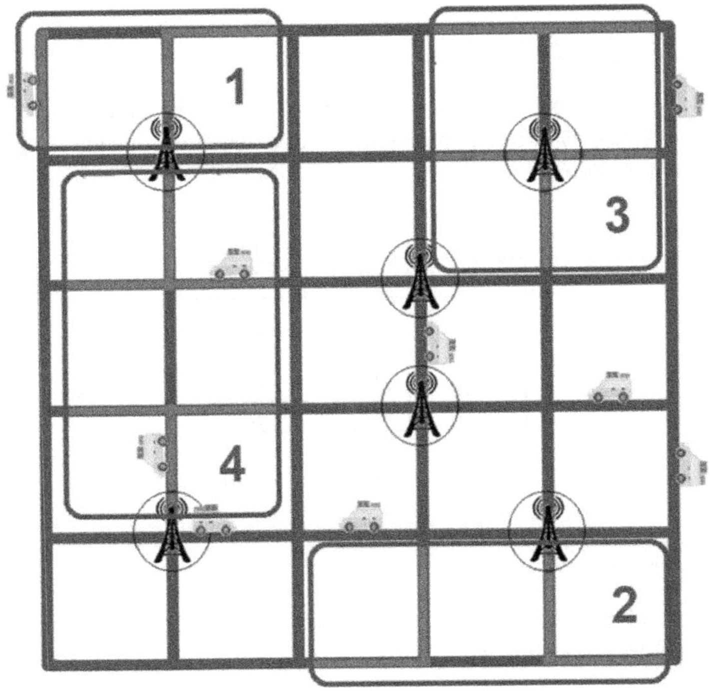

Figure 7.13. *Example of segment configurations. For a color version of this figure, see www.iste.co.uk/mendiboure/transportsystems.zip*

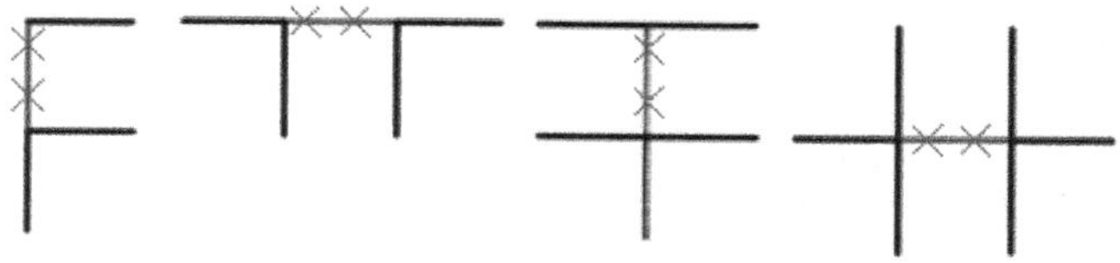

Figure 7.14. *Schematic representations of segment configurations. For a color version of this figure, see www.iste.co.uk/mendiboure/transportsystems.zip*

7.4.4.4. *Dissemination*

For each configuration *X*, *HX* expresses the maximum number of hops needed to cover a segment, *MX* expresses the minimum number of hops and *HMX* expresses the average number. For each configuration, we obtain the values *H*, *M* and *HM* as follows:

	H_X	M_X	HM_X
Config. 1	$\left\lceil \frac{L}{P_V - \gamma} \right\rceil$	$\left\lceil \frac{L}{P_V} \right\rceil$	$\left(\frac{H_1 - M_1}{2}\right)$
Config. 2	$\left\lceil \frac{L - P_{RSU}}{P_V - \gamma} \right\rceil$	$\left\lceil \frac{L - P_{RSU}}{P_V} \right\rceil$	$\left(\frac{H_2 - M_2}{2}\right)$
Config. 3	$\left\lceil \frac{L - 2P_{RSU}}{2(P_V - \gamma)} \right\rceil$	$\frac{L - 2P_{RSU}}{2P_V}$	$\left(\frac{H_3 - M_3}{2}\right)$

Table 7.7. *Number of hops*

We wish to compute N_1, N_2, ..., N_n, with N_1 being the number of segments without any RSU that are connected to at least one segment with one RSU. *N*2 is the number of segments without an RSU that are connected to at least one segment ∈ N_1, without being connected to a segment ∈ N_0, etc. To have the number of segments at level *n*, we separate the computation according to the number of connected roads and then sum them. A segment can be connected to $x = 3$ to 6 segments, and N_{nx} is the probability that a segment is connected to *x* segments, while being connected to at least one segment of level $n - 1$. We deduce the following expression:

$$N_n = \sum_{x=3}^{6} O_x \, N_{n_x} = 1$$

and N_{n_x} is expressed as follows:

$$N_{n_x} = \sum_{y=1}^{x} C_x^y \left(\frac{A * B}{C}\right)$$

with:

$$A = \left(\prod_{w=0}^{y-1} N_{n-1} - w\right)$$

$$B = \left(\prod_{m=0}^{x-y-1} N_s - N_{RSU} - \sum_{k=0}^{n-1} N_k - m\right)$$

$$C = \left(\prod_{m=0}^{x-y-1} N_s - N_{RSU} - z - \sum_{k=0}^{n-1} N_k\right)$$

The variable x refers to the number of roads connected to a segment. This is given by the global formula [1.17] and therefore takes a value between 3 and 6. y refers to the number of these roads that are of level $n - 1$ and takes a value between 1 and x.

Finally, N_{n_x} is computed by summing the probability of having a segment linked to a road of level $n - 1$ with that connected to two roads of level $n - 1$, three roads of level $n - 1$, etc. up to x roads.

The result of computing the probability of having y roads of level $n - 1$ needs to be multiplied by the number of combinations of y among x, because the road, or road of level $n - 1$ can have different positions. For example, if a segment of 05 that is connected to 5 roads is adjacent to a road of level $n - 1$, the latter can have five different positions.

The formula A expresses the probability of being connected to w segments of levels $n - 1$, from which we subtract, for each step, the segments already taken into account, which explains why w is used. We then obtain the following result:

$$(N_{n-1} - 0) * (N_{n-1} - 1) * \ldots * (N_{n-1} - w)$$

We multiply the result of this expression with the probability, for the other segments, of not being of a level lower than n. This is expressed by B.

C is used to express the number of segments of a level higher than n, meaning the total number of segments in the network from which we subtract the segments having two RSUs and the segments of a level lower than n. We must also subtract the segments of a level higher than n already taken into account, which are represented by m. m can take a value of between 0 *et* $x - y - 1$; and if we wished to expand, we would obtain an expression that takes the following form:

$$(N_s - N_{RSU} - \sum_{k=0}^{n-1} N_k) * (N_s - N_{RSU} - \sum_{k=0}^{n-1} N_k - m) * \ldots * (N_s - N_{RSU} - \sum_{k=0}^{n-1} N_k - m)$$

Finally, we divide it by the number of segments of a level higher than $n - 1$. This number is obtained by subtracting the number of segments with two RSUs and the number of segments of a level lower than or equal to $n - 2$ ($\sum_{k=0}^{n-2} N_k$) from the total number of segments N_S. In the same way, we must remove the segments already taken into account at each stage of multiplication, hence z is used, which has a value between 1 and x. We start at 1, as we have to subtract the segment that we have computed (shown in red in Figure 7.14).

7.4.5. *Approaches*

To compare the network load (NL) and the data freshness (FR) with each approach, we propose an analytical study involving the different variables of Tables 7.2–7.4. For each approach, we define the expression of the respective NL and FR. We thus obtain, for each approach, a formula for computing network load and data freshness.

DEFINITION OF *NL*.–

To measure the additional load generated by ICN exchanges, on the G5 channel, we define the network load as the sum of the network load caused by the ICN interests (NL_i) and that caused by the data (NL_d):

$$NL = NL_i + NL_d$$

DEFINITION OF *FR*.–

Data freshness, FR, is defined as the percentage of vehicles receiving the data before expiration:

$$FR = \frac{\alpha_D}{\alpha} * 100$$

7.4.5.1. *Expression of network load and data freshness for NDN*

Figure 7.15 shows the different exchanges that are carried out with the NDN approach in the proposed environment. Firstly, a vehicle sends one interest per sensor that interests it (1). Then, the traffic light equipped with an RSU receives the interests. As a sensor, it responds with its data and, as a Resolution Handler (RH), routes the other interests (2). Each sensor questioned responds to the RH with its own data (3). Finally, the RH may transmit the data from the other traffic lights to the vehicle (4). These operations are repeated regularly.

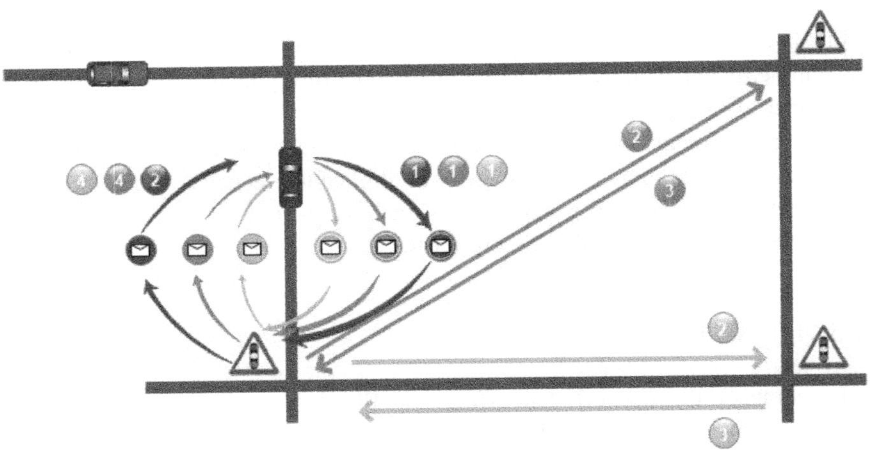

Figure 7.15. *NDN exchanges. For a color version of this figure, see www.iste.co.uk/mendiboure/transportsystems.zip*

DEFINITION OF NL.–

We define NL as the sum of NL_i and NL_d; therefore, for NDN:

$$NL = PO_i * I_s * I_{RSU} * PO_d$$

Thus, we obtain:

$$NL = (S + P_{if}) * \beta\left(\frac{\alpha}{100} * \delta\right) * F_i + \beta\left(\frac{\alpha}{100} * \delta\right) * F_i * (S + P_{df} + P_{if})$$

DEFINITION OF α_D.–

With NDN, the number of α_D vehicles represents the number of vehicles connected to an RSU, to which we add the number of vehicles connecting to RSU between two data updates:

$$\alpha_D = \left(\frac{\alpha}{100} * \delta\right) + \frac{\lambda * \varphi * \beta}{\sigma}$$

DEFINITION OF FR.–

α_D is applied to the broad definition of freshness, which gives us:

$$FR = \frac{\left(\frac{\alpha}{100} * \delta\right) + \frac{\lambda * \varphi * \beta}{\sigma}}{\alpha} * 100$$

7.4.5.2. *Expression of network load and data freshness for RENE*

Figure 7.16 shows the exchanges that occur in the RENE approach in the proposed environment. Firstly (1), a vehicle subscribes to the server for the sensors that interest it. A sensor then generates new data (2). This sensor makes a request to the server to obtain the list of subscribers and how to reach them (3). The server provides the list and the requested information (4). Finally, the sensor is capable of transmitting the data to the ad hoc router (5), which in turn will transmit it to the vehicle concerned (6).

Without the repetition of interests, RENE drastically reduces bandwidth consumption. With its publisher/subscriber approach, RENE also reduces the data latency. However, if a vehicle is not within the range of a router when the data is generated, it will never receive them.

DEFINITION OF NL.–

The network load dedicated to ICN exchanges in the RENE approach is given by the following expression:

$$NL = PO_i * I_s + \frac{\alpha}{100} * \delta_{RENE} * \sigma * PO_d$$

This is equal to:

$$NL = (S + P_{if}) * \omega * F_\lambda + \frac{\alpha}{100} * (\delta + \delta_{RSU})/2 * \sigma * (S + P_{if} + P_{df})$$

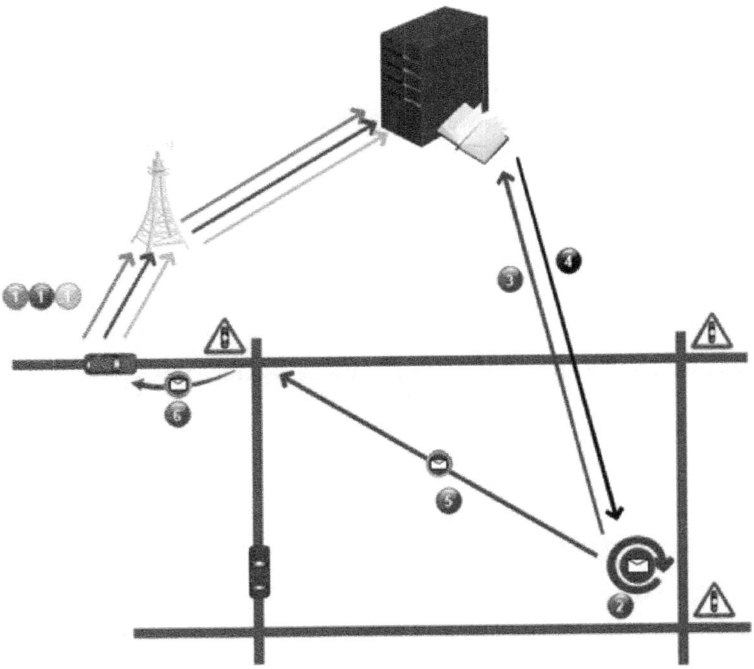

Figure 7.16. *Exchanges in RENE. For a color version of this figure, see www.iste.co.uk/mendiboure/transportsystems.zip*

DEFINITION OF α_D.–

With this approach, α_D is the number of vehicles in the coverage area at the instant when the data is generated; we obtain the following expression:

$$\alpha_D = \left(\frac{\alpha}{100} * \delta_{RENE}\right)$$

DEFINITION OF FR.–

We apply α_D to the general definition of network freshness and obtain:

$$FR = \frac{\left(\frac{\alpha}{100} * \delta\right)}{\alpha} * 100$$

7.4.5.3. *Expression of network load and data freshness for VCDN*

The VCDN approach consists of grouping together data from multiple sensors into a metadata and adding a version number.

DEFINITION OF NL.–

$$NL = PO_i * I_s + \frac{\alpha}{100} * \delta_V * \sigma * PO_d + PO_d * F_\lambda$$

DEFINITION OF α_D.–

With this approach, α_D is obtained by summing the number of vehicles in the coverage area at the moment the data is generated and the number of vehicles entering the coverage area before expiration of the data.

$$\alpha_D = \left(\frac{\alpha}{100} * \delta_V\right) + \frac{F_\lambda}{\sigma}$$

DEFINITION OF *FR*.–

We apply the definition of α_D to the general expression of network freshness:

$$FR = \frac{\left(\frac{\alpha}{100} * \delta_V\right) + \frac{F_\lambda}{\sigma}}{\alpha} * 100$$

7.4.5.4. *Expression of network load and data freshness for VCDNoCAM*

By including messages in the CAMs, we no longer need security data dedicated to ICN messages, and we no longer send a message to a vehicle; however, we broadcast it to all vehicles in an area.

DEFINITION OF *NL*.–

The overall network load generated by the VCDNoCAM approach is expressed as follows:

$$NL = \left(\frac{\alpha}{100} * \delta_{RSU}\right) * F_{CAM} * PO_i + \beta * PO_d * (\vartheta(F_{CAM}, \sigma) + \vartheta(F_{CAM}, F_v))$$

DEFINITION OF α_D.–

For this approach, α_D is computed by summing the number of vehicles in the coverage area at the moment the data is generated and the number of vehicles entering the coverage area before expiration of the data. We thus obtain:

$$\alpha_D = \left(\frac{\alpha}{100} * \delta_{RSU}\right) + \frac{F_\lambda}{\vartheta(\sigma, F_{CAM})}$$

DEFINITION OF FR.–

We apply this expression to the general definition of network freshness:

$$FR = \frac{\left(\frac{\alpha}{100} * \delta_{RSU}\right) + \frac{F_\lambda}{\vartheta(\sigma, F_{CAM})}}{\alpha} * 100$$

7.4.5.5. *Expression of network load and data freshness for GeoVCDN*

With GeoVCDN, we keep the same mechanisms as VCDNoCAM, and we add V2V exchanges. This new functionality requires taking account of the maximum number of hops that data can make in V2V.

DEFINITION OF NL.–

Therefore, we obtain the network load that is generated by our approach: $NL = NL_i + NL_{d0} + NL_{d1} + NL_{d2}$.

With NL, which expresses the network load caused by the segments with 0, 1 and 2 RSUs. The result obtained is an upper limit of the expected result, as we maximize the number of segments without an RSU that are contaminated at each step.

DEFINITION OF α_D.–

For freshness, the worst case is where segments without an RSU do not receive the information. Therefore, we pose the following hypothesis: the length of an intersection is greater than a vehicle's range multiplied by HL, i.e.:

$$L - P_{RSU} > P_V * HL$$

Freshness is given by the number of vehicles in the coverage area when the data is generated, to which we add the number of vehicles entering that area and those receiving the information from other vehicles. This depends on the relationship between the inter-distances and the range of the vehicles. Therefore, in the first case, if the inter-distance is higher than the vehicles' transmission range, α_D is expressed as follows:

$$\left(\frac{\alpha}{100} * \delta_{RSU}\right) + \frac{F_\lambda}{\vartheta(\sigma, F_{CAM})}$$

In the second case, if the inter-distance is lower than the vehicles' transmission range, α_D is obtained with the following expression:

$$\left(\frac{\alpha}{100} * \delta_{RSU}\right) + \frac{\alpha}{N_S} * POR_3 * N_{RSU} * \vartheta\left(\frac{HM_1}{HL}, 1\right) + \frac{\alpha}{N_S} * POR_2 * N_0 * \vartheta\left(\frac{HM_2}{HL}, 1\right)$$

DEFINITION OF *FR*.–

Finally, if we return to the general definition of freshness, if the inter-distance is higher than the vehicles' range, then $FR = \left(\left(\frac{\alpha}{100} * \delta_{RSU}\right) + \frac{F_\lambda}{\vartheta(\sigma, F_{CAM})}\right) / \alpha * 100$; if not:

$$FR = \left(\left(\frac{\alpha}{100} * \delta_{RSU}\right) + \frac{\alpha}{N_S} * POR_3 * N_{RSU} * \vartheta\left(\frac{HM_1}{HL}, 1\right) + \frac{\alpha}{N_S} * POR_2 * N_0 * \vartheta\left(\frac{HM_2}{HL}, 1\right)\right) / \alpha * 100$$

7.4.6. *Discussion*

Firstly, we presented a model in which we can express the behavior of different ICN approaches. This model takes into account the network topology and vehicle mobility. We also proposed a model of message propagation in a vehicular network. We then focused on two performance indicators, namely the network load generated by ICN exchanges and the network freshness.

The definitions of our model allow us to obtain an analytical expression of any ICN approach, as we did with the two "lighter" versions of our approach (VCDN and VCDNoCAM). In the following section, we compare the different approaches based on the analytical model and the results obtained with our simulator.

7.5. Evaluation

In this section, we compare the NDN, RENE, VCDN, VCDNoCAM and GeoVCDN approaches based on the models defined in the previous section. These comparisons aim to confirm and evaluate the benefit of our approach, compared to existing methods. In addition, to validate our analytical model, we develop an ad hoc simulator that takes into account certain aspects of vehicular communication and vehicle mobility. Finally, we test NDN, RENE and GeoVCDN on shortest pathfinding in a connected city.

7.5.1. *Simulator description*

In order to validate our model, we develop a simulator from scratch, with Figure 7.17 illustrating it. This is a discrete-time simulator with a step of 100 ms. It takes into account the vehicle and RSU communication ranges and vehicle mobility. Each vehicle emulates the entire communication stack standardized by ETSI (GeoNetworking, security layer and Cooperative Awareness Messages (CAM)) and simulates NDN, RENE and our approaches. In accordance with the working hypotheses introduced earlier, it uses a Manhattan topology and the vehicles are distributed homogeneously.

At the beginning of a scenario, the RSUs are placed at the intersections randomly and the vehicles are stationary. After all components have been put in place, the simulator core is activated, placing the vehicles in motion at constant speed and proceeding with exchanges between C-ITSs (vehicles and RSUs). After a certain time, the RSUs start transmitting at regular intervals for 10 seconds. At the end of the 10 seconds, we note the results concerning network load and freshness. These operations are repeated 20 times from the placement of the RSUs. Finally, we average the results obtained.

The simulations were run on a computer with the configuration described in Table 7.8. As a result of its performance, we had to limit the scale of the simulations; the default values for each variable are given in Table 7.9.

Nevertheless, despite these limitations, we can distinguish the trends of each approach. Note that the values chosen are realistic and based on observations made in the European C-ITS deployment projects (Scoop, InterCor and C-Roads).

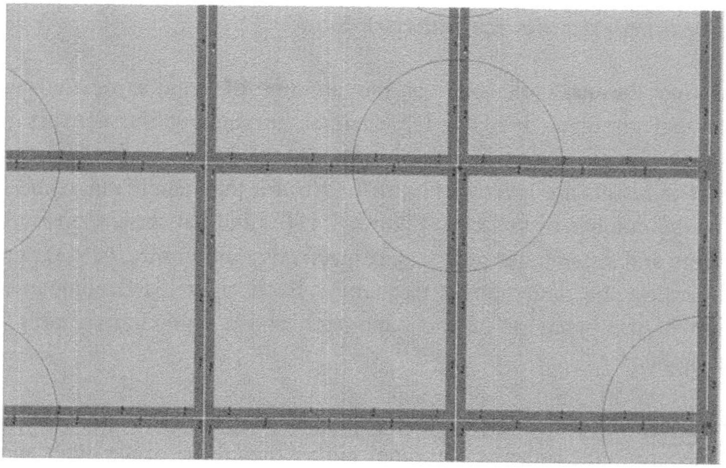

Figure 7.17. *Simulation extract. For a color version of this figure, see www.iste.co.uk/mendiboure/transportsystems.zip*

Component	Model
CPU	i7 @3.6 GHZ
RAM	32 Gb
OS	Windows 10 x64
GPU	Intel HD Graphics 630
IDE	Eclipse neon 3
Programming language	Java 8

Table 7.8. *Computer configuration*

Designation	Description	Value
R_I	Intersection size	300 m
P_V	Vehicle range	200 m
P_{RSU}	RSU range	250 m
D	Grid size	4
A	Number of vehicles	267
B	Number of RSUs	12
Σ	Data update frequency	1 Hz
Ω	Number of sensors	12

Table 7.9. *Variables and default values*

7.5.2. *Simulation results for network load*

We define the network load as the amount of data associated with ICN exchanges that circulate over the G5 channel, throughout the network. We will compare the behavior of each approach further to the variations in the parameters of Table 7.9. The results are given in Figure 7.19(b) for the vehicle range and in Figure 7.19(a) for the number of vehicles. Figures 7.19(c) and (d) show the results related to RSU range and the number of RSUs, respectively, and Figure 7.19(e) presents the results regarding the data update frequency. Each figure is accompanied by an explanation of the behavior of each approach and a comparative table with the analytical model.

For Figures 7.18 and 7.19, we do not present the results of the NDN approach, because they are in the order of 100 times higher than those of the other approaches, making it impossible to read a comparison of the latter. We illustrate this in Figure 7.18, where, for 20,000 vehicles, NDN generates 1.2 GB/s of data; whereas, VCDN generates only 12 MB.

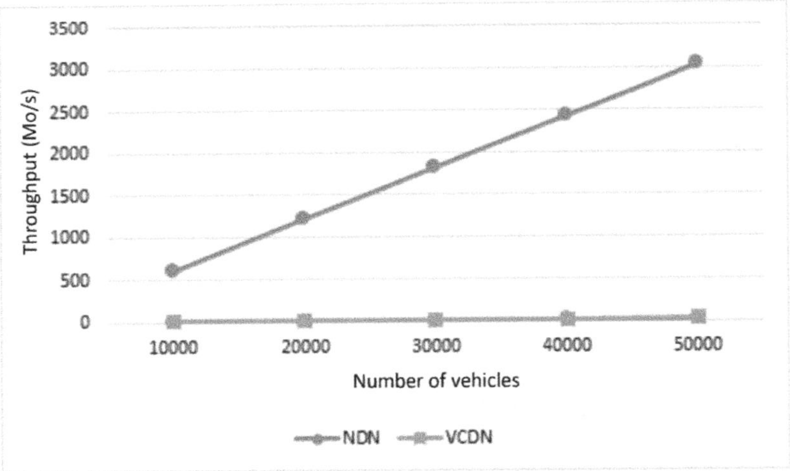

Figure 7.18. *Network load – number of vehicles – NDN versus VCDN. For a color version of this figure, see www.iste.co.uk/mendiboure/transportsystems.zip*

– *NDN*: the amount of data and interests increases in proportion to the number of vehicles in the network. On the one hand, the data is sent once per vehicle and each time a vehicle requests it. On the other hand, there are as many security headers as there are sensors.

– *VCDN*: only the number of interests increases, and it increases more slowly; the data is sent only when updated. On the contrary, the approach benefits from a gain in security data. Indeed, there is only one security header for all sensors. We can see that the metadata and version-number mechanisms make a significant difference by reducing the network load caused by ICN exchanges and making VCDN more scalable.

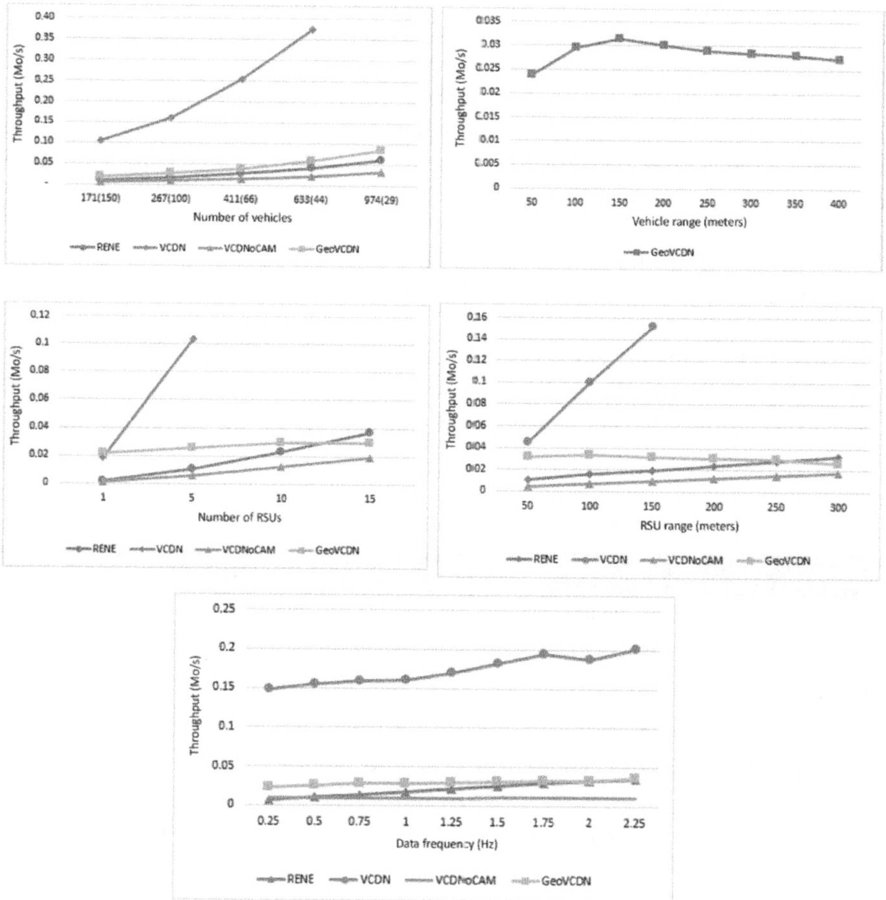

Figure 7.19. *Comparison of network load. For a color version of this figure, see www.iste.co.uk/mendiboure/transportsystems.zip*

7.5.2.1. *Number of vehicles (Figure 7.19(a))*

Approach	Expected value	Observed value	Difference
NDN	3,507,200.00	3,712,913.93	+5.86%
RENE	27,021.88	27,453.20	+1.5%
VCDN	252,934.88	254,684.26	+0.69%
VCDNoCAM	14,596.58	14,678.86	+0.56%
GeoVCDN	40,220.45	39,822.73	-0.99%

Table 7.10. *Network load – 411 vehicles*

– *RENE*: vehicles send more interests than VCDN (one per sensor), but only data from sensors whose status have changed is sent. Since data size is larger than that of an interest, the RENE performance is better than VCDN when the number of vehicles increases.

– *VCDNoCAM*: the number of vehicles has no effect on the network load caused by the data sending. Only the number of interests increases.

– *GeoVCDN*: with all vehicles emitting interests to benefit from V2V exchanges, the resulting network load is higher than for other approaches. However, despite this mechanism causing a greater network load, this approach is more scalable than with RENE.

7.5.2.2. *Vehicle range (Figure 7.19(b))*

Approach	Expected value	Observed value	Difference
NDN	5,126,400.00	5,422,778.06	+6%
RENE	28,483.56	28,486.66	+0.01%
VCDN	251,951.88	253,434.53	+0.59%
VCDNoCAM	15,049.23	15,079.13	+0.20%
GeoVCDN	31,614.90	31,383	-0.73%

Table 7.11. *Comparison of network load – vehicle range 150 m*

– *GeoVCDN*: we see a slight increase in network load with a range of 50–100 m, which corresponds to the phase where the vehicles' range is not sufficient to communicate with their neighbors. In this simulation, we had an inter-vehicle distance of 107 m. After this threshold, we see a decrease in network load as the vehicle range increases, since the number of hops, to cover a segment of road, decreases.

– *Other approaches*: considering that the range of vehicles is less than the range of the RSUs (which is a realistic consideration), this only impacts the GeoVCDN approach, since it is the only approach that supports V2V exchanges.

7.5.2.3. *Number of RSUs (Figure 7.19(c))*

In intelligent transport systems networks, RSUs are the portals between vehicles and the rest of the world. Increasing the number or range of RSUs will automatically increase the number of data access points. Therefore, in most cases, the network load will increase. For each approach, we show the consequences of variations in RSU number or range.

The coverage area is also affected by these variations. The larger the area, the greater the number of vehicles concerned by ICN messages.

Approach	Expected value	Observed value	Difference
NDN	890,000.00	956,702.13	+7.49%
RENE	10,298.97	10,469.80	+1.69%
VCDN	100,984.96	103,212.86	+2.2%
VCDNoCAM	5,889.28	6,001.46	+1.90%
GeoVCDN	27,812.43	25,939.06	-6.73%

Table 7.12. *Comparison of network load – five RSUs*

– *RENE and VCDN*: increasing the number of RSUs leads to a significant increase in network load. For RENE, the network load generated by the interests is proportional to the number of entry points of the coverage area; while, for VCDN, it is the surface area of the coverage area that counts.

– *GeoVCDN*: increasing these parameters results in a reduction in bandwidth consumption. Indeed, the more RSUs there are, the rarer the exchanges between vehicles.

7.6. Simulation results for data utility

7.6.1. *Simulation results for data validity*

7.6.1.1. *Range of RSUs (Figure 7.19(d))*

Approach	Expected value	Observed value	Difference
NDN	3,075,840.00	3,240,097.20	+5.34%
RENE	19,672.56	19,734.53	+0.32%
VCDN	151,559.88	151,847.26	+0.19%
VCDNoCAM	9,709.23	9,680.13	-0.30%
GeoVCDN	30,725.85	31,360.93	+2.07%

Table 7.13. *Comparison of network load – RSU range 150 m*

– *VCDN*: the range of the RSUs modifies the coverage area and, consequently, the number of vehicles receiving the data.

The result is an increase in the number of interests and data sent.

– *RENE*: the coverage area does not affect the network load related to interests; these represent the bulk of the load related to this approach. However, we send the data to more vehicles, which is why we can observe a slight increase.

– *VCDNoCAM*: changes in the coverage area have no effect on the data-related load. They only affect the interest-related load.

– *GeoVCDN*: it is the only approach whose network load tends to decrease when the RSU range increases; the V2V exchanges necessary to disseminate the data are less numerous.

7.6.1.2. *Data validity period (Figure 7.19(e))*

Approach	Expected value	Observed value	Difference
NDN	2,278,400	2,338,049.13	+2.6%
RENE	17,554.36	17,281.2667	-1.55%
VCDN	164,315.36	160,374.2	-2.39%
VCDNoCAM	9,684.26	9,462.53	-2.28%
GeoVCDN	28,900.20	28,399.6	-1.73%

Table 7.14. *Network load comparison – data update frequency*

– *VCDN, RENE and VCDNoCAM*: the increase in the update frequency leads to an increase in the data passing through the network.

– *VCDNoCAM*: the maximum frequency of the CAMs is 10 Hz; beyond this stage, the network load is constant.

– *GeoVCDN*: the higher the frequency, the fewer exchanges there are between vehicles. Beyond 10 Hz, exchanges between vehicles decrease; the hop number limit is computed according to the update frequency. For this reason, from 10 Hz onwards, the higher the update frequency, the less the network bandwidth is occupied.

7.6.2. *Simulation results for data freshness*

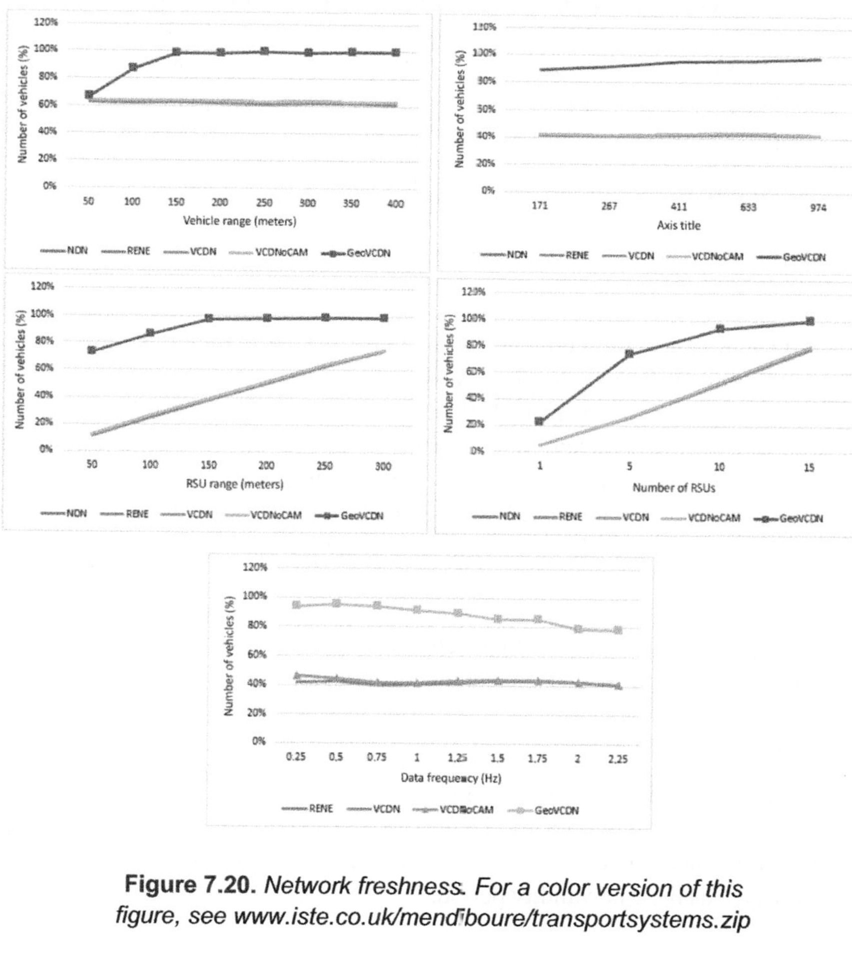

Figure 7.20. *Network freshness. For a color version of this figure, see www.iste.co.uk/mendiboure/transportsystems.zip*

Network freshness is defined as the percentage of vehicles receiving data before expiration. The data expires when updated data has been generated. In our environment, the data update frequency (σ) also gives the validity period of the data. We will compare the behavior of each approach by varying the parameters of Table 7.9. The results are given in Figure 7.20(a) for the vehicle range and in Figure 7.20(b) for the number of vehicles. Figures 7.20(c) and (d) show the results related to the range of the RSUs and the number of RSUs, respectively, and Figure 7.20(e) presents the results regarding the data update frequency. Each figure is accompanied by an explanation of the behavior of each approach and a comparative table with the analytical model.

7.6.2.1. *Vehicle range*

Approach	Expected value	Observed value	Difference
NDN	64.13%	64%	-0.13 pt
RENE	62.50%	63%	+0.5 pt
VCDN	64.13%	64%	-0.13 pt
VCDNoCAM	64.13%	64%	-0.13 pt
GeoVCDN	95%	98%	+ 3pt

Table 7.15. *Comparison of network freshness – vehicle range 150 m*

– *GeoVCDN*: when the communication range of the vehicles is sufficient to reach neighboring vehicles, the data freshness increases rapidly until it reaches 100%. For example, with a range of 150 m, we reach a freshness of 100% compared to 60% for the other approaches. With a range lower than the inter-distance, we still do better, with almost 16% more vehicles receiving the data; thanks to the two-way traffic. The more the range is increased, the more the broadcast from client to client is optimized, ultimately reaching the optimal freshness if the other parameters allow it.

– *Other approaches*: the vehicle communication range has no effect on approaches that do not manage V2V communication.

7.6.2.2. *Number of vehicles*

– *GeoVCDN*: the increase in the number of vehicles has a positive effect on network freshness as it optimizes the use of the maximum transmission range, allowing a larger area to be covered. Note that network freshness is limited by other parameters such as the data validity period.

– *Other approaches*: increasing the number of vehicles has no effect on the approaches that do not handle the V2V exchanges. As expected, the percentage of vehicles in the RSU coverage area and entering this coverage area before the expiration of the data remains the same, since the vehicle distribution is homogeneous and their speed is constant.

Approach	Expected value	Observed value	Difference
NDN	42.75%	42%	-0.75 pt
RENE	41.67%	41%	-0.67 pt
VCDN	42.75%	42%	-0.75 pt
VCDNoCAM	42.75%	42%	-0.75 pt
GeoVCDN	76.67%	91%	+24.33 pt

Table 7.16. *Comparison of network freshness – 267 vehicles*

7.6.2.3. *RSU range*

Approach	Expected value	Observed value	Difference
NDN	26.63%	26%	-0.63 pt
RENE	25%	25%	0 pt
VCDN	26.63%	26%	-0.63 pt
VCDNoCAM	26.63%	39%	-0.63 pt
GeoVCDN	95%	98%	+3 pt

Table 7.17. *Comparison of network freshness – RSU range 100 m*

– *GeoVCDN*: the more the RSU range is increased, the more the number of V2V hops needed to disseminate the data is reduced. This makes it possible to reach 100% network freshness.

– *Other approaches*: increasing the range of the RSUs means increasing the number of vehicles within their range; hence, we observe this improvement in network freshness.

7.6.2.4. *Number of RSUs*

Approach	Expected value	Observed value	Difference
NDN	26.72%	27%	+0.28 pt
RENE	25.04%	26%	-0.4 pt
VCDN	26.72%	27%	-0.28 pt
VCDNoCAM	26.72%	27%	-0.28 pt
GeoVCDN	54.17%	74%	+19.83 pt

Table 7.18. *Comparison of network freshness – eight RSUs*

– *All approaches*: in Figure 7.20(d), we can see that increasing the number of RSUs has a positive effect on network freshness for all approaches. This increases the number of entry points of the coverage area and consequently increases the number of vehicles passing through it with the expiration of data. For GeoVCDN, increasing the number of RSUs also means increasing the number of starting points for propagation of the data in V2V, which explains why the curve grows much faster than those of the other approaches (Figure 7.20(d)). A comparison of the expected values and observed values is given in Table 7.18.

7.6.2.5. *Data validity period*

– *VCDN, NDN and VCDNoCAM*: we observe a slight decrease caused by the decrease in the number of vehicles entering the coverage area between two versions of the data.

– *RENE*: regardless of the frequency, the number of vehicles in the coverage area is constant (as many vehicles leave as enter); therefore, the network freshness remains constant.

– *GeoVCDN*: by increasing the data update frequency, the number of possible V2V exchanges, and therefore, its penetration into the network is reduced. However, even with a very high frequency, GeoVCDN remains more efficient.

7.6.2.6. *Conclusion*

Tables 7.15–7.18 highlight the small gap between our analytical model and the simulations. In the case of GeoVCDN, our analytical model is pessimistic with respect to the network freshness obtained, since it is considered that messages only propagate on segments with an RSU, so the simulation will always give better results.

7.6.3. *Discussion of the simulation*

The NDN approach gives disastrous results in this environment where the data is large in number and changes often, resulting in the need to frequently repeat each request for each sensor in order to keep updated. NDN has been designed to make the load lighter across the entire network but without taking into account the final link to the client. However, in vehicular networks, the G5 channel has a low bandwidth capacity. In order to authenticate and ensure the integrity of all messages, the latter is quickly overloaded.

The RENE approach is more suited to this type of environment, since the data is distributed to the interested parties when it changes and is only distributed when it changes. In addition, the subscription is less frequent than its equivalent in NDN. However, as with NDN, the final link with the client is not taken into account and all vehicles within range of the same RSU will subscribe individually and receive the information individually.

The VCDN results, with the version number and metadata mechanism, are not entirely satisfactory. VCDNoCAM provides the same level of service as existing approaches with respect to freshness. By exploiting mechanisms of technologies already in place (ITS-G5 and CAM), we managed to substantially reduce the cost related to ICN exchanges on the bandwidth.

For GeoVCDN, the addition of V2V transfers to disseminate the data has improved network freshness but with a cost on the load; this cost remains lower than what is proposed elsewhere and V2V transfers are not an obligation. In addition, we have highlighted that our proposal behaves better with respect to scalability.

7.7. Use case study

In this section, we propose applying our data exchange method, in a context cloud, to the pathfinding problem.

The simulator used is the same as that shown in Figure 7.17: a Manhattan-like city and the vehicles uniformly distributed in it. Firstly, we place traffic lights and RSUs in five fixed intersections. A traffic light has two possible configurations: fast or slow. The fast state means that the light changes phase every 5 seconds; with the slow state, it is 15 seconds per phase. Then, we choose at random one or more intersections that will also be equipped with RSUs.

At the beginning of the simulation, we insert an additional three vehicles. One implements the NDN approach, the other RENE and the last implements GeoVCDN. In their initial state, each vehicle has knowledge of the topology of the city and the state of each light; the three vehicles will choose to take the same path. After 2 seconds, all traffic lights switch state from slow to fast or from fast to slow, which naturally changes the fastest road. An ICN packet is transmitted by the RSUs to inform of change; all vehicles within the coverage area receive the data. Among vehicles outside the coverage area, those implementing the NDN approach will receive the data when entering the area, those implementing GeoVCDN will receive it from the other vehicles, and for those implementing RENE, it will never be received.

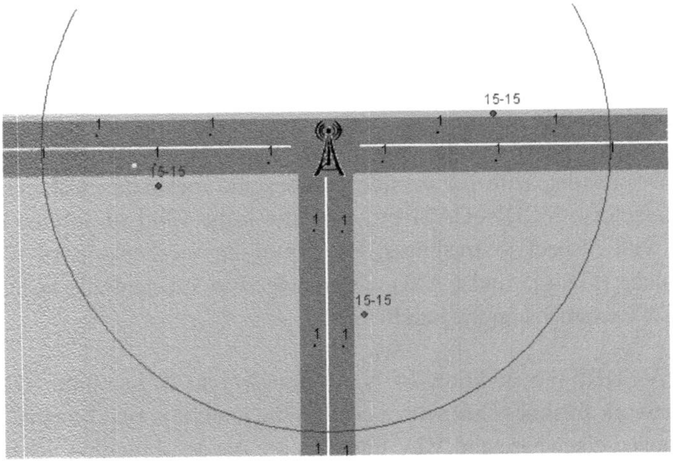

Figure 7.21. *Illustration of an intersection of the simulator. For a color version of this figure, see www.iste.co.uk/mendiboure/transportsystems.zip*

In Figure 7.21, an RSU is placed in the center of the intersection, as well as a traffic light at each entry. The red circle represents the transmission range of the RSU. The yellow square represents a vehicle with GeoVCDN.

7.7.1. *Scenario*

We ran four scenarios with different numbers of random RSUs, which are added to the five fixed ones:

– Scenario 1: one random RSU.

– Scenario 2: two random RSUs.

– Scenario 3: eight random RSUs.

– Scenario 4: all the intersections are equipped with RSUs.

For each scenario, we ran 100 simulations (the position of the RSUs changing each time). At the end, we measured the time a vehicle spent in the city for each approach.

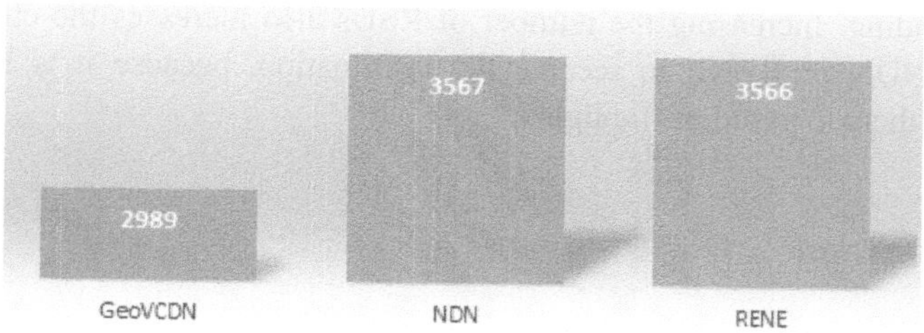

Figure 7.22. *Best pathfinding with one random RSU. For a color version of this figure, see www.iste.co.uk/mendiboure/transportsystems.zip*

Figure 7.22 shows the results for the first scenario. Vehicles using GeoVCDN spent 2,989 seconds on the road, while the others (with NDN and RENE) spent 3,566 seconds on the road, i.e. a 17% improvement with our approach.

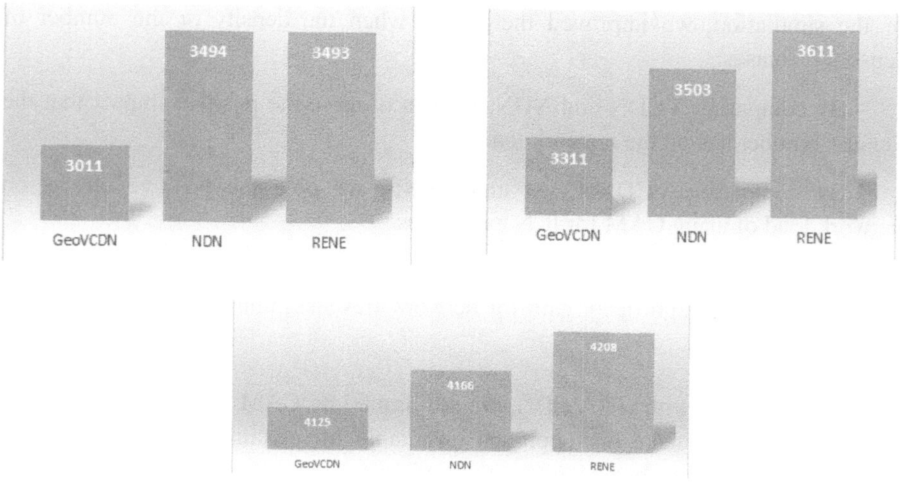

Figure 7.23. *Best pathfinding for different numbers of RSUs (2, 8 and 11)*

With Figure 7.23(a)–(c), which is the results of scenarios 2, 3 and 4, respectively, we observe that the more RSUs there are, the closer the results are. With RSUs at each intersection, the difference is only 1–2%.

With few RSUs, the initial best pathfinding (without dynamic information) had the effect of avoiding traffic lights to save time. Ironically, by avoiding traffic lights, the vehicles could not receive dynamic information and therefore could not update their pathfinding. Increasing the number of RSUs also increases the chances for a vehicle in NDN or RENE to receive the information, because it is increasingly difficult for them to avoid traffic lights.

7.7.2. *Discussion*

For the analytical model, we made two choices: (1) for network load, message dissemination was maximized; and (2) data freshness was minimized. Therefore, the simulation was expected to give better results than the analytical model for our approach. The results obtained are consistent with those expected. The analytical expression of our approach could be improved, but we were nevertheless able to confirm that it is better than the others, even in the worst case.

Table 7.19 presents an evaluation of the results obtained for each approach on the two performance indicators that we examined. Whether for the analytical model or the simulation, we improved the results when the density or the number of sensors increases:

– By comparing VCDN and NDN, we can observe the positive impact that the version number has on the network load.

– By comparing VCDNoCAM and RENE, we note the positive effects on network load of using CAM for ICN exchanges.

– By comparing GeoVCDN with VCDNoCAM, we can observe the effects of V2V exchanges: a significant gain for network freshness and a low additional cost for the network load.

Let us add that with GeoVCDN, we are limited by CAM frequency, which can affect latency and make our approach unsuitable for certain "critical" use cases. Nevertheless, it allows bandwidth to be freed up for the said use cases. As we have shown with pathfinding, we obtain both better quality of service and lower bandwidth consumption.

	Parameters	NDN	RENE	VCDN	VCDNoCAM	GeoVCDN
Network load	Number of vehicles	Very high	Medium	High	Very low	Low
	Size of a segment	Very high	Low	High	Very low	Medium
	RSU range	Very high	Medium	High	Very low	Medium
	Coverage area	Very high	Medium	High	Very low	Low
	Vehicle range	Very high	Low	High	Very low	Low
	Number of RSUs	Very high	Medium	High	Very low	Very low
	Data update frequency	Very high	Low	High	Very low	Very low
Data freshness	Number of vehicles	Poor	Poor	Poor	Poor	Very good
	RSU range	Medium	Medium	Medium	Medium	Very good
	Vehicle range	Poor	Poor	Poor	Poor	Very good
	Number of RSUs	Medium	Medium	Medium	Medium	Good
	Data update frequency	Poor	Poor	Poor	Poor	Medium

Table 7.19. *Performance of approaches. For a color version of this table, see www.iste.co.uk/mendiboure/transportsystems.zip*

7.8. Conclusion

In this chapter, we presented our approach based on a context cloud architecture formed by C-ITS stations. The connected city was considered as an environment for our study, but it can be easily adapted to any type of environment: a country, a shopping center, a region, a stretch of road, etc. We used the ICN paradigm for data exchange within the cloud. We chose this approach because it is suitable for vehicular networks (data-oriented, rather than host-oriented) and facilitates mobility management. We included ICN exchanges (interest and data) in the application beacon (CAM) standardized by ETSI, meaning that our approach can be easily implemented in the currently deployed C-ITSs, without any modification of the standards. We also used a central C-ITS station to gather data from multiple sensors and add a version number to avoid rebroadcasting data that had not been updated.

We compared our approach to two existing solutions, one push-oriented (RENE) and the other pull-oriented (NDN). The comparison was made on two performance indicators: network load and data freshness. For this, we proposed a probabilistic model that takes into account vehicle mobility, the number and random layout of RSUs and the dissemination of vehicle-to-vehicle messages. In order to confirm the results of this model, we developed a simulator that implements the ETSI standardized communication stack, mobility and ICN exchanges. Slight variations were observed between the simulation and the analytical results (generally less than

2%), which can be explained by the type of simulator (discrete time) and the working hypotheses posed for the analytical model.

Our results were promising, as shown in Table 7.19; both network load and data freshness had been improved. In addition, we showed that our approach adapts better to scale up (lots of data, lots of vehicles and lots of RSUs).

7.9. Acknowledgment

This work was made possible by EC Grant No. INEA/CEF/TRAN/A2014/1042281 from the ENEA agency for the SCOOP project. The statements made here are the sole responsibility of the authors.

7.10. References

Ahlgren, B., Dannewitz, C., Imbrenda, C., Kutscher, D., Ohlman, B. (2012). A survey of information-centric networking. *IEEE Communications Magazine*, 50, 26–36.

Alhowaidi, M., Nadig, D., Hu, B., Ramamurthy, B., Bockelman, B. (2021). Cache management for large data transfers and multipath forwarding strategies in Named Data Networking. *Computer Networks*, 199, 108437.

Amadeo, M., Campolo, C., Quevedo, J., Corujo, D., Molinaro, A., Iera, A., Vasilakos, A.V. (2016a). Information-centric networking for the internet of things: Challenges and opportunities. *IEEE Network*, 30, 92–100.

Amadeo, M., Campolo, C., Molinaro, A. (2016b). Information-centric networking for connected vehicles: A survey and future perspectives. *IEEE Communications Magazine*, 54, 98–104.

Amadeo, M., Ruggeri, G., Campolo, C., Molinaro, A. (2021). Diversity-improved caching of popular transient contents in vehicular named data networking. *Computer Networks*, 184, 107625.

ICN2020: Horizon 2020 Project (2020). Information Centric Networking [Online]. Available at: www.icn2020.org/information-centric-networking/.

Bitam, S. and Mellouk, A. (2012). ITS-cloud: Cloud computing for Intelligent Transportation System. In *2012 IEEE Global Communications Conference (GLOBECOM)*, Anaheim, CA, 2054–2059.

Bo Yu, C.X. (2010). Vehicular ad-hoc networks: An information-centric perspective. *ZTE Communications*, 8, 42.

Bouk, S.H., Ahmed, S.H., Kim, D., Song, H. (2017). Named-data-networking-based ITS for smart cities. *IEEE Communications Magazine*, 55, 105–111.

Cheriton, D. and Gritter, M. (2000). TRIAD: A new next-generation internet architecture. Technical Report, Stanford University, Stanford, CA, January.

Dannewitz, C. (2009). NetInf: An information-centric design for the future internet. In *Proc. 3rd GI/ITG KuVS Workshop on The Future Internet*, Munich.

Djama, A., Djamaa, B., Senouci, M.R. (2020). Information-centric networking solutions for the internet of things: A systematic mapping review. *Computer Communications*, 159, 37–59.

ETSI (2014). EN 302 636-3 V1.2.1; Intelligent Transport Systems (ITS); Vehicular communications; GeoNetworking; Part 3: Network architecture. ETSI Standard.

ETSI (2015). TS 103 175 V1.1.1; Intelligent Transport Systems (ITS); Cross layer DCC management entity for operation in the ITS G5A and ITS G5B medium. ETSI Standard.

ETSI (2019a). EN 302 637-2 V1.4.1; Intelligent Transport Systems (ITS); Vehicular communications; Basic set of applications; Part 2: Specification of cooperative awareness basic service. ETSI Standard.

ETSI (2019b). EN 302 637-3 V1.3.1; Intelligent Transport Systems (ITS); Vehicular communications; Basic set of applications; Part 3: Specifications of decentralized environmental notification basic service. ETSI Standard.

Feng, C., Han, P., Zhang, X., Yang, B., Liu, Y., Guo, L. (2022). Computation offloading in mobile edge computing networks: A survey. *Journal of Network and Computer Applications*, 103366.

Festag, A. (2014). Cooperative intelligent transport systems standards in europe. *IEEE Communications Magazine*, 52, 166–172.

Gerla, M. (2012). Vehicular cloud computing. In *The 11th Annual Mediterranean Ad Hoc Networking Workshop (Med-Hoc-Net)*, Ayia Napa, 152–155.

Ghafoor, K., Abu Bakar, K., Mohammed, M., Lloret, J. (2013). *Vehicular Cloud Computing: Trends and Challenges*. IGI Global, Hershey, PA.

Grewe, D., Wagner, M., Frey, H. (2018). A domain-specific comparison of information-centric networking architectures for connected vehicles. *IEEE Communications Surveys Tutorials*, 20, 2372–2388.

Hamdani, S.E., Benamar, N., Younis, M. (2020). Pedestrian support in intelligent transportation systems: Challenges, solutions and open issues. *Transportation Research Part C: Emerging Technologies*, 121, 102856.

Ijemaru, G.K., Ang, L.M., Seng, K.P. (2022). Transformation from IoT to IoV for waste management in smart cities. *Journal of Network and Computer Applications*, 204, 103393.

ISO (2019). TS 19091:2019; Intelligent Transport Systems – Cooperative ITS – Using V2I and I2V communications for applications related to signalized intersections. ISO Standard.

Khan, W.Z., Ahmed, E., Hakak, S., Yaqoob, I., Ahmed, A. (2019). Edge computing: A survey. *Future Generation Computer Systems*, 97, 219–235.

Liu, X., Li, Z., Yang, P., Dong, Y. (2017). Information-centric mobile ad hoc networks and content routing. *Ad Hoc Network*, 58, 255–268.

Morello, R., Mukhopadhyay, S.C., Liu, Z., Slomovitz, D., Samantaray, S.R. (2017). Advances on sensing technologies for smart cities and power grids: A review. *IEEE Sensors Journal*, 17, 7596–7610.

Musaddiq, A., Ali, R., Bajracharya, R., Qadri, Y.A., Al-Turjman, F., Kim, S.W. (2020). Trends, issues, and challenges in the domain of IoT-based vehicular cloud network. In *Unmanned Aerial Vehicles in Smart Cities*, Al-Turjman, F. (ed.). Springer International Publishing, Cham, 49–64.

Petrolo, R., Loscrì, V., Mitton, N. (2015). Towards a smart city based on cloud of things, a survey on the smart city vision and paradigms. *Transactions on Emerging Telecommunications Technologies*, 1–11.

Qiu, T., Liu, X., Li, K., Hu, Q., Sangaiah, A.K., Chen, N. (2018). Community-aware data propagation with small world feature for internet of vehicles. *IEEE Communications Magazine*, 56, 86–91.

Rao, B.P., Saluia, P., Sharma, N., Mittal, A., Sharma, S.V. (2012). Cloud computing for Internet of Things & sensing based applications. In *2012 Sixth International Conference on Sensing Technology (ICST)*, 374–380.

Suciu, G., Vulpe, A., Halunga, S., Fratu, O., Todoran, G., Suciu, V. (2013). Smart cities built on resilient cloud computing and secure internet of things. In *2013 19th International Conference on Control Systems and Computer Science (CSCS)*, 513–518.

Thomas, Y., Fotiou, N., Toumpis, S., Polyzos, G. (2020). Improving mobile ad hoc networks using hybrid IP-Information centric networking. *Computer Communications*, 156.

Whaiduzzaman, M., Sookhak, M., Gani, A., Buyya, R. (2014). A survey on vehicular cloud computing. *Journal of Network and Computer Applications*, 40, 325–344.

Xylomenos, G., Ververidis, C.N., Siris, V.A., Fotiou, N., Tsilopoulos, C., Vasilakos, X., Katsaros, K.V., Polyzos, G.C. (2014). A survey of information-centric networking research. *IEEE Communications Surveys Tutorials*, 16, 1024–1049.

Yi, B., Wang, X., Huang, M. (2021). Content delivery enhancement in vehicular social network with better routing and caching mechanism. *Journal of Network and Computer Applications*, 177, 102952.

Zaslavsky, A., Perera, C., Georgakopoulos, D. (2013). Sensing as a service and big data. arXiv preprint arXiv:1301.0159.

8

Optimal Placement of Edge Servers in C-ITS Systems

Sabri KHAMARI, Toufik AHMED and Mohamed MOSBAH
LaBRI, Bordeaux INP, France

8.1. Introduction

Vehicle networks are considered a significant element for the future of Intelligent Transport Systems (ITS). Vehicular applications offer safer and more comfortable driving, improved traffic efficiency and a variety of value-added applications (Lin 2017). Vehicular applications are emerging and becoming computationally intensive, but they require complex data processing and significant storage space (Liu 2021). Generally, vehicles generate and transmit data to remote cloud computing infrastructures that have abundant resources for processing. Cloud computing enables on-demand resource delivery, cost reduction, scalability and rapid development of applications (Apostu 2013). However, cloud computing resources are generally centralized in large data centers. As a result, data transmission delays between users and the cloud can be long and unpredictable, making them unsuitable for time-sensitive vehicular applications (Javadzadeh 2020).

To overcome these limitations, the concept of edge computing was developed. Indeed, placing storage and computing resources at the network frontier, as close as possible to end users, reduces latency and achieves more responsive real-time operation (Cao 2020). The concept of edge computing consists of deploying a

Cooperative Intelligent Transport Systems,
coordinated by Léo MENDIBOURE. © ISTE Ltd 2024.

collection of nodes or edge servers within a geographic region. Processing and storage resources are thus accessible to end users, generally in a single hop.

Deploying edge servers in vehicle networks is a major challenge. The increase in the number of edge servers in the metropolitan area reduces latency for the automotive applications envisaged, but increases costs. The physical location of edge servers is crucial to ensuring high performance. Inefficient edge server placement results in increased latency and uneven workloads between edge servers, meaning that certain edge servers are overloaded while others are underused (Li 2021). This raises issues regarding the efficient placement of an optimal number of edge servers while taking into account deployment cost and network performance constraints. This problem becomes even more difficult to solve in metropolitan environments, where massive volumes of data are generated due to the high user density.

This chapter introduces an optimization model and a scientific methodology for placing edge servers effectively. More specifically, it proposes an integer linear programming approach to minimize the edge server deployment cost while respecting the desired latency threshold and balancing workloads between edge servers. The proposed solution is evaluated by in-depth simulations with open data from real vehicle traffic on the roads of the city of Bordeaux, France (OpenDataBordeaux 2021). The results obtained show that our solution can solve the edge server placement problem in a real vehicular environment effectively. It surpasses existing edge server placement strategies such as Random, Top-K, K-means and Wang et al. (2019). In particular, the main contributions of this chapter can be summarized as follows:

– We formulate the edge server placement problem in the form of an integer linear programming (ILP) model to identify the most appropriate locations to deploy edge servers. The model aims to minimize cost, respect a specific latency and ensure workload balance between servers.

– We propose a methodology based on open data from the real automotive traffic of the city of Bordeaux in order to provide a realistic solution for edge server placement. This methodology can be extended to other traffic sources and is valid for any other data.

– We evaluate the performance of our proposed solution through simulations and compare it with other placement strategies. The results obtained demonstrate the effectiveness of our solution compared to other strategies.

8.2. Context

8.2.1. *Vehicular applications*

Within the context of automotive IoT, V2X (Vehicle-to-Everything), which covers Vehicle-to-Vehicle (V2V), Vehicle-to-Infrastructure (V2I) and Vehicle-to-Pedestrian (V2P), requires a critical communication infrastructure where reliability and ultra-low latency are decisive factors (Zakaria 2017). Intelligent Transport System (ITS) applications can be classified into three categories: road traffic safety, road traffic efficiency and value-added services (Festag 2014; Singh 2019). The use cases in these categories include collision warnings, assistance (movement at intersections, left turn, cooperative merging), congestion warnings, etc. (Singh 2019). In order to exploit an efficient and reliable vehicular application that could include real-time traffic monitoring, continuous detection and cooperative and augmented vision, communication and computing capabilities need to be improved (Porambage 2018). However, the current cloud-based network and infrastructure are unable to fully satisfy the application requirements. With this in mind, multi-access edge computing architecture should offer a higher level of flexibility and provide a viable solution that can accelerate the development of ITS applications (Datta 2017).

8.2.2. *Multi-access edge computing (MEC)*

Mobile edge computing was introduced by the Industrial Specification Group (ISG) of the European Telecommunications Standards Institute (ETSI) as a way to extend intelligence to the network edge, as well as higher processing and storage capabilities (Hu 2015). In 2017, the ETSI industry group renamed it as multi-access edge computing (MEC), as the advantages of MEC technology go beyond the framework of mobile to extend to Wi-Fi and fixed access technologies. The principle of MEC is to extend cloud computing capabilities to the edge of networks. This aims to minimize network congestion and improve resource optimization, user experience and overall network performance. Several technologies are identified as "enabling technologies" for the realization of MEC, notably software-defined networking (SDN), network function virtualization (NFV), information-centric networking (ICN) and network slicing.

8.2.3. *Deployment of MEC systems*

The primary motivation for MEC is to bring cloud computing capacity close to the network edges in order to mitigate cloud computing limitations such as high latency and throughput bottleneck. However, choosing the site for edge server deployment represents a significant challenge. The server placement strategy is very different from the conventional base station site selection problem. Optimal edge server placement is linked to the provision of computing resources and is limited by the deployment budget (Mach 2017) and vehicular traffic on the roads. The choice of the location of MECs, in particular edge servers, is the first step in the implementation of the MEC system and the deployment of vehicular applications. MEC suppliers must take account of the computing requirements associated with the arrival of tasks and deployment costs, such as site rental, for the MEC server placement strategy to be cost-effective. In general, given the system deployment budget, more MEC servers should be installed in regions with the highest computing demands. The use of existing telecommunications infrastructure, such as base stations, could reduce costs, but would not solve all problems. Indeed, due to the continuous increase in the computing needs of vehicles and various smart devices (IoT), application requirements cannot be guaranteed. Therefore, edge servers must be deployed with a more distributed strategy that is closer to the user.

8.3. State of the art

The majority of existing studies on vehicle networks have addressed the cost-effective placement of roadside units, mainly from a communication point of view (Balouchzahi 2015; Zhang 2017; Nikookaran 2017). The majority of works targeting edge computing architecture have focused on resource allocation and scheduling (Sorkhoh 2019; Zhao 2019). While several solutions exist that are explicitly aimed at edge server placement in vehicle networks, there is no established methodology to solve the problem. The article presented in Premsankar (2018) developed an optimization problem and methodology that infrastructure providers can use to deploy edge computing devices in a smart city. The authors proposed a mixed-integer linear programming formulation that minimizes the cost of deploying edge computing devices by jointly satisfying a target level of network coverage and computing demand. However, this work considers the deployment of edge servers alongside roadside units, which leads to challenges in RSU deployment to ensure a good network coverage ratio. Although these works aim to assure a certain level of computing demand, they do not take into account the latency requirements of vehicular applications. Wang et al. (2019) studied the edge server placement problem in mobile edge computing environments for smart cities. They formulated the problem as a multi-objective constraint optimization problem that places edge servers in certain strategic locations in order to balance the workload between edge servers and minimize the edge server access delay. They then adopted mixed-integer programming to find the optimal solution. While these works are not specifically aimed at the vehicle network, they are the most closely connected to the problem discussed in this chapter. We will take this into account for comparison with our proposed solution. Xu et al. (2015) addressed the cloudlet placement problem in a large-scale wireless metropolitan area network (WMAN) composed of many wireless access points (APs). This work aims to place *K* cloudlets at strategic locations in the WMAN in order to minimize the average access delay between mobile users and the cloudlets serving their requests. The authors presented an exact solution to the problem by formulating it as ILP and proposed a scalable heuristic solution for large-scale scenarios. However, the main objective of this placement strategy is to minimize the cloudlet access delay, without studying the cost-effectiveness of the solution. Laha et al. (2020) studied the placement of the edge server in 5G-compatible vehicle urban networks. They discussed the efficient deployment of a limited number of edge servers in an urban scenario with a limited budget. They took into consideration the structural properties of the road network using centrality measures based on complex networks and vehicle traffic distribution to rank candidate sites for edge servers. The problem is transposed to the Knapsack 0-1 problem in order to maximize the total benefit based on the parameters mentioned above. However, this work focuses on the coverage ratio – the ratio of the number of vehicles covered by deployed edge servers to the total number of vehicles in the network – and does not take into account the latency and workload balancing constraints.

Our work aims to provide an effective approach to edge server placement to minimize the server deployment cost by jointly satisfying a target latency threshold and ensuring workload balancing between edge servers.

8.4. OptPlacement: efficient edge server placement

Before continuing, let us present the reference architecture shown in Figure 8.1. We consider a smart city with vehicles moving within an urban area. These vehicles run applications based on artificial intelligence and rely on edge servers for advanced learning capabilities based on data collected in the environment. For example, vehicles communicate wirelessly with road-side units (RSUs) in application of ITS-G5/IEEE802.11p (IEEE 2016), while edge servers communicate with RSUs via a broadband IP connection, generally over fiber optics.

Multiple RSUs can be connected to a single edge server. Edge servers process data offloaded by vehicles that are moving in the vicinity of these RSUs. In addition, these edge servers are connected to the Internet, which provides additional storage and processing resources to enable the cloud–edge continuum. Edge servers are also connected to the road operator platform that provides V2X services and security mechanisms (e.g. AAA and PKI). In the next section, we propose "OptPlacement", our solution for efficient edge server placement under real traffic and server capacity constraints, followed by the scientific methodology used for the evaluation.

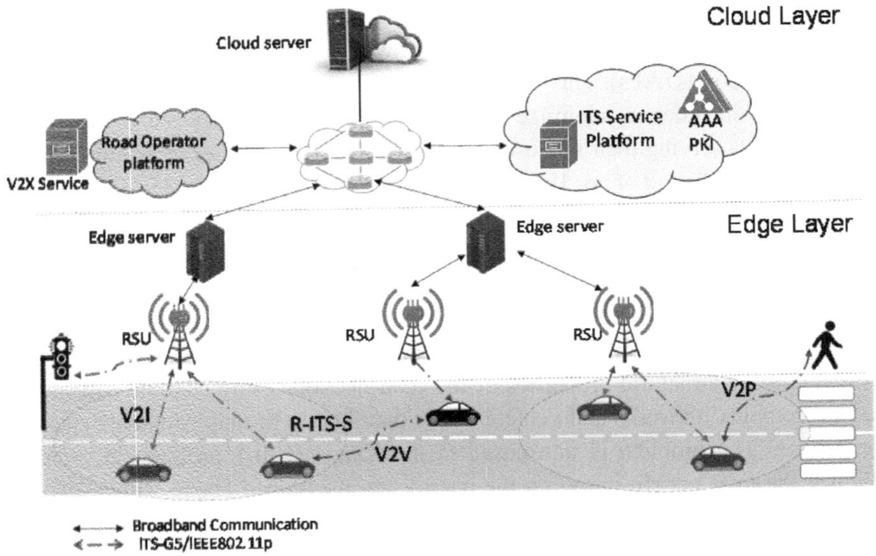

Figure 8.1. *Reference architecture for vehicular edge computing. For a color version of this figure, see www.iste.co.uk/mendiboure/transportsystems.zip*

8.4.1. *System modeling*

The system is composed of a set, $J = \{S1, S2, \ldots, Sm\}$, of potential locations for edge servers and a set, $I = \{R1, R2, \ldots, Rn\}$, of RSUs. The edge server placement problem can be considered in the form of an undirected graph, $G = (V, E)$, where $V = I \cup J$ and E represents the set of links between roadside units and edge servers at a location in J. The system components are shown in Figure 8.2. It should be noted that the connections between vehicles and roadside units are omitted because they are highly dynamic due to the mobility of vehicles and are not at the forefront of our concerns. Vehicles transmit data requests or messages to an RSU, which delegates processing to the edge server and receives responses in return. The total number of requests received by an RSU will be used to compute its requests. The RSU request represents the amount of processing, memory and storage resources that should be available on the edge server to process all requests received by this RSU. We assume that the vehicles send their requests with the same frequency.

We will assume that K edge servers are allocated to K different locations, where K is a constant (in Figure 8.2, $K = 3$). We assume that each edge server has the same limited computing resource (storage, processing and memory) to process the requests transmitted by the vehicles. Each roadside unit is connected to an edge server to offload vehicle requests. As a result, each edge server must process multiple requests from different RSUs, without exceeding its capacity. The latency between the vehicle and the edge server is assumed to be proportional to their distance.

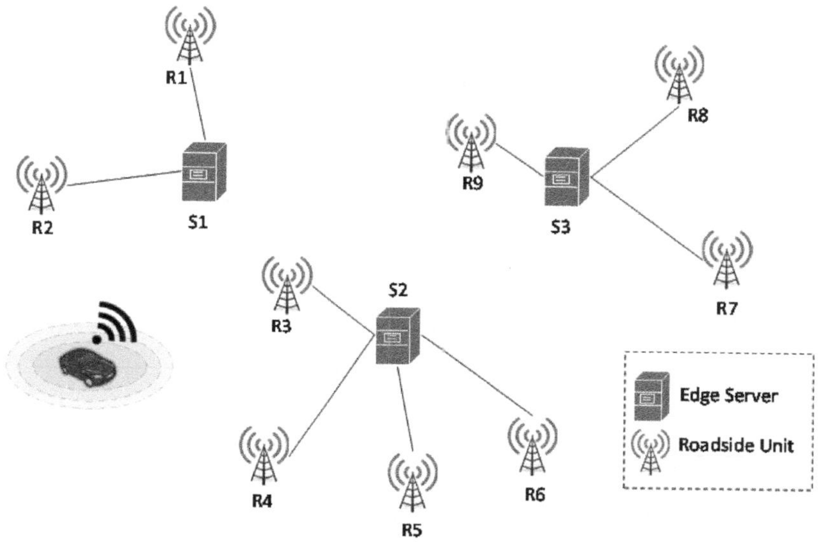

Figure 8.2. *Example of edge server placement. For a color version of this figure, see www.iste.co.uk/mendiboure/transportsystems.zip*

Based on the notation in Table 8.1, our objective is to minimize the edge server deployment cost while ensuring a target latency requirement and load balancing to prevent the edge servers from being overloaded. A number of K edge servers must be placed among m potential locations. The following constraints are taken into account in our formulation:

– Each roadside unit is connected to a single edge server.

– Each edge server processes vehicle requests from the RSUs.

– Each RSU has a request that represents the sum of the requests from vehicles communicating with that RSU. The requests represent the amount of resources required to perform a vehicle task in the edge server.

– Each edge server has its own computing and storage capacity (considered equal in this work).

Symbol	Description
$V = I \cup J$	*I*: a set of RSUs *J*: a set of potential locations of edge servers
$G = (V,E)$	*E*: a set of links between road units and edge servers
n	Number of roadside units
m	Number of potential locations for edge servers
K	Number of edge servers deployed
c_j	Cost of deploying an edge server at location *j*
L_{max}	Maximum latency (proportional to distance)
L_{ij}	Latency between RSU *i* and the edge server at location *j*
Pr_i	RSU *i* processing request
Mr_i	RSU *i* memory request
Sr_i	RSU *i* storage request
Ps_j	Processing capacity of server *j*
Ms_j	Memory capacity of server *j*
Ss_j	Storage capacity of server *j*
x_j	Binary variable for the edge server placed at location *j*
y_{ij}	Binary variable for RSU *i* linked to the edge server at location *j*
Z	Total cost of edge server deployment

Table 8.1. *Notation*

Accordingly, we formulate an ILP model that takes into account the characteristics of the model under study. We begin by defining the decision variables of the model:

$$x_j = \begin{cases} 1, if\ the\ server\ is\ deployed\ at\ location\ j \\ 0, \quad Otherwise \end{cases} \quad [8.1]$$

$$y_{ij} = \begin{cases} 1, if\ RSU\ i\ is\ linked\ to\ server\ j \\ 0, \quad Otherwise \end{cases} \quad [8.2]$$

$$Z = \sum_{j \in J} c_j x_j\text{: total cost} \quad [8.3]$$

The objective function is defined as follows:

$$Min\ Z\ , i.e.\ \ Min \sum_{j \in J} c_j x_j \quad [8.4]$$

s.t.

$$\sum_{j \in J} y_{ij}\, x_j = 1 , \qquad \forall\, i \in I \qquad [8.5]$$

$$\sum_{j \in J} L_{ij}\, y_{ij}\, x_j \leq L_{max} , \qquad \forall\, i \in I \qquad [8.6]$$

$$\sum_{i \in I} Pr_i\; y_{ij} \leq Ps_j\, x_j , \qquad \forall\, j \in J \qquad [8.7]$$

$$\sum_{i \in I} Mr_i\; y_{ij} \leq Ms_j\, x_j \qquad \forall\, j \in J \qquad [8.8]$$

$$\sum_{i \in I} Sr_i\; y_{ij} \leq Ss_j\, x_j \qquad \forall\, j \in J \qquad [8.9]$$

$$x_j \in \{0,1\} \qquad \forall\, j \in J \qquad [8.10]$$

$$y_{ij} \in \{0,1\} \qquad \forall\, j \in J, \forall\, i \in I \qquad [8.11]$$

We explain each expression in our model, starting with the decision variables in [8.1] and [8.2], x_j designates a Boolean variable indicating whether a candidate position is selected to deploy the edge server. y_{ij} is a Boolean variable indicating to which edge server each road unit is connected. For example, $y_{34} = 1$ indicates that RSU unit "3" is connected to the edge server deployed at position "4". The total cost function, which depends on the number of edge servers and their deployment costs, is presented in equation [8.3]. The objective function in [8.4] aims to minimize deployment cost by decreasing the number of edge servers to be deployed, where c_j designates the cost of deploying an edge server at position j among J. Next, we shall describe the constraints. Equation [8.5] ensures that each roadside unit is connected to one of the deployed servers. Therefore, no RSU is left without a connection or is connected to more than one server. Constraint [8.6] ensures that each RSU is connected to an edge server without exceeding the maximum allowed latency L_{max}, which is defined by the distance. Therefore, this constraint ensures that each RSU is within the coverage area of the deployed edge servers. Constraints [8.7]–[8.9] ensure that the memory, storage and processor requirements are met. They ensure that the total number of requests submitted to the edge server by the road units does not exceed its capabilities. Finally, constraints [8.10] and [8.11] assure the completeness requirements of the decision variables.

8.4.2. *Methodology and simulation*

This section describes the preliminary processing needed to generate the input data for our model. Next, the evaluation methodology and simulations are described.

8.4.2.1. *Preliminaries*

We consider the city of Bordeaux, France, as the target area where edge servers are to be placed. A map of central Bordeaux, including road and building information, is acquired on OpenStreetMap (2021). The map delineates an area of 15.10 km². The area in question is shown in Figure 8.3.

We will begin by examining the placement of roadside units in this map using a uniform strategy. We used this approach because there is no data available on the actual deployment of RSUs. At road intersections, roadside units are spaced approximately 800 meters apart. A total of 27 roadside units were installed to achieve optimal network coverage.

The next step consists of identifying potential sites for edge servers. We chose these candidate locations using a simple technique. A location is allocated between every three or four RSUs. This ensures that each RSU is associated with a possible edge server without exceeding the maximum acceptable latency L_{max} (defined by the distance). A set of $m = 19$ candidate locations for edge servers is introduced into an optimization model that determines the most strategic locations for deploying edge servers.

In order to perform simulations that most accurately represent the real environment, we must generate traffic traces identical to the real traffic on Bordeaux roads. By specifying the vehicle number, speed and trajectory, the traffic trace defines how vehicles travel on the roads shown on the map. To do this, we used the OpenDataBordeaux (2021) open dataset, which provided the number and speed of vehicles at a geographical point. These statistics are provided by a network of sensors installed on the roads of Bordeaux. Based on this data, we identified three road classes: high-traffic roads, moderate-traffic roads and low-traffic roads. Next, we generated the road traffic trace using the RandomTrips tool provided by simulation of urban mobility (SUMO) (Lopez 2018) with the classification of roads as input (by integrating the RandomTrips Customized Weights option).

Figure 8.3. *RSU locations and candidate locations for edge servers on the map of Bordeaux. For a color version of this figure, see www.iste.co.uk/mendiboure/transportsystems.zip*

Once the traffic trace for our simulation had been created, we determined the demands for each roadside unit, included in our model. RSU requests are the number of messages or tasks transmitted to the RSU by the vehicles. Assuming that all vehicles transmit messages at the same rate, RSU requests are computed based on the number of vehicles in the region covered by each RSU. Then, we ran a simulation to record the number of messages received by each RSU. Based on the results presented in Figure 8.4, road units are gathered into three categories: high demand, moderate demand and low demand. Table 8.2 presents the results obtained (Ri representing the roadside unit at location i).

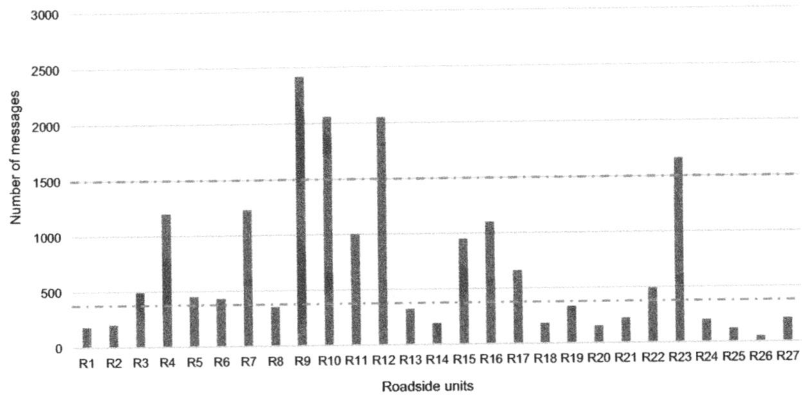

Figure 8.4. *Roadside unit requests on the roads of Bordeaux. For a color version of this figure, see www.iste.co.uk/mendiboure/transportsystems.zip*

Category	RSU
High demand (>= 1,500)	R9, R10, R12, R23
Moderate demand (400 =< x < 1,500)	R3, R4, R5, R6, R7, R11, R15, R16, R17, R22
Low demand (< 400)	R1, R2, R8, R13, R14, R18, R19, R20, R21, R24, R25, R26, R27

Table 8.2. *Classification of RSUs*

8.4.2.2. *Simulation methodology and tools*

Figure 8.5 provides a complete overview of the simulation methodology and configuration used to evaluate our placement strategy. The solution consists of two main parts. The first part is devoted to solving optimization problems using AIMMS (Advanced Interactive Multi-dimensional Modeling System) (AIMMS 2020) in conjunction with the CPLEX12.10 solver. AIMMS is a prescriptive analytics platform for the modeling and development of optimization-based applications. An optimal solution is determined using the following inputs: the geographical coordinates of the roadside units, the geographical coordinates of the edge server candidate locations, the latency threshold (measured as a distance), the capacity of the edge servers and the RSU requirements (in terms of processing, memory and storage). As a result, the solver provides the number of edge servers to be deployed and the locations of the servers among the candidate locations. In addition, the total cost of deployment is determined. Finally, one of the edge servers is allocated to each RSU. The connection between the RSU and the server respects the latency constraint, and the set of RSU requests linked to a single server must not exceed the server's capacity. The computer used for this optimization problem has the following configuration: Intel processor i7-10610U CPU @ 1.80 GHz 2.30 GHz and RAM 16 GB.

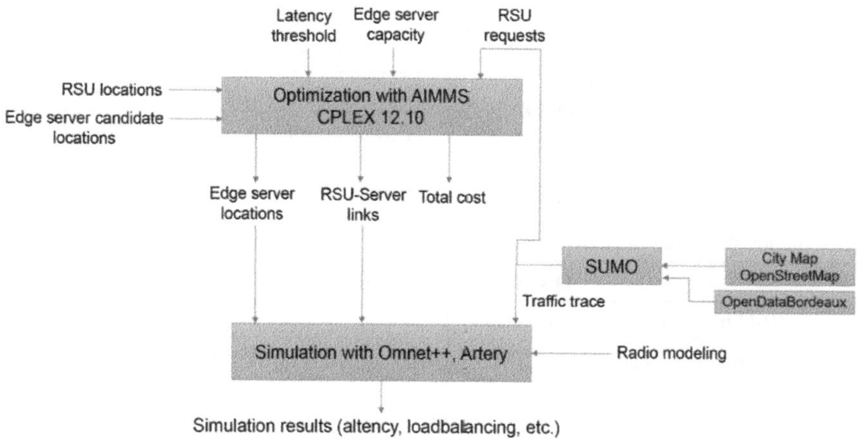

Figure 8.5. *Overview of the methodology used to evaluate "OptPlacement". For a color version of this figure, see www.iste.co.uk/mendiboure/transportsystems.zip*

The second part of this solution uses the results obtained in the first part to perform simulations using OMNeT++ (Varga 2020) and Artery (Riebl 2019). In addition, the traffic trace generated by SUMO (Lopez 2018) is also required as an input for the simulation. Finally, the simulation also integrates the radio model, which comprises the following characteristics: propagation model, obstacle loss model, path loss model and background noise model, as presented in Table 8.3. The simulations were carried out on a computer with 16 GB of RAM and Intel processor i7-9700 @ 3.00 GHz x8.

Table 8.3 shows the different parameters used to find the optimal solution and to perform the simulations. We assume that all candidate locations have the same deployment cost, which is equal to 1. We assume that all edge servers have the same capacity. The RSU processing, memory and storage requirements depend on the road unit group (high requirement, moderate requirement and low requirement). RSU request values are given per unit, which represents the number of messages received per second. Server capacity is expressed in units, which represents the processing, memory and storage capacity required to process received messages per second.

AIMMS parameters		
Cost of edge server deployment at location i	1 unit	
RSU requirements (processing/memory/storage)	High requirement	30 units
	Moderate requirement	20 units
	Low requirement	10 units
Edge server capacity (processing/memory/storage)	150 unit	
Latency threshold (as a distance)	1,500 m	
OMNeT/artery parameters		
RSU transmission power	47.9 mW	
RSU antenna height	5 m	
Vehicle transmission power	200 mW	
Frequency band	5.9 GHz	
Propagation model	ConstantSpeedPropagation (speed of light)	
Path loss model	GEMv2	
Obstacle loss model	DielectricObstacleLoss	
Background model	IsotropicScalarBackgroundNoise (–110 dBm)	
Message size (payload)	1,000 bytes	
Message frequency	1 Hz	

Table 8.3. *Simulation parameters*

Figure 8.6 shows the variation in vehicle density over the course of the simulation based on the traffic trace generated from the OpenDataBordeaux dataset. The maximum density at the end of the simulation reached 729 vehicles.

Our methodology and solution are generic and can be used with any map or traffic data source. According to the different inputs, the model provides the optimal solution for efficient edge server placement.

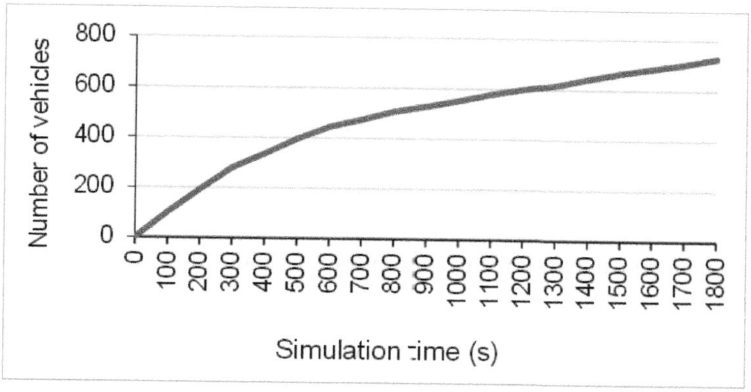

Figure 8.6. *The variation in vehicle density across the duration of the simulation. For a color version of this figure, see www.iste.co.uk/mendiboure/transportsystems.zip*

8.4.3. *Performance evaluation*

We compared the performance of our proposed OptPlacement approach with other placement strategies in terms of latency and workload balance. We chose the following strategies: Random, Top-K and K-means, as baseline comparisons because they are the most widely used strategies in this type of problem. In addition, the approach by Wang et al. (2019) was also used for comparison as it is the most relevant work identified in the literature. The different strategies are described as follows:

– Random: this approach places the K edge servers randomly among the candidate locations. We evaluated two variants of this approach based on how the roadside units are connected to the edge servers. The first is Random-Random, where RSUs are connected to servers at random. The second is Random-Nearest, where each RSU is linked to the nearest edge server.

– Top-K: based on traffic requests (Figure 8.4), the K RSUs with the highest requests are chosen. Next, the closest locations (K locations) of each RSU are chosen to deploy the edge servers. Each RSU is connected to the nearest edge server.

– K-means: this approach is commonly used to automatically gather a dataset into K groups. We use the K-means clustering algorithm to find K groups of road units and then place K edge servers in their centers.

– Approach by Wang et al. (2019): we implemented the solution proposed by Wang et al. (2019), which takes into account edge server placement in edge mobile computing environments for smart cities. They adopted mixed-integer programming to find the optimal solution. Their strategy aims to balance the workload between edge servers and minimize the edge server access delay. We compared the results to our approach.

For each strategy, we measured two critical metrics: latency and workload balance. The latency measurement represents the time it takes to send a message from the vehicle to the edge server. Workload balance is represented by the percentage of dropped tasks for each approach. To do this, we assume that each edge server maintains a queue (of size N) for incoming tasks and that each edge server is capable of processing M tasks simultaneously. If the server queue is full, the next task received is abandoned. Workload balancing will depend primarily on the number of road units connected to each server and their associated requests.

8.4.3.1. *Placement results*

According to our scenario, the results obtained by the AIMMS solver indicate that four (K = 4) edge servers should be placed in locations 1, 6, 12 and 19 (see Figures 8.3 and 8.7). This means that the total cost of edge server deployment is equal to four units (the optimal solution). The link between each road unit and the edge server is shown in Table 8.4, as well as the latency between entities, expressed in terms of distance. For example, server 1, located at position 1, is used to connect R1 at a distance of 591 meters, R2 at a distance of 1,088 meters etc. In addition, we show the workload of the edge servers, which is the aggregation of requests from the roadside units connected to this server. It is 110 units for server 1.

Edge server	RSU ID (distance with respect to the edge server)
Server 1 (at location 1)	R1 (591 m), R2 (1,088 m), R3 (458 m), R5 (736 m), R7 (1,210 m), R9 (1,465 m)
Server 2 (at location 6)	R4 (193 m), R6 (1,272 m), R8 (975 m), R11 (841 m), R13 (1,148 m), R15 (1,332 m), R16 (1,384 m)
Server 3 (at location 12)	R10 (1,441 m), R12 (983 m), R14 (293 m), R17 (967 m), R18 (439 m), R19 (1,036 m), R20 (1,120 m)
Server 4 (at location 19)	R21 (1,462 m), R22 (1,090 m), R23 (1,152 m), R24 (563 m), R25 (367 m), R26 (550 m), R27 (1,275 m)

Table 8.4. *Results of edge server placement*

This optimal solution is achieved in 42.11 seconds, with 129.1 MB of memory used. The results obtained satisfy the latency and capacity constraints of the server. All RSUs (R1 to R27) are connected to one of the edge servers without exceeding the maximum latency (expressed as a distance) of 1,500 m, which is used as an input to the model. With regard to the capacity constraint, for example, the sum of the RSUs connected to server 1 is 110 units, which is less than the capacity of server 1 (150 units).

Table 8.5 summarizes the results obtained using other placement approaches. For comparison purposes, in the Random, Top-K and K-means approaches, we set K = 4 as the number of edge servers to be deployed. This enables these approaches to align with the exact solution in relation to our approach. Therefore, we consider that they have the exact deployment cost (K = 4). In the approach by Wang et al. (2019), two servers are deployed according to the optimal solution of the proposed model. In this approach, the maximum latency threshold is not taken into account; therefore, two servers are deployed to reduce the cost. The results of the random approach are not presented in this table since they are random. However, we did run several simulations with different random placements of edge servers to evaluate their performance.

Strategy	Edge server	RSU ID (distance with respect to the edge server)
Top-K	Server 1 (at location 7)	R2, R7, R8, R9, R12, R13
	Server 2 (at location 8)	R1, R3, R4, R5, R6, R10, R11, R16
	Server 3 (at location 10)	R14, R15, R17, R18, R19, R20
	Server 4 (at location 17)	R21, R22, R23, R24, R25, R26, R27
K-means	Server 1 (at location 6)	R1, R2, R3, R4, R7, R8, R9, R10, R12, R13
	Server 2 (at location 8)	R5, R6, R11, R16
	Server 3 (at location 13)	R14, R15, R17, R18, R19, R20, R21
	Server 4 (at location 17)	R22, R23, R24, R25, R26, R27
Wang et al. (2019)	Server 1 (colocated with RSU 3)	R3, R6, R10, R11, R12, R13, R14, R15, R16, R17, R20, R21, R22, R25
	Server 2 (colocated with RSU 15)	R1, R2, R4, R5, R7, R8, R9, R18, R19, R23, R24, R26, R27

Table 8.5. *Placement results of other approaches*

Figure 8.7 summarizes and shows the locations chosen for deploying edge servers for each placement strategy. The Random-Random and Random-Nearest approaches are also not presented here as their results are random.

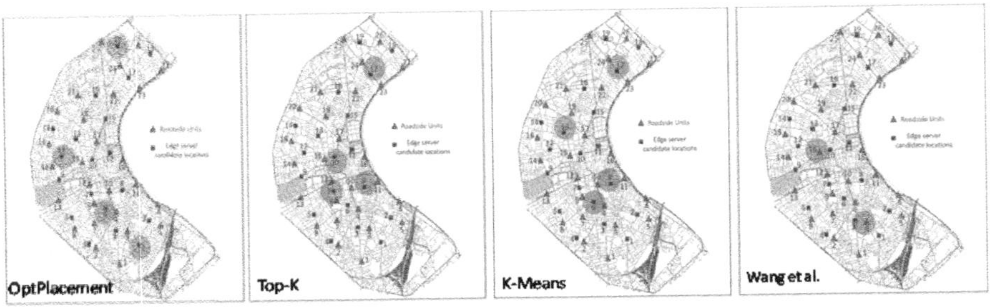

Figure 8.7. *Locations of edge servers for different approaches: OptPlacement, Top-K, K-means and Wang et al. For a color version of this figure, see www.iste.co.uk/mendiboure/transportsystems.zip*

8.4.3.2. *Performance results*

As mentioned earlier, the measurements considered are the latency and percentage of rejected tasks to evaluate workload balance between servers.

Figures 8.8 and 8.9 show the latency curves generated using the Artery simulation tool for the above-mentioned approaches. The results show the average latency of messages received by the servers. Our OptPlacement approach achieves latency varying between 1.51 ms in low density and 1.61 ms in high density. The other approaches offer latency in approximately the same interval.

When we examine the results more closely, we note that OptPlacement outperforms Random-Random and Wang et al. (2019). In addition, Random-Nearest and OptPlacement have overlapping curves and provide similar results. Furthermore, our solution presents a higher latency of 0.01 ms (maximum) compared to the Top-K and K-means approaches. This can be explained by the fact that these approaches use the nearest-server criterion, i.e. each RSU is connected to the nearest server, while not taking into account other constraints such as server capacity and workload balancing. For example, in the Top-K approach, server 2, which has a capacity of 150 units, cannot process all of the tasks of the road units connected to it (160 units of requests in total). Therefore, this minor increase in latency is acceptable to our solution to ensure that server capacity is not exceeded. On the contrary, the proposed model guarantees that the constraint of the latency threshold (L_{max}) is always respected.

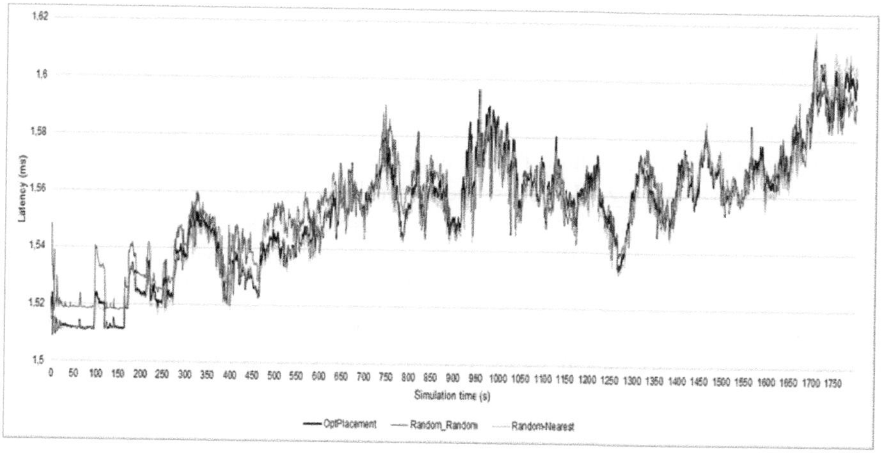

Figure 8.8. *Results of the "latency" simulations: OptPlacement, Random-Random and Random-Nearest. For a color version of this figure, see www.iste.co.uk/mendiboure/transportsystems.zip*

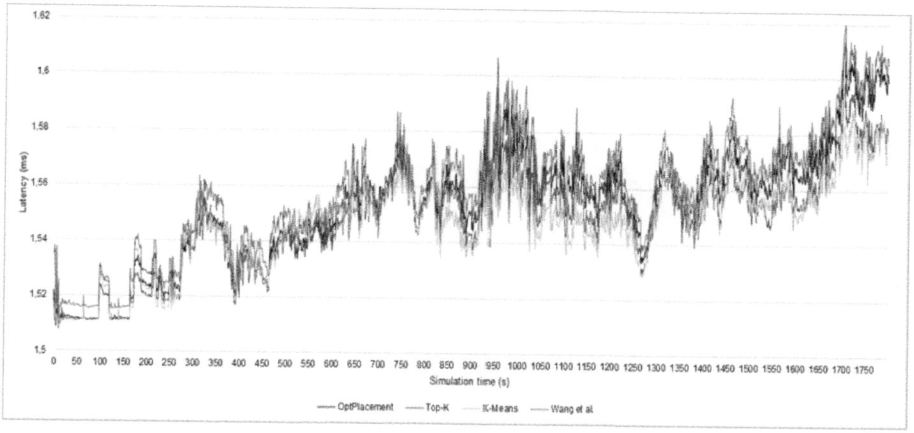

Figure 8.9. *Results of the "latency" simulations: OptPlacement, Top-K, K-means and Wang et al. (2019). For a color version of this figure, see www.iste.co.uk/mendiboure/transportsystems.zip*

To examine in greater depth the performance of placement methods, we studied workload balancing between edge servers to determine whether one of them was overloaded. To do this, we set the queue of each server to 150 tasks (N = 150) and each server can process six tasks simultaneously (M = 6). If a server's queue is full, the next task that arrives is abandoned. Figure 8.10 shows the percentage of abandoned tasks for each approach. The results show that our approach can provide the best performance since 2.90% of tasks are abandoned. Specifically, the percentage of tasks abandoned by OptPlacement is lower than the percentage of tasks abandoned by Random-Random and Random-Nearest by 48% and 69%, respectively. The K-means and Top-K approaches demonstrated relatively better results in terms of latency. However, their workload balancing performance is not promising. The Top-K approach has 5.67% abandoned tasks, which is 95.5% higher than our approach, and the K-means strategy has 13.06% abandoned tasks. The percentage of tasks abandoned in the approach by Wang et al. (2019) is 394% higher compared to our approach. These results confirm that our proposed model can provide a better workload balance compared to other edge server placement strategies while assuring the latency constraint.

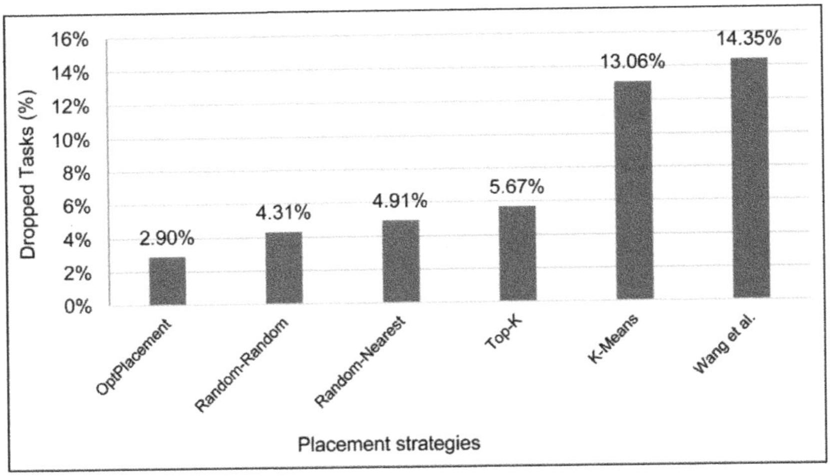

Figure 8.10. *Results of abandoned tasks. For a color version of this figure, see www.iste.co.uk/mendiboure/transportsystems.zip*

8.5. Conclusion

Edge computing is a significant emerging technology that can be used to extend computing and storage capabilities by moving processing workloads from the cloud to the edge, thus improving performance for a wide variety of applications, particularly vehicular applications. Up until now, little attention has been paid to edge server placement. This chapter proposes an effective strategy for edge server placement that minimizes deployment costs and balances workload between edge servers while respecting the maximum authorized latency. We formulated the edge server placement problem in the form of an ILP model. To give the edge servers a realistic location, we used a methodology based on the set of real traffic data from Bordeaux. Our model is generic and applicable to other maps and different traffic datasets. Using the AIMMS analytical platform, we obtained the optimal solution to the placement problem. Next, we performed simulations to evaluate the performance of the solution. The results obtained showed that the proposed solution is effective and outperforms existing approaches by ensuring load balancing between edge servers and minimizing deployment costs. As a result, our solution reduces the number of abandoned tasks by an average of 56% compared to other placement strategies. In addition, it satisfies the maximum latency constraint that cannot be exceeded. As a result, the strategy we propose for edge server placement provides an optimal compromise between deployment cost, workload balancing and latency. As part of future works, we intend to explore the migration of services between edge servers to ensure service continuity and reduce application delays.

8.6. References

AIMMS (2020). Download AIMMS developer [Online]. Available at: https://www.aimms.com/english/developers/downloads/download-aimms/ [Accessed 11 June 2021].

Apostu, A., Puican, F., Ularu, G., Suciu, G., Todoran, G. (2013). Study on advantages and disadvantages of Cloud Computing – The advantages of telemetry applications in the Cloud. In *Recent Advances in Applied Computer Science and Digital Services*, 17668067.

Balouchzahi, N.M., Fathy, M., Akbari, A. (2015). Optimal road side units placement model based on binary integer programming for efficient traffic information advertisement and discovery in vehicular environment. *IET Intelligent Transport Systems*, 9(9), 851–861.

Cao, K., Liu, Y., Meng, G., Sun, Q. (2020). An overview on edge computing research. *IEEE Access*, 8, 85714–85728.

Datta, S.K., Haerri, J., Bonnet, C., Ferreira Da Costa, R. (2017). Vehicles as connected resources: Opportunities and challenges for the future. *IEEE Vehicular Technology Magazine*, 12(2), 26–35.

Festag, A. (2014). Cooperative intelligent transport systems standards in Europe. *IEEE Communications Magazine*, 52(12), 166–172.

Hu, Y.C., Patel, M., Sabella, D., Sprecher, N., Young, V. (2015). Mobile edge computing – A key technology towards 5G. ETSI White Paper, 11(11), 1–16.

IEEE (2016). 802.11.2016 – IEEE Standard for Information technology – Telecommunications and information exchange between systems Local and metropolitan area networks – Specific requirements – Part 11: Wireless LAN Medium Access Control (MAC) and Physical Layer (PHY) Specifications.

Javadzadeh, G. and Rahmani, A.M. (2020). Fog computing applications in smart cities: A systematic survey. *Wireless Networks*, 26(2), 1433–1457.

Laha, M., Kamble, S., Datta, R. (2020). Edge nodes placement in 5g enabled urban vehicular networks: A centrality-based approach. In *2020 National Conference on Communications (NCC)*, 1–6. IEEE, Kharagpur.

Li, X., Zeng, F., Fang, G., Huang, Y., Tao, X. (2021). Load balancing edge server placement method with QoS requirements in wireless metropolitan area networks. *IET Communications*, 14(21), 3907–3916.

Lin, Y., Wang, P., Meng, M.A. (2017). Intelligent transportation system (ITS): Concept, challenge and opportunity. In *2017 IEEE 3rd International Conference on Big Data Security on Cloud (Bigdatasecurity), IEEE International Conference on High Performance and Smart Computing (Hpsc), and IEEE International Conference on Intelligent Data and Security (IDS)*, Beijing, 167–172.

Liu, L., Chen, C., Pei, Q., Maharjan, S. and Zhang, Y. (2021). Vehicular edge computing and networking: A survey. *Mobile Networks and Applications*, 26(3), 1145–1168.

Lopez, P.A., Behrisch, M., Bieker-Walz, L., Erdmann, J., Flötteröd, Y.-P., Hilbrich, R., Lücken, L., Rummel, J., Wagner, P, Wießner, E. (2018). Microscopic traffic simulation using sumo. In *2018 21st International Conference on Intelligent Transportation Systems (ITSC)*, 2575–2582. IEEE, Maui.

Mach, P. and Becvar, Z. (2017). Mobile edge computing: A survey on architecture and computation offloading. *IEEE Communications Surveys & Tutorials*, 19(3), 1628–1656.

Nikookaran, N., Karakostas, G., Todd, T.D. (2017). Combining capital and operating expenditure costs in vehicular roadside unit placement. *IEEE Transactions on Vehicular Technology*, 66(8), 7317–7331.

OpenDataBordeaux (2021). Bordeaux Métropole [Online]. Available at: https://opendata.bordeaux-metropole.fr/" [Accessed 11 June 2021].

Open Street Map France (2021). [Online]. Available at: https://www.openstreetmap.fr/" [Accessed 11 June 2021].

Porambage, P., Okwuibe, J., Liyanage, M., Ylianttila, M., Taleb, T. (2018). Survey on multi-access edge computing for internet of things realization. *IEEE Communications Surveys & Tutorials*, 20(4), 2961–2991.

Premsankar, G., Ghaddar, B., Di Francesco, M., Verago, R. (2018). Efficient placement of edge computing devices for vehicular applications in smart cities. In *NOMS 2018–2018 IEEE/IFIP Network Operations and Management Symposium*, 1–9. IEEE, Taipei.

Riebl, R., Obermaier, C., Gunther, H.-J. (2019). Artery: Large scale simulation environment for ITS applications. In *Recent Advances in Network Simulation*, Virdis, A. and Kirsche, M. (eds). Springer, Cham, 365–406.

Singh, P.K., Nandi, S.K., Nandi, S. (2019). A tutorial survey on vehicular communication state of the art, and future research directions. *Vehicular Communications*, 18, 100164.

Sorkhoh, I., Ebrahimi, D., Atallah, R., Assi, C. (2019). Workload scheduling in vehicular networks with edge cloud capabilities. *IEEE Transactions on Vehicular Technology*, 68(9), 8472–8486.

Varga (2020). OMNeT++ Discrete Event Simulation System, Release 5.6.

Wang, S., Zhao, Y., Xu, J, Yuan, J., Hsu, C.-H. (2019). Edge server placement in mobile edge computing. *Journal of Parallel and Distributed Computing*, 127, 160–168.

Xu, Z., Liang, W., Xu, W., Jia, M., Guo, S. (2015). Efficient algorithms for capacitated cloudlet placements. *IEEE Transactions on Parallel and Distributed Systems*, 27(10), 2866–2880.

Zakaria, O., Britt, J., Forood, H. (2017). U.S. Patent No. 9,717,012, U.S. Patent and Trademark Office, Washington, DC.

Zhang, R., Yan, F., Xia, W., Xing, S., Wu, Y., Shen, L. (2017). An optimal roadside unit placement method for vanet localization. In *GLOBECOM 2017–2017 IEEE Global Communications Conference*, Singapore, 1–6.

Zhao, J., Li, Q., Gong, Y., Zhang, K. (2019). Computation offloading and resource allocation for cloud assisted mobile edge computing in vehicular networks. *IEEE Transactions on Vehicular Technology*, 68(8), 7944–7956.

9

Risk Estimation: A Necessity for the Connected Autonomous Vehicle

Dominique GRUYER[1], Sio-Song IENG[1], Sébastien GLASER[2], Sébastien DEMMEL[2], Charles TATKEU[3], Sabrine BELMEKKI[3]

[1]*PICS-L, Gustave Eiffel University, Marne-la-Vallée, France*

[2]*Queensland University of Technology, Brisbane, Australia*

[3]*COSYS-LEOST, Gustave Eiffel University, Villeneuve-d'Ascq, France*

9.1. Context and objectives

In recent years, numerous research works and projects have aimed to propose new and innovative mobility solutions addressing current societal challenges: safety, sustainability and accessibility of mobility. The proposed solutions generally consist of using public transport, sharing individual or private mobility means and developing connected and automated mobilities. In the near future, the large-scale deployment and use of fleets of connected and automated vehicles (levels 3 and 4) in a variety of traffic conditions (highway, urban, peri-urban, rural and closed site) could complement the conceivable solutions. Nevertheless, questions remain as to the impact of these systems and their true benefit. Therefore, studies seek to assess, at best, this impact and the performance of these new mobilities on the criteria of safety, energy and pollution, traffic management and comfort and health

For a color version of all figures in this chapter, see www.iste.co.uk/mendiboure/transportsystems.zip.

Cooperative Intelligent Transport Systems,
coordinated by Léo MENDIBOURE. © ISTE Ltd 2024.

(Gruyer et al. 2021; Cui et al. 2022). These studies are carried out both locally and globally, notably with the use of digital and numerical infrastructures. From a safety point of view, questions persist regarding the impact and appropriateness of using decision-making and autonomous driving systems based on AI (Artificial Intelligence). Given that human operators will be increasingly absent from the task of driving, it is imperative to guarantee that autonomous driving systems will take into account strong safety constraints, both for passengers of the vehicle and for other road users. This safety is reflected both by local risk estimation and by the prediction and anticipation of risk beyond the perception range of the ego-vehicle sensors. This anticipation is important in order to interact with the driver. The latter must have enough time to react efficiently to a request to regain control in the event of a future accidentogenic or near-accident situation. In this context, the use of high-quality means of communication is essential. Indeed, current perception solutions are not able to satisfy this constraint of regaining control. This safety constraint and anticipation of risky situations is critical to ensure efficient and safe traffic management in mixed traffic conditions (autonomous vehicles and conventional vehicles). Being able to develop these new mobilities offers public authorities the benefit of optimizing global objectives including safety, reduction of consumption and pollutant emissions and optimization of traffic conditions. Moreover, for reasons of acceptability and trust in these systems, it is also necessary to guarantee a high level of comfort. This high level of comfort will include, among other aspects, the ability to anticipate risky situations while having time to make the best decision, while not arriving at situations requiring extreme dynamic and emergency behavior of the vehicle. In fact, we want to remain as much as possible within the "normal driving" area shown in Figure 9.1 and limit entry into the next three areas.

To meet these new challenges in terms of safe and efficient automated mobility, cooperative systems appear as a promising, and indeed indispensable, solution. Indeed, in addition to being able to predict and anticipate risky situations, physical and digital infrastructures, including the deployment of means of communication, will allow access to a real-time global perception of road infrastructure in order to better manage traffic by extending perception, anticipating risky situations and reducing energy consumption. These improvements will then be applicable not only on a local (vehicle) scale but also on a more global (infrastructure manager) scale. This will also make it possible to adapt speed profiles and behaviors by taking into account a succession of remote road situations and configurations. The autonomous car, the ultimate level of vehicle automation (levels 4+ and 5), will then need to optimize the criteria mentioned above and will need to integrate and use a virtual co-driver (decision, planning and action) in order to be eco-responsible and much more cautious and efficient than a human driver. The deployment of connected and autonomous vehicles should also allow more efficient traffic regulation by using them as adjustment "variables". It is therefore evident that cooperative systems will have a positive impact as a traffic regulation tool to homogenize vehicle flows and thus minimize accelerations, decelerations, stop-and-go waves and other factors of fuel overconsumption and near-accident generation.

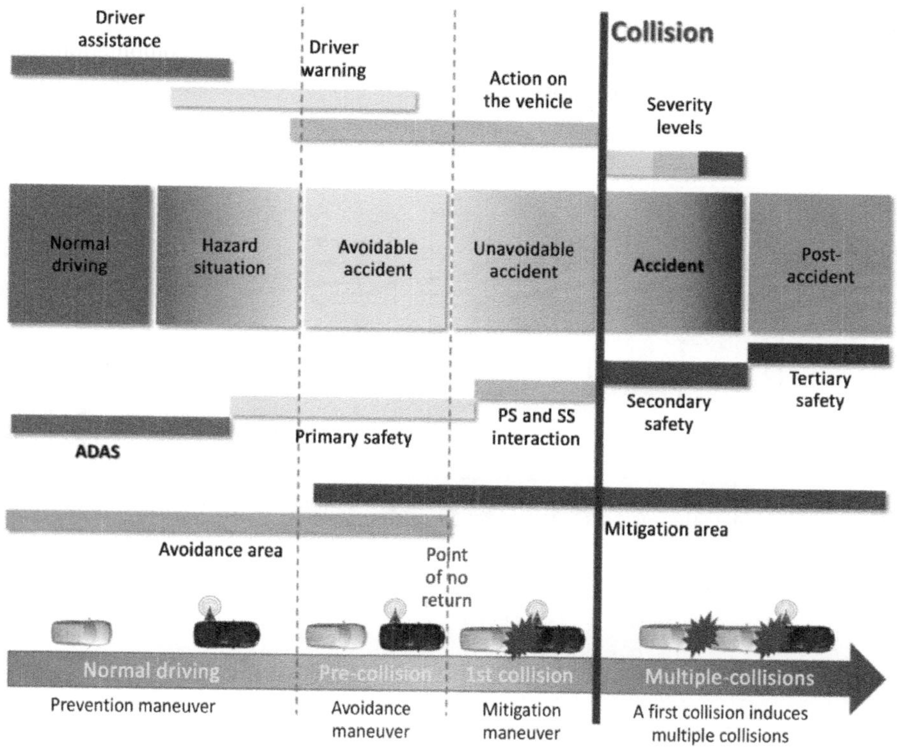

Figure 9.1. *Classification of driving situations with their criticality level and on-board system action levels*

In this chapter, we will discuss the contribution of cooperative systems for the estimation of local risk at vehicle level, extended local at the level of a vehicle and its close environment and the global risk with several vehicles at a link road. This notion of risk is essential to ensure increased safety and better traffic management. Indeed, risk is a pivotal factor as it forms the interface between perception of the environment and the decision-making necessary for the selection of maneuvers, trajectories and commands controlling the dynamics of the vehicle. The introduction of new systems in driving must aim to improve road safety and thus reduce risks.

Figure 9.2 demonstrates the strategic benefit of risk computation in a complete architecture of driving automation well. In the next section, the basic concepts of local risk computation will be presented. As we will see, there are a great number of "local" indicators. These indicators are necessary but unfortunately insufficient. We will then proceed with another section proposing the implementation of a communication strategy to dynamically build clusters of communication points with the constraints of allowing for continuity and high quality services. The objective of this communication strategy will be to propose new levels of risk computation with extended local, and then global, risk estimators.

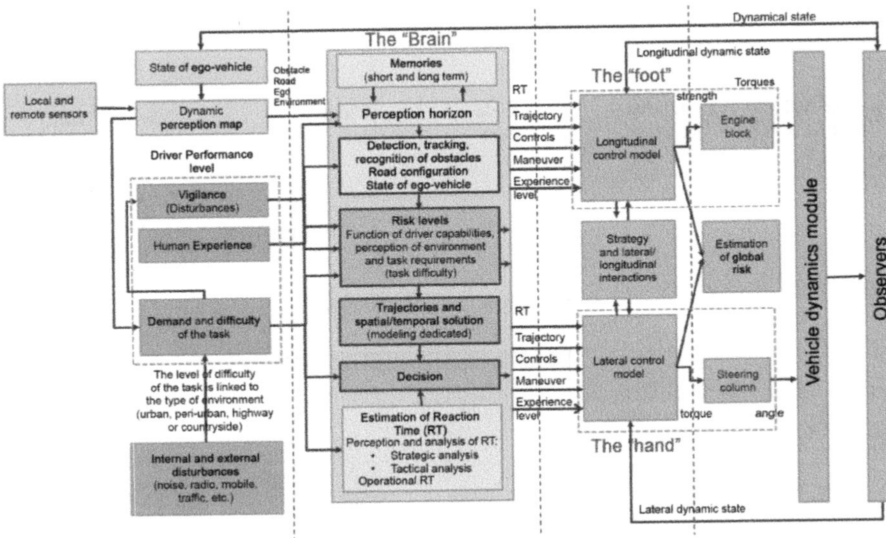

Figure 9.2. *Modeling of an autonomous driving system drawing a parallel between the computer modules and the human functions that constitute it*

9.2. Estimation of risk local to the ego-vehicle: some existing metrics

From the point of view of the ego-vehicle, in the event of an imminent critical situation that could lead to a collision (see Figure 9.1), it is necessary to detect this situation (or exposure to risk) and its severity (or proven result) to react as rapidly as possible and avoid collision or possibly mitigate the severity of the collision and thus limit the damage. This type of function relies solely on the use of on-board sensors to estimate this risk. Nevertheless, the possibility of estimating the potential severity of a collision and the ability to do so are strongly limited by the capabilities and performance of the on-board sensors. Generally, the perception range is limited to 100 or even 200 meters under favorable visibility and road geometry conditions.

In this configuration, the risk to be estimated is local and will only allow an event to be reacted to, and not a critical situation to be anticipated, in the sense defined previously with a human capable of regaining control of driving.

Monkhouse (2020) suggested a generic definition of this risk for driving automation systems. In this work, an enhanced vehicle control model is proposed based on the definition provided by the ISO 26262 standard and by SOTIF (ISO 21448). The automotive risk, R, is described as: $R = F(f, S, C)$, where f is the frequency of occurrence of a hazardous event, C is the controllability of the system, and S is the resulting severity of the event. For the risk of road situations, the frequency of occurrence of a hazardous event corresponds to the collision probability: controllability is the existence of avoidance maneuvers, and severity is the mortality rate estimated from the equivalent energy speed (EES) at the moment of impact between the ego-vehicle and an obstacle.

In more concrete terms, existing risk estimators can be broken down into several categories (spatial, temporal and other) but will always consist of building a metric that will take into account the information on position, orientation, speed, and potentially the acceleration and mass of the objects generating the near-accident or accident event. We can interpret this as the need to estimate the distance between obstacles or to estimate the collision probability. In any case, a problem or research question to be solved consists of knowing how to model imperfections on the data we manipulate, namely imprecision, uncertainty, reliability and trust.

Regarding recent works on this subject, it is interesting to note that they focus mainly on taking account of the attributes of the two actors, "ego-vehicle" and "obstacle". The most commonly used approach is based on the use of the severity, EES and collision probability functions. The collision probability function is often derived from time-to-collision (TTC), the ratio between inter-distance and speed differential. However, in many works, the functions used to generate this probability as a function of TTC give more functions of "possibility" of collision. To obtain a true collision probability, we can either use an adaptation of the works of Lambert et al. (2008) or Monte Carlo experiments as proposed in Glaser et al. (2011). For the G(EES) severity function, several templates have been proposed in the literature, but generally it is the function proposed by the LAB that is used.

The metrics obtained from TTC are generally seen as specific indicators. For example, TTC and THW (Time HeadWay) are clearly "ego-vehicle"-oriented. Other operators will be more "road-user" oriented. Initially, TTC was only applied to longitudinal configurations in the evolution direction of the ego-vehicle. In the past decade, works have proposed to extend this concept to several dimensions. This is the case in Jiménez et al. (2013), which proposes an improved TTC for two-dimensional collisions. Ward et al. (2015) also considered extending TTC to 2D configurations using probabilistic reasoning. St-Aubin et al. (2015) made a comparison between TTC-based predictions and aggregation methods for alternative safety measures. In-depth reviews of the literature on recent "proximity" risk indicators are available in Mahmud et al. (2017), Pinnow et al. (2021) and Westhofen et al. (2023). The latter reference proposes the comparison of more than 40 critical metrics for the autonomous vehicle. It also provides a relevant diagram of the links between these metrics. In order to better describe and above all detect accidentogenic situations, numerous indicators derived from TTC have been proposed. Some of the best known include the following:

– *Time-Exposed-to-TTC (TET)*: gives the ego vehicle's exposure time below a critical threshold (TTC critical value).

– *Time-Integrated TTC (TIT)*: gives the integral of the TTC function below a "critical" threshold.

– *Modified TTC (MTTC)*: gives a TTC that takes into account the relative acceleration, in addition to the distance and speed differential.

– *Collision Index (CI)*: metric that includes the influence of speed on the kinetic energy involved in a collision. This metric reflects the severity of a potential accident. Very close to MTTC.

– *Time-To-Accident (TTA)*: gives the time remaining for the ego-vehicle to perform an avoidance maneuver.

– *Encroachment Time (ET)*: gives the length of time taken by an ego-vehicle maneuvering to encroach on a conflict area (area containing an obstacle).

– *Post-Encroachment Time (PET)*: gives the time between the end of the encroachment of the maneuvering ego-vehicle and the moment when the "obstacle" vehicle actually arrives at the potential collision point. The PET gives the time interval between the departure of one vehicle and the entry of another vehicle into a conflict area. Assuming that the ego-vehicle passes the conflict area before the second vehicle.

– *Time-To-Brake (TTB)*: gives the braking time to be applied to avoid a collision.

– *Time-Headway (THW) or Gap Time (GT)*: gives the time interval that separates the ego-vehicle from the position of an obstacle vehicle with the constraint that it is at the same speed and on the same path.

– *Proportion of Stopping Distance (PSD)*: gives the ratio between the inter-vehicle distance available for maneuvering and the remaining distance to the projected collision location (the minimum acceptable stopping distance).

– *Initially-Attempted Post-Encroachment Time (IAPET)*: gives the time period between the start of encroachment by the maneuvering ego-vehicle plus the time expected for the "obstacle" vehicle to reach the collision point and the time the ego-vehicle takes to complete encroachment.

– *Potential Index (PI)*: index used for collision situations with urgent deceleration.

– *Difference between Space distance and Stopping distance (DSS)*: gives the safe stopping distance. An improved version is given in Gürbüz and Buyruk (2019).

– *Time-Integrated DSS (TIDSS)*: improved version of DSS with time integration and a possibly negative DSS.

– *Unsafe Density (UD)*: gives the critical density value for which the global collision risk is above a threshold.

– *Deceleration Rate to Avoid Collision (DRAC)*: gives the deceleration rate to be applied to avoid a collision.

– *Crash Potential Index (CPI)*: gives an index based on the time during which the DRAC is greater than a maximum available deceleration rate (MADR).

– *Critical Index Function (CIF)*: gives a function combining relative speed as the severity factor and TTC as the probability of exposure.

– *Integrated Conflict Risk Index (ICRI)*: gives the quadratic distance between TTC and DRAC.

– *Potential Index for Collision with Urgent Deceleration (PICUD)*: gives a potential collision index in a situation requiring urgent deceleration. This metric is compared to five other metrics (TTC, S-PDRF, DRAC, PFS and CFS) (Lu et al. 2021).

– *Margin-To-Collision (MTC)*: gives a metric quite similar to the stopping distance. The index also represents the possibility of a collision in the event that the preceding vehicle and the following vehicle suddenly decelerate at the same time.

– *Time-to-Line-Crossing (TLC)*: gives the time remaining before a line is crossed (Mammar et al. 2004).

– *Lane-Changing Risk Index (LCRI)*: gives a method of estimating the accident risk while an "obstacle" vehicle is changing lanes. The LCRI integrates the duration of exposure to a potential accident and the expected severity level of the accident by applying a fault-tree analysis (FTA). This metric merges SSD, SDI, REL (Risk Exposure Level) and RSL (Risk Severity Level) (Park et al. 2018).

Other less-cited additional indicators include the following (Mahmud et al. 2017):

– *Jerks*: studying the jerk profile helps to identify collision avoidance attempts.

– *J-value*: cumulative risk value computed in a platoon situation.

– *Standard Deviation of Lateral Position (SDLP)*: lateral indicator for keeping in lane and the driver's level of alertness and vehicle control. It is used as an additional indicator of longitudinal risk, as well as operators of critical deviation, deviation time, time to intersection, time to stop line, expected minimum distance and potential energy.

In Lambert et al. (2008), a multidimensional probabilistic method is proposed. This method uses a generalized Gaussian modeling to estimate the probability of collision between an ego-vehicle and an obstacle. From this collision probability, the collision risk is presented as the product of the collision probability function and the collision cost, approximated by the EES. This modeling has been integrated into many recent works.

Dahl et al. (2019) also proposed an in-depth literature review of the most recent collision avoidance techniques and threat assessment techniques. Batsch et al. (2020) presented a set of longitudinal indicators based on distance or time to collision, supplementing previous studies. This work investigates their use for assessing autonomous driving in various microscopic-simulation projects.

Recently, more global accident risk indicators have been successfully developed. Among these indicators, we can cite a global indicator of accident risk based on the statistical description and analysis of the ODD (Gelder et al. 2019). Philipp and Goehring (2019) also proposed using an octagonal representation of the collision areas and the space surrounding a vehicle. This method can be seen as an analytical approach that will use two probabilistic methods to compute risks. The first method is collision state probability (CSP), which represents the probability of spatial overlap of two objects at a certain moment in time. The second is the density of collision event probability (CEP) at a certain moment in time. CEP, which is the probability of a collision during a period of time, is computed by integrating CEP density over time. The concept developed in this work then provides an indicator of spatial safety. In Katrakazas et al. (2017), the collision risk model is obtained from an object interaction-aware motion model and is based on the use of dynamic Bayesian networks (DBN). In this method, the authors take into account a more global view of the current situation (network and traffic level) in order to estimate a "collision prediction at network level". This prediction capability makes it possible to have a risk anticipation that significantly improves the risk assessment in complex and large traffic scenarios.

It is then impossible to talk about risk without addressing a set of models derived from responsibility-sensitive safety (RSS). RSS was proposed by Intel and Mobileye to ensure the "safe state" of autonomous vehicles by defining the concept of a safe distance in order to avoid any liability in the event of an accident.

Studies such as Shalev-Shwartz et al. (2017), Chai et al. (2019) and Koopman et al. (2019) proved that the RSS model can significantly improve safety in car-following maneuvers compared to integrated adaptive cruise control (ACC). RSS is of particular interest because it takes into consideration additional sets of parameters to represent the ego-vehicle state, road geometry and environmental parameters. This operator is clearly the first operator taking into account three of the five actors of the road scene. Kuang et al. (2015) proposed a modeling to rank the collision risk in tree form. In fact, by producing a possible disturbance on the "leader" vehicle, the action of the ego-vehicle is represented as a probabilistic causal model. From this information, a tree is built to describe the eight possible types of conflict exploitable by the model. This risk measurement is called ACI (Aggregated Crash Index).

In order to assess the risk between two vehicles, in 2021, Leroy et al. (2021) proposed and developed a new indicator of multidimensional risk under uncertainties. This indicator is defined as follows:

– It is scaled between 0 and 1, with 0 signifying that there is no risk and 1 signifying that the risk of fatal accident is at a maximum.

– It is the product of two components: the collision probability and the severity function.

– The collision probability is the normalized function applied to the distance of Gruyer et al. (2019) for multidimensional configurations with uncertainties.

– The severity function is the probability of death according to the extended EES computed using the scalar speed differential.

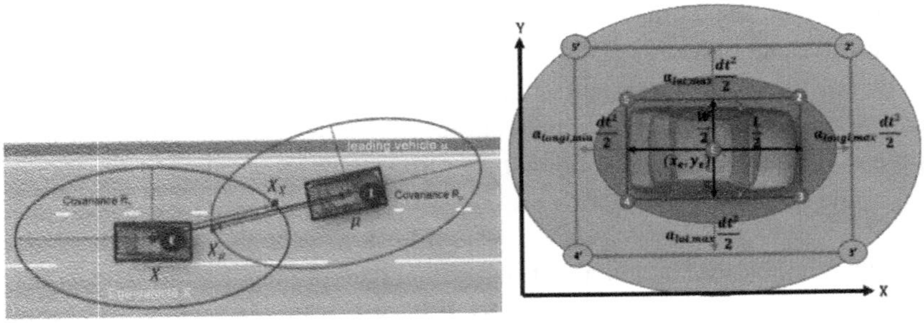

Figure 9.3. *Uncertainty modeling and computation of the distance of Gruyer ($D_{i,j}$) (Gruyer et al. 2019) based on the inter-class distance normalized by the sum of the intra-class distances*

Therefore, this risk estimator is appropriate for representing a risk by taking into account the collision probability due to the proximity of the two vehicles with the uncertainties modeling these objects, and the collision severity in the event of a collision, with respect to the EES of the vehicles. This risk metric is written in the form:

$$\text{RIMU } M_{i,j} = F\ (D_{i,j}) \times G\ (EES(DV_{scal,ij}))$$

with $D_{I,j}$ being the distance of Gruyer, $DV_{scal,ij}$ being the scalar speed differential, and F being the distance normalization function and G being the severity function.

Nevertheless, all of these metrics, although widely used to estimate local collision risks, are insufficient to extend the scope of knowledge of a risky situation and thus allow the human driver enough time to resume the driving task or allow for a maneuver to be performed, via the virtual co-driver, that is safe and comfortable for users.

9.3. Development of communication strategy to extend risk: CBL and CBL-G

In late 2015, as part of a collaboration between the LEOST and LIVIC laboratories, Gustave Eiffel University proposed innovative work on this communications issue, and more specifically quality of service routing protocols in VANETs adapted to distributed positioning issues in a constellation of vehicles and to the estimation of global risk. This work is based on the strong hypothesis of an absence of infrastructure and of a global positioning service and that each vehicle has only a local perception (obstacles, road and ego-vehicle). A V2V communication organization model called CBL (Chain Branch Leaf) (Rivoirard et al. 2017) was proposed to create and maintain a structure supporting close and direct exchanges (no multi-hops) between neighboring vehicles and remotely through the traffic (Figure 9.4). This structure creates a hierarchy where disjoint groups of neighboring vehicles (leaves) are connected via a leader vehicle (branch). The set of branches are connected to build a backbone (chain) formed along the road. The CBL structure is self-organizing and dynamic. It only requires knowledge of the one-hop neighborhood (direct communication) and can be integrated into all existing routing protocols. In addition, this communication structure is perfectly adapted to the deployment of cooperative positioning methods (Rivoirard et al. 2018), as well as those developed in partnership with the University of Sherbrooke: Cooperative Map Matching and Dynamic DGPS (Rohani et al. 2016).

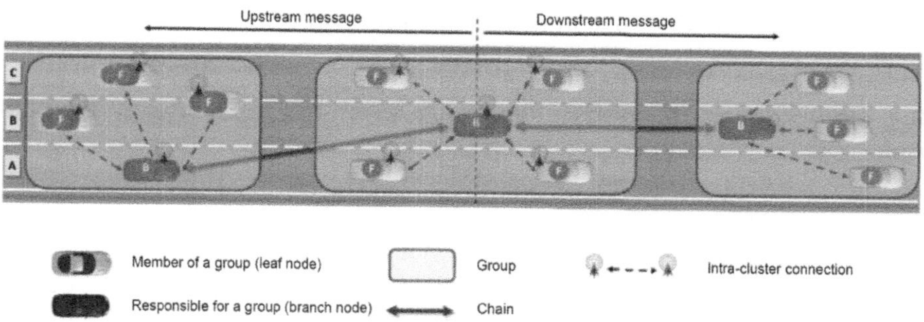

Figure 9.4. *Principle of the CBL strategy for road communication node clustering*

Since 2019, an evolution of the CBL method has been under development, proposing to integrate RSUs (Road-Side Units) in the form of a gateway to the CBL method to guarantee a high quality of communication in case of degraded configurations, such as the uncovered area, low vehicle density, high vehicle speed, presence of a tunnel, etc. (Belmekki et al. 2022). The initial V2V strategy is now extended to V2I configurations in highway areas where RSUs are present. The objective of CBL-G consists of developing a dynamic communication node clustering strategy with a distributed and dedicated RSU topology for optimal communication coverage (Figure 9.5). The proposed solution improves the quality of service for critical functions (risk assessment, precise and distributed positioning) by redirecting isolated vehicle nodes and clusters to RSUs. CBLG is currently assessed in different vehicle density configurations with an implementation in the physical-realistic simulation platform SiVIC-MobiCoop (NS3, Pro-SiVIC and RTMaps).

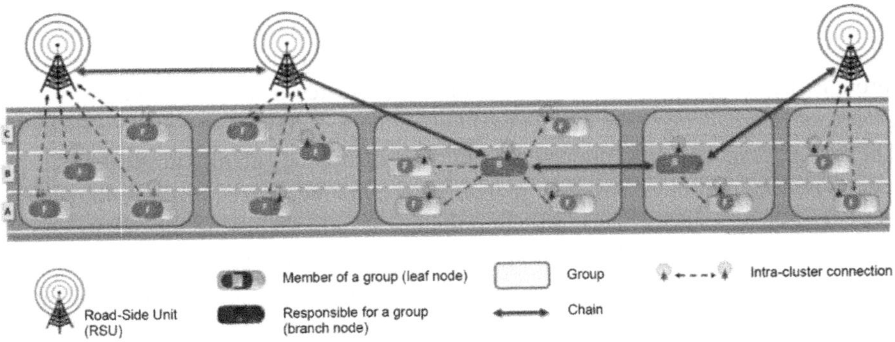

Figure 9.5. *CBL-G, an extension of the CBL strategy by integrating RSUs as gateways to ensure continuity of service*

9.4. Computation of cooperative risks: extended local risk and global risk

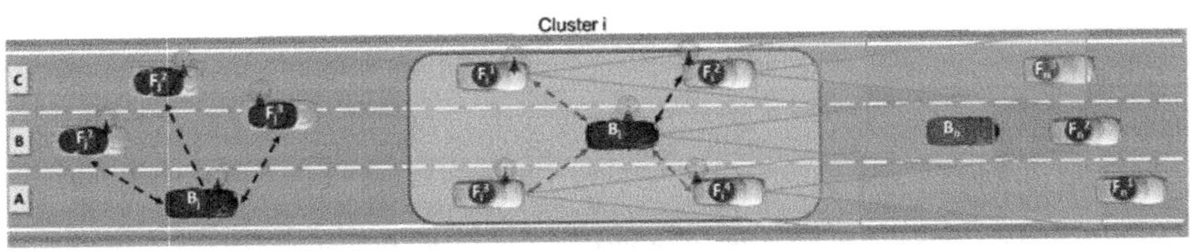

Figure 9.6. *Dynamic clustering of a fleet of vehicles equipped with means of communication and front perception*

With the help of the strategy for dynamic clustering of CBL-G communication nodes, it is now possible to implement three classes of collision risk metrics. The first class corresponds to the local metrics discussed in section 9.2 of this chapter. The second class concerns the development of new indicators built using information collected by a cluster "branch" through its "leaves". The third class includes global risk metrics. Global risks will be built through the cooperation and merging of the different extended local risks computed by the "branches" of the different clusters present on an area of interest. As mentioned in the previous section, the CBL-G strategy is not only adapted to the implementation of these different risk levels but also allows distributed and cooperative methods of estimating a vehicle's positioning to be proposed. In order to respond optimally to this second type of application, the optimal size of a cluster must be more than five vehicles (one branch and four leaves) on a range in accordance with the local perception range, the direct communication range and access to the same satellite constellation.

Therefore, for this type of "optimal" configuration (see Figure 9.6), we will be able to use the information perceived by vehicles F_i^1 and F_i^3 to perceive the presence of the two vehicles, F_i^2 and F_i^4 (see Figure 9.7). In this example, we limit perception to the front area of vehicles (i.e. RADAR) and apply a modeling of uncertainties on both the positioning of vehicles, obtained with a GNSS (Global Navigation Satellite System), and the detection of frontal obstacles obtained with the RADAR. This perception hypothesis also allows vehicles F_i^2 and F_i^4 to perceive the vehicles of the cluster downstream without being able to obtain their GPS positioning since there is no communication between the leaves of two different clusters. In addition, branch B_i is able to perceive branch B_n. Since the adjacent branches communicate to form a chain, branch B_i will also be able to receive the GPS position of branch B_n. We thus have the building of a local dynamic perception map, for the partial leaves.

By collecting all of the information of the leaves of cluster i by branch i, it will then be possible to estimate an extended dynamic perception map, which can then be used to build a map of extended local risks. The building of this map by B_i will be performed by applying a method of multi-source merging and multi-object tracking by belief theory (Gruyer et al. 2019) or using the BPM (Belief Plot Map) method (Magnier et al. 2021). The risk map will be built from the local dynamic perception map (estimation of positions, headings, speeds and uncertainties) and the RIMU metric.

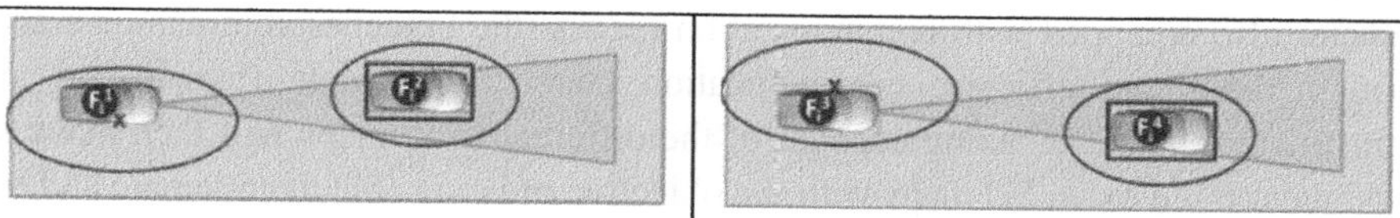

Perception from leaf F_i^1 | Perception from leaf F_i^3

Figure 9.7. *Frontal dynamic local perception taking into account positioning uncertainties; elaboration of Figure 9.6 on cluster i*

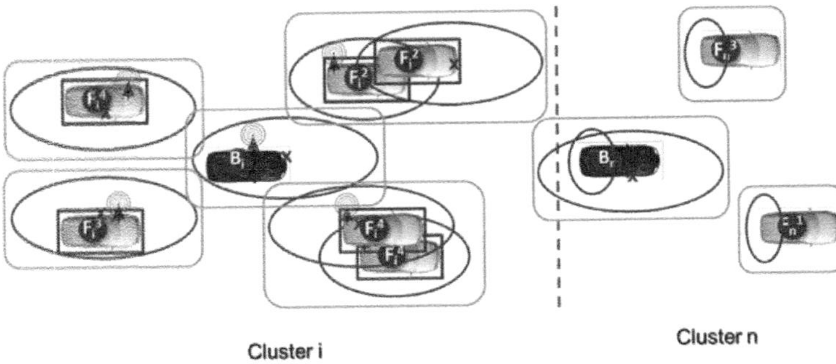

Figure 9.8. *Local dynamic perception map from the point of view of branch i, detail of Figure 9.6*

9.5. Impact of global risk and anticipation of risky situations

Previous studies (Giang et al. 2015) highlighted that the use of communication means that 802.11p could have an impact on reducing collisions. This result was an important first step in encouraging us to continue along this path of developing cooperative systems. In the next step, we have extended our reasoning by seeking not only to limit the number of collisions but also to anticipate risky situations and to know whether we could guarantee a maximum level of safety or a minimum level of risk. In this case, the transmission of a warning message is no longer sufficient; there is a need to handle more-complex information on the current situation. This information corresponds to all or part of the data of the local dynamic perception map of an ego-vehicle (the branch in CBL-G). Each vehicle equipped with on-board perception and communication means will then transmit its local knowledge to other vehicles also equipped with communication means (the branches communicate and form a rapid and efficient communication chain). A vehicle (branch) or an infrastructure manager taking account of the different local maps will then make it possible to build extended maps with the objective of being able to anticipate and be informed of remote events and of the future road configuration. This extended perception can be reflected both in a greater perception scope and in an enrichment of the description of the actors of the scene (extended local perception and extended perception). Increasing our perception range will be critical to anticipate and predict events that may lead to accidentogenic scenarios. Of course, the interpretation of these future situations is very useful to enable a decision to be made that can lead to a modification in the behavior of our ego-vehicle or the vehicles of a cluster (in CBL-G). In all cases, the objective is to minimize a situation's risk level. Consequently, in this new section, we will present our works showing that it is possible, and above all appropriate, to estimate a level of global risk from this extended perception. Moreover, it seemed logical to quantify the contribution, if any, of this global risk with respect to the local risk (Demmel et al. 2018).

To conduct this study, we must either choose or build a single risk criterion. In this section, we propose using a conventional, simple risk metric based on TTC, i.e. the ratio between the inter-distance and the speed differential between two vehicles if they do not alter their current behaviors. The risk is then obtained by applying the probability of the accident occurring, multiplied by the expected severity of the collision if it occurs. The collision probability is estimated with respect to TTC or (Glaser et al. 2010). The severity is obtained using the EES, as we saw earlier in this chapter. The EES is obtained by assuming an elastic collision between the two vehicles. Indeed, when we consider the collision of two vehicles, the EES will give us an indication of the kinetic energy that was dissipated by the collision. This value is then linked to the probability of the level of injuries sustained by the occupants of the vehicle(s) (using the case of frontal collisions).

Start of the journey with a homogeneous queue	Emergency braking of the first vehicle with transmission of an alert

Figure 9.9. *Generation of a queue of vehicles in the Pro-SiVIC platform for an "Emergency braking" scenario involving front perception of vehicles, reaction time, warning-type communications and a variable perception/communication equipment rate*

To highlight the benefit of computing a global risk, we consider a conventional scenario of a queue of n vehicles, in which the vehicles are equipped with exteroceptive sensors allowing them to perceive the frontal scene with a field of view limited to the traffic lane. The first step is to compute the TTC from the point of view of an ego-vehicle with its predecessor. This indicator provides a local risk assessment based on a single-vehicle system and local perception (see Figure 9.9).

At the same time, we also assess the extended local risk using the extended perception map. Therefore, with the local and extended maps, all TTC indicators are computed between an ego-vehicle and all of the moving objects that have been perceived.

Once all of these values are known, we apply the function given by Glaser et al. (2010) to compute a collision probability, $P_{i,n}$, for each vehicle $i \in [1, n-1]$ with respect to the ego-vehicle. Then, the risk indicators, $R_{i,n}$, for each vehicle with respect to the ego-vehicle can be computed by applying the function g, computing the severity of the potential collision as a function of the EES and the transfer function, G. G is the transfer function linking the EES to the probability of injury, ranging from mild injury to driver death. We then choose the maximum out of the two computed values to represent two variants of the severity of the same accident. $g(V_n, V_i)$ represents the severity of a collision when the two vehicles have not altered their speeds relative to the current time. Furthermore, $g(V_n, V_i - \gamma TTC_{i,n})$ represents the collision severity in the case, where vehicle i would carry out emergency braking with high deceleration γ ($\gamma = 0.8$ g in our implementation).

In the scenario presented in this section, we took only four "obstacle" vehicles and one ego-vehicle. Our ego-vehicle is placed at the end of the queue in position $n = 5$. Therefore, in a noncooperative application, we will only perceive vehicle 4 since the queue configuration prevents us the other vehicles from being perceived. This also implies that we will only be able to compute risk $R_{4,5}$. In contrast, in a cooperative application where all vehicles are equipped with means of communication, we will be able to compute the series of risks: $\{R_{1,5}, R_{2,5}, R_{3,5}, R_{4,5}\}$. In fact, in the cooperative application, each vehicle can compute similar risk vectors, which implies that for each vehicle, x, we will have a risk vector $\{R_{i,x}, \ldots R_{n,x}\}$ with $i \in [1, n], i \neq x$. This reasoning can be applied to CBL-G. In this case, each vehicle providing a risk is a branch that centralizes the information of its group. Therefore, $R^c_{1,5}$ will represent the risk between branch 5 and branch 1. Of course, each branch will reflect a cluster of vehicles involving a more complex and extended situation. Therefore, a branch, x, will have a risk vector $\{R_{i,x}, \ldots R_{n,x}\}$.

In order to compare the performance of the local risk estimation with respect to the global risk based on the use of extended perception, we found it more relevant and appropriate to simply create a single risk value describing the dangerousness of the situation, rather than a set of risks determined for each vehicle. Within the framework of CBL-G, this assertion translates into a single extended local risk value for a group. Therefore, for each test over time, we will have two risk values that will be calculated. Initially, we will have the global risk, $R_{g,x}$, computed for a single vehicle, x. Then, we will have the global risk, R_g, perceived for all vehicles. The higher the value of R_g, the more hazardous the driving situation. The risk $R_{g,x}$ is computed by applying the maximum of all of the risks computed 2-by-2 with vehicle x. This risk then provides an upper limit of the global risk perceived by a vehicle with respect to the other vehicles. In fact, this global risk, $R_{g,x}$, is very close to the local risk, computed from local perception, of vehicle x, due to the vehicle immediately in front being likely to be the most significant source of risk generation. However, in the formulation employed, the other vehicles are also taken into account, which may be useful in certain driving scenarios (Demmel et al. 2018).

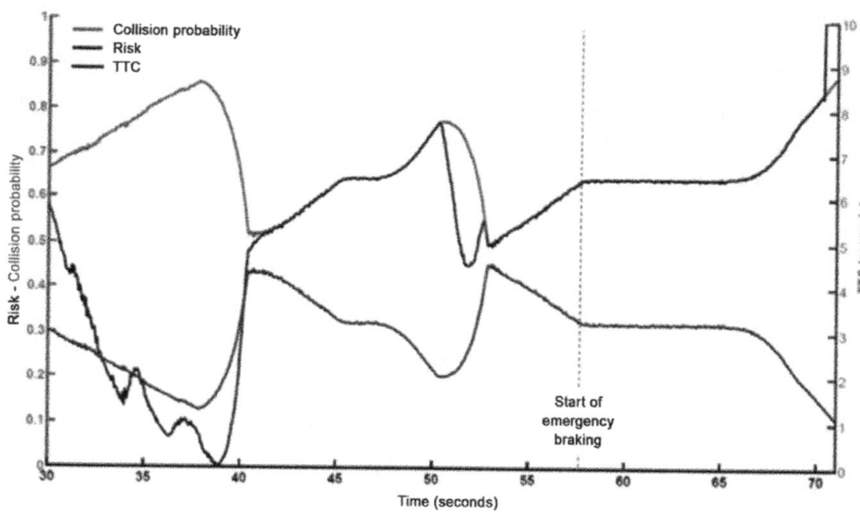

Figure 9.10. *Local risk, collision probability and TTC in the vehicle queue scenario for vehicles 4 and 5 (ego-vehicle)*

The global risk, R_g, for its part, can be seen as the average of the risks centered on each vehicle, or more explicitly as the total risk of a driving situation over a defined area of interest. In the case of CBL-G, R_g would be the average of the extended local risks perceived by each branch. This approach is similar to the work presented in Fitzgerald and Landfeldt (2012), which computes a weighted average risk (the value of the risk perceived by other vehicles is weighted). However, we do not apply weighting that we consider to be equally probable to the risks received.

In the scenarios, the first four vehicles will transmit their position at 2 Hz with a maximum latency of 5 ms. The ego-vehicle is positioned in the last place of the queue and updates its own positioning and its local perception map at a frequency of

10 Hz and 20 Hz, respectively. The extended map is updated at a frequency of 10 Hz. The local perception map of the ego-vehicle is obtained using a RADAR or a LiDAR. The positioning simply uses a GPS, an inertial unit and an odometer that are merged using a nonlinear Kalman filter. In Figure 9.10, we can observe the TTC, the collision probability and the local risk, $R_{4,5}$, for vehicles 4 and 5 (ego-vehicle). We also observe that $TTC_{4,5}$ begins to decrease at the instant when vehicle 4 begins to brake, while vehicle 5 is still at the same speed. In this scenario, a reference speed is transmitted from RSUs present along the infrastructure. Due to the position of the vehicles and the RSU transmission range, the application of the reference speed setpoint (maximum speed) is received at different instants by the different vehicles. This then results in a regulation of the inter-distances, which will vary the TTC and therefore the risk level. From emergency braking, we can clearly see that the risk will increase after a stabilization plateau step of about 10 seconds (stable inter-vehicular distance). The starting point of this increase in risk at $t = 67\ s$ means that vehicle 4 has just braked in order to avoid a collision with vehicle 3.

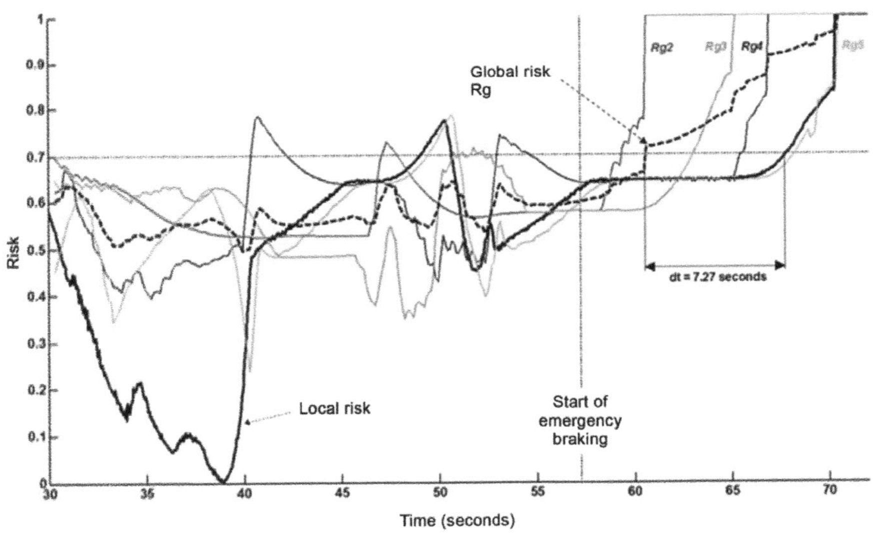

Figure 9.11. *Local risk and global risk in the vehicle queue scenario*

If we set a critical danger threshold to the risk value 0.7 (meaning that the driving situation is hazardous), then the automated driving system (or driver) will be warned 5 seconds prior to collision. This warning time is short and will only allow the driver to react (emergency braking or avoidance maneuver) and not to anticipate the situation to implement a maneuver minimizing the risk and not risking causing a collision with a rear vehicle. If we now observe the results presented in Figure 9.11, obtained with the use of communications, we can observe significant changes to the events. In fact, halfway through the scenario, the local risk between vehicles 4 and 5 is close to the danger threshold, while the global risk remains well below this threshold.

In this situation, vehicle 5 considers that it is in a very risky situation, whereas this is not the case for the other vehicles. This therefore mitigates a very local perception of the risk by a much more global perception. The two risk indicators (local and global) are therefore in contradiction, but they nevertheless perfectly reflect the situation from two different points of view. At the moment the first vehicle performs emergency braking, we can observe that the risk level for vehicle 2 increases almost immediately. Then, the other vehicles also see their respective risk increase in cascade, which is reflected by a series of emergency braking maneuvers. However, by observing the passing of the local and global risk indicators above the danger threshold, we can observe that the global risk crosses this threshold approximately 7 seconds before the local risk. This observation is very significant as it tells us that vehicle 5 no longer has only 7 seconds to react, but double this, i.e. 14 seconds. This result is highly significant for two reasons. The first concerns vehicles with a human driver who will have more time to make better decisions and engage in maneuvers that minimize the global and local risk both for them and for other road users. The second reason is linked to new mobilities and more specifically to the deployment of autonomous vehicles. Indeed, this global risk and the time saving obtained to prevent the autonomous driving system from arriving in a high-risk area will allow it to choose the best strategy without having to resort to emergency braking or another maneuver that may be equally hazardous. An adapted visualization of the local and global risk levels will enable the driver to improve their understanding of the environment. In addition, another solution can consist of the virtual co-driver warning the human driver and requesting that they resume the driving task (co-driver disengagement). In this case, with the execution of a secondary task, a human driver will need from 6 (average) to 21 (maximum) seconds to perceive and understand the situation so as to best react (Eriksson and Stanton 2017). In Merat et al. (2014), the study shows that it takes 40 seconds for the driver to ensure stable and appropriate control of the vehicle. In our example, the global risk gives us 14 seconds, which is more than enough for the driver to become aware of the situation, analyze and understand it and make the best decision.

9.6. Discussion

With the development and deployment of on-board advanced driver-assist system (ADAS) technologies and more recently, connected and autonomous driving systems, it is necessary to detect and analyze the road environment in all its complexity. The complexity of interactions between vehicles in mixed traffic implies being able to assess the level of safety associated with a road situation, through the use of local dynamic perception maps. However, these perception maps involving five key players (obstacles, road, ego-vehicle, environment and driver) and their attributes are not sufficient. Based on this perception data and these key elements, it is necessary to build new risk indicators that are more advanced and above all multi-actor, multidimensional, under uncertainties and under constraints.

A first ontology of the risk layer was proposed in Leroy et al. (2020). This ontology is significant as it identifies the different attributes, variables and parameters to be taken into account in future risk metrics that are more advanced and more representative of the real situation and the ability of the vehicle and the driver to respond to them. Figure 9.12 presents a non-exhaustive list of the different variables that can be taken into account in the computation of new multi-actor risk indicators. As we can observe from Figure 9.12, the risks will be computed per actor.

Reviewing the literature, the majority of articles dealing with risks mainly relate to the road configuration, the presence of obstacles and the current state of the driver, without carrying out real merging of this information. With regard to the "*road*" actor, we decided to break it down into four domains: characteristics of the roadsides, road markings, road geometry and road surface, which are the main attributes of this actor. The "*obstacles*" actor was broken down into three main categories: distance, dynamic and type. The "*driving*" actor was separated into four areas: somatic, psychological, experience-related and perception-related risks. The "*environment*" actor risks relate to the weather and visibility conditions. The last actor, "*ego-vehicle*", is more dedicated to the current state of the vehicle (positioning and dynamic state), faults in on-board components and the physical limits of the system (grip, braking and steering capacity, etc.).

Nevertheless, while we have proposed five risk classes based on the five actors, it is obvious that these classes are not really independent of each other. Indeed, regardless of the taxonomy and ontology used, there is a side effect to classification that masks the interactions between the different risk categories identified. For example, grip performance can be found in three different key components: ego-vehicle, road and environment. This is due to the choice made to separate the risks according to the scene into five key components. However, the same side effect would appear in any other type of ontology. Nevertheless, this finding is not really a weakness but rather a strength of the method. Indeed, with the existence of the same type of parameters in different key components, it is possible to use this redundancy in order to improve the certainty, reliability and robustness of the risk indicators using them. Therefore, it will be necessary to generate not five risk indicators (one per actor) but many more single- and multi-actor indicators.

In complement to our previous remarks, different attributes of the actors have a simultaneous influence on the same risk indicator. For example, those most affected are as follows:

– grip performance with data from the ego-vehicle, the road and the environment;

– lane and marking crossing time using road, ego-vehicle and obstacle data;

– visibility distance based on information from the environment, the road and the ego-vehicle;

– marking detection using attributes of the road, the ego-vehicle and the environment;

– the driver's reaction time, taking account of the environment.

These interactions must be taken into account, particularly when working on an indicator. Should a risk indicator impacted by several actors be estimated only from one of them, this could imply biases leading to a fault or inconsistency in the computed risk. This ontology, a first draft of which is presented in Figure 9.13, can then be used to verify that the risk indicators built include all of the actors involved as well as all of the necessary attributes.

However, this type of classification does not particularly highlight the interactions between actors. Therefore, future work will need to extend this representation to reveal all of these interactions. In addition, as shown in Figure 9.14, the processing operations of each actor makes it possible to estimate a set of specific attributes used to compute the different metrics. This figure highlights the interactions and interdependencies of the different significant variables computed from the processing modules presented in Figure 9.13.

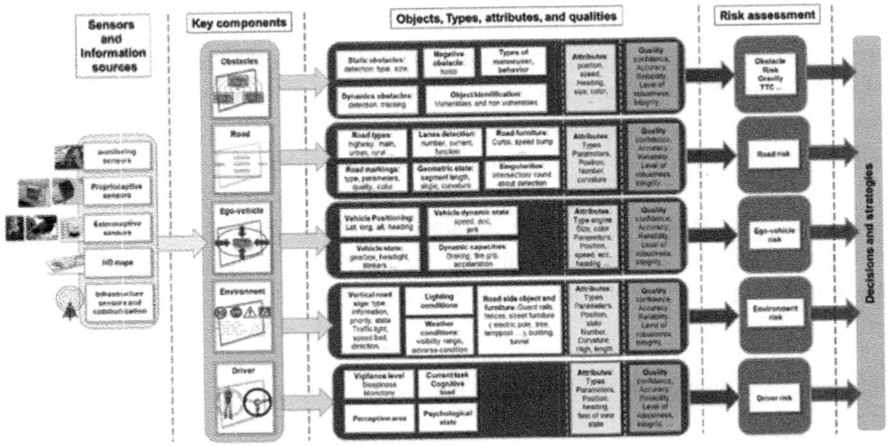

Figure 9.12. *All the actors and attributes involved in the construction of new indicators of multi-actor risks, multidimensional risks and risks under uncertainty*

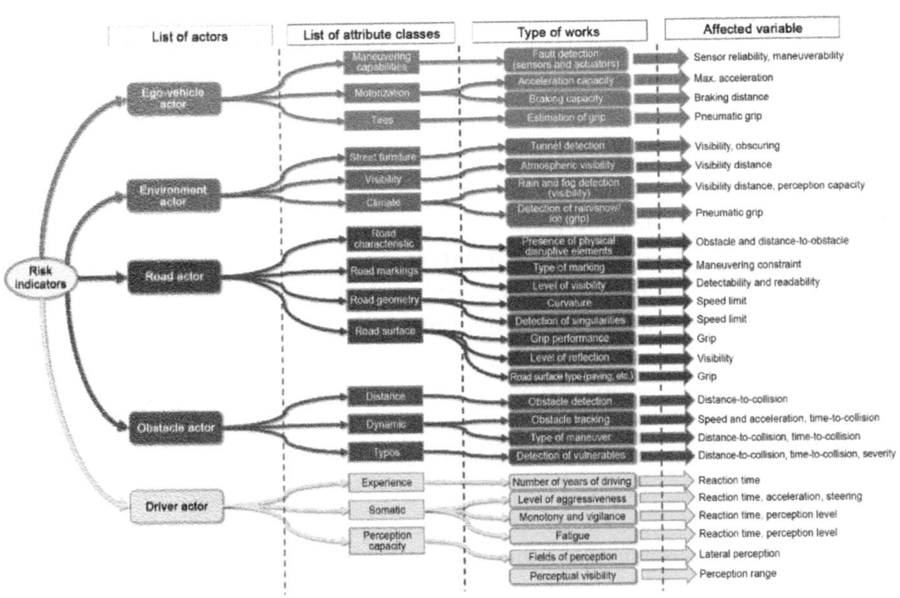

Figure 9.13. *Proposal of an ontology of the risk and the components/elements/actors involved*

It is also then evident that the different risk indicators will not necessarily have the same semantics for each actor, and the impact of an attribute can be different from one indicator to another. As a result, a mechanism needs to be found for homogenizing risk indicators in order to then merge them. It is imperative for the merging step to take into account not only the risk values but also their imperfections (imprecision, uncertainty, reliability and trust). To achieve this merging, the use of specialized sources in belief theory seems relevant. Another solution may consist of putting this risk data in matrix or vector form and then feeding either a CNN (Convolutional Neural Network) or an ANN (Artificial Neural Network).

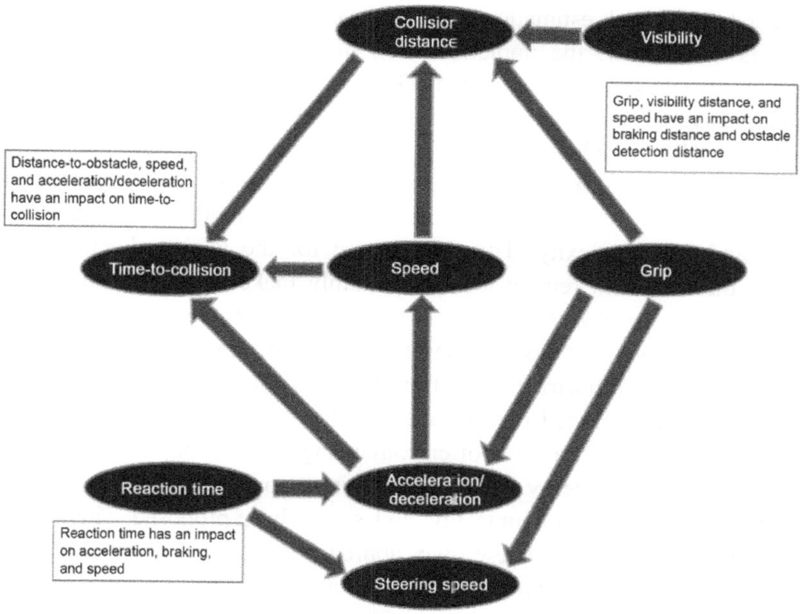

Figure 9.14. *Different input variables in the computation of risk metrics with their interdependencies and interactions*

In this second proposal relating to the use of supervised AI and deep learning, it will be necessary to have access to a training database with the associated ground truths. The outputs of these networks can be an enumeration of different risk types (collision, road exit, discomfort, etc.). This quantification of the multi-modal risk(s) will likely be very useful to feed driving models closer to the driving behavior of a human.

9.7. Conclusion and prospects

The deployment of new automated mobilities requires the development and use of new hardware and software technologies that allow the environment to be perceived in all situations that may also initially be managed by a human driver. However, like for human drivers, it is necessary to make the best driving decisions with respect to often complex situations and degraded information. This means we need to interpret the current driving situation and, above all, understand it. For humans, this understanding then leads to the estimation of a perceived risk. The driver will make their decision and act on the actuators according to this risk. In the design of autonomous vehicles, we find the same steps: perception, interpretation and understanding, risk estimation, decision-making and planning of trajectory and maneuvers and finally, the implementation of an action. As with humans, risk estimation by the system is essential. Generally, the risk is computed with the information available locally. The metrics employed are generally derived from TTC, TIV or HW, to which we apply a collision probability, and a severity level dependent on EES. However, in recent years, new operators have appeared and are seeking to be more exhaustive and above all more representative of the road situation and its complexity. This is the case for the RSS and RIMU metrics. However, these metrics remain local and mainly make it possible to react to an event, while not having the ability to predict and anticipate it. To allow this anticipation of risky situations, it is necessary to have access to remote information sent either from the infrastructure or from other road users. The first research and methods proposed consisted of implementing broadcasting and multi-hop routing strategies. While these strategies offer convincing results, they are not optimal in terms of resource and energy use. It is for this reason that strategies for dynamic clustering of single-hop communication nodes have been proposed. These strategies include CBL and CBLG. These two approaches are appropriate, and indeed efficient, for distributed processing such as extended dynamic perception, cooperative and distributed positioning (DDGPS and CMM) and estimation of extended local and global risk metrics. In this context, the extended local risk will be obtained by the "branch" of a cluster by merging the information provided by the "leaves" of this cluster. The branch will then transmit its information to adjacent branches until it forms a communication chain several kilometers long. This approach is efficient and much more optimal in terms of resource and energy use. Moreover, this generic architecture is easily adaptable to the different local metrics presented in the second section of this chapter and ensures a continuity of service essential for the deployment of autonomous vehicles. The application of this communication strategy and the computation of the global risk make it possible to then predict and anticipate remote critical situations with enough time to react appropriately and above all safely. Therefore, the example presented in section 9.5

highlights the time saving obtained to detect a remote risk. With only five vehicles, we double the detection time range (14 seconds instead of 7 seconds). Knowing that an efficient and safe regaining of control of the driving task by a human driver requires 8 to 10 seconds, the 14 seconds obtained are consistent with a possible request for intervention requested by the virtual co-driver.

Lastly, in the final section of this chapter, we address new research questions requiring future development of cooperative-risk metrics and more adapted/representative taking into account the complexity of a road scene. As we were able to highlight, the road scene includes a great number of "actors" that we must consider in order to be more comprehensive and efficient. These actors include the obstacles, the road, the ego-vehicle, the environment and the driver. In order to, at best, estimate the risk of a situation, it will be necessary to propose merging operators taking into account the uncertainties and constraints on each actor. The resolution pathways are likely to involve the use of AI-based architecture and evidence theory enabling the quantification and propagation of uncertainties.

9.8. References

Batsch, F., Kanarachos, S., Cheah, M., Ponticelli, R., Blundell, M. (2020). A taxonomy of validation strategies to ensure the safe operation of highly automated vehicles. *Journal of Intelligent Transportation Systems*, 1–20. doi: 10.1080/15472450.2020.1738231.

Belmekki, S., Gruyer, D., Tatekeu, C. (2022). Chain branch leaf-gateway, a strategy for dynamic clustering and optimal coverage of communication for automated mobility. In *Accepted in the 2nd IEEE International Conference on Digital Twins and Parallel Intelligence (IEEE DTPI 2022)*, 26–28 October, Boston, 1–5.

Chai, C., Zeng, X., Wu, X., Wang, X. (2019). Safety evaluation of responsibility-sensitive safety (RSS) on autonomous car-following maneuvers based on surrogate safety measurements. In *2019 IEEE Intelligent Transportation Systems Conference (ITSC)*, Auckland, 175–180.

Cui, G., Zhang, W., Xiao, Y., Yao, L., Fang, Z. (2022). Cooperative perception technology of autonomous driving in the internet of vehicles environment: A review. *Sensors*, 22, 5535. doi: 10.3390/s22155535.

Dahl, J., de Campos, G.R., Olsson, C., Fredriksson, J. (2019). Collision avoidance: A literature review on threat-assessment techniques. *IEEE Transactions on Intelligent Vehicles*, 4(1), 101–113. doi: 10.1109/TIV.2018.2886682.

Demmel, S., Gruyer, D., Burkhardt, J.-M. Glaser, S., Larue, G., Orfila, O., Rakotonirainy, A. (2018). Global risk assessment in an autonomous driving context: Impact on both the car and the driver, in the special session "Human-Machine Interactions in Autonomous Driving". In *2nd IFAC Cyber-Physical & Human Systems (CPHS 2018)*, 14–15 December, Miami.

Eriksson, A. and Stanton, N.A. (2017) Take-over time in highly automated vehicles: Non-critical transitions to and from manual control. *Human Factors*, 59(4), 689–705. doi: 10.1177/0018720816685832.

Fitzgerald, E. and Landfeldt, B. (2012). A system for coupled road traffic utility maximisation and risk management using VANET. In *15th International IEEE Conference on Intelligent Transportation Systems (ITSC)*, Anchorage, 1880–1887.

de Gelder, E., Khabbaz Saberi, A., Elrofai, H. (2019). A method for scenario risk quantification for automated driving systems. In *26th International Technical Conference on the Enhanced Safety of Vehicles (ESV)*. NHTSA, Eindhoven.

Giang, A.T., Busson, A., Lambert, A., Gruyer, D. (2015). Spatial capacity of IEEE 802.11p based VANET: Models, simulations and experimentations. *IEEE Transaction on Vehicular Technology*, PP(99). doi: 10.1109/TVT.2015.2474156.

Glaser, S., Vanholme, B., Mammar, S., Gruyer, D., Nouveliere, L. (2010). Maneuver-based trajectory planning for highly autonomous vehicles on real road with traffic and driver interaction. *IEEE Transactions on Intelligent Transportation Systems*, 11, 589–606.

Glaser, S., Vanholme, B., Gruyer, D., Mammar, S. (2011). Probability and risk-based manoeuvre planning for collision avoidance. In *Proceeding of First International Symposium on Future Active Safety Technology*. Toward Zero-Traffic-Accident, Tokyo.

Gruyer, D. and Rahal, M. (2019). Multi-layer laser scanner strategy for obstacle detection and tracking. In *2019 IEEE/ACS 16th International Conference on Computer Systems and Applications (AICCSA)*, 3–7 November. IEEE, Abu Dhabi.

Gruyer, D., Demmel, S., d'Andrea-Novel, B., Larue, G., Rakotonirainy, A. (2013). Simulating cooperative systems applications: A New complete architecture, *International Journal of Advanced Computer Science and Applications (IJACSA)*, 4(12), 171–180.

Gruyer, D., Orfila, O., Glaser, S., Hedhli, A., Hautière, N., Rakotonirain, A. (2021). Are connected and autonomous vehicles the silver bullet for future transportation issues? Benefits and weaknesses on safety, consumption, and traffic congestion. *Frontiers in Sustainable Cities 2, Special Collection "Advances in Road Safety Planning"*. doi: 10.3389/frsc.2020.607054.

Gürbüz, H. and Buyruk, S. (2019). Improvement of safe stopping distance and accident risk coefficient based on active driver sight field on real road conditions. *IET Intelligent Transport Systems*, 13(12), 1843–1850. doi: 10.1049/iet-its.2019.0322.

Jiménez, F., Naranjo, J.E., Garcia, F. (2013). An improved method to calculate the time-to-collision of two vehicles. *International Journal of Intelligent Transportation Systems Research*, 11(1), 34–42. doi: 10.1007/s13177-012-0054-4.

Katrakazas, C. (2017). Developing an advanced collision risk model for autonomous vehicles. PhD Thesis, Loughborough University, Loughborough.

Koopman, P., Osyk, B., Weast, J. (2019). Autonomous vehicles meet the physical world: RSS, variability, uncertainty, and proving safety. In *International Conference on Computer Safety, Reliability, and Security*, 245–253. Springer, Cham.

Kuang, Y., Qu, X., Wang, S. (2015). A tree-structured crash surrogate measure for freeways. *Accident Analysis and Prevention*, 77, 137–148.

Lambert, A., Gruyer, D., Pierre, G.S. (2008). A fast Monte Carlo algorithm for collision probability estimation. In *Proceedings of International Conference on Control, Automation, Robotics and Vision 2008 (ICARCV 2008)*, December. IEEE, Hanoi.

Leroy, J., Gruyer, D., Orfila, O., El Faouzi, N.-E. (2020). Five key components-based risk indicators ontology for the modelling and identification of critical interaction between human driven and automated vehicles. In *3rd IFAC Conference on Cyber-Physical & Human-Systems (CPHS)*, 3–5 December, Shanghai.

Leroy, J., Gruyer, D., Orfila, O., El Faouzi, N.-E. (2021). Adapted risk indicator for autonomous driving system with uncertainties and multi-dimensional configurations modeling. In *24th IEEE International Conference on Intelligent Transportation (ITSC2021)*, 19–22 September, Indianapolis, 2034–2041.

Lu, C., He, X., van Lint, H., Tu, H., Happee, R., Wang, M. (2021). Performance evaluation of surrogate measures of safety with naturalistic driving data. *Accident Analysis and Prevention*, 162, 106403. doi: 10.1016/j.aap.2021.106403.

Magnier, V., Gruyer, D., Godelle, J. (2021). Belief Plot Map – An efficient way to model static objects and environments for Autonomous Driving Systems. In *24th IEEE International Conference on Intelligent Transportation (ITSC2021)*, 19–22 September, Indianapolis.

Mahmud, S.S., Ferreira, L., Hoque, M.S., Tavassoli, A. (2017). Application of proximal surrogate indicators for safety evaluation: A review of recent developments and research needs. *IATSS Research*, 41(5), 153–163.

Mammar, S., Glaser, S., Netto, M., Blosseville, J.-M. (2004). Time-to-line crossing and vehicle dynamics for lane departure avoidance. In *7th International IEEE Conference on Intelligent Transportation Systems (IEEE Cat. No. 04TH8749)*, Washington, DC, 618–623.

Merat, N., Jamson, A.H., Lai, F.C.H., Daly, M., Carsten, O.M.J. (2014). Transition to manual: Driver behaviour when resuming control from a highly automated vehicle. *Transportation Research Part F, Traffic Psychology and Behaviour*, 27, 274–282. doi: 10.1016/j.trf.2014.09.005.

Monkhouse, H.E., Habli, I., McDermid, J. (2020). An enhanced vehicle control model for assessing highly automated driving safety. *Reliability Engineering and System Safety*, 202(8). doi: 10.1016/j.ress.2020.107061.

Park, H., Oh, C., Moon, J., Kim, S. (2018). Development of a lane change risk index using vehicle trajectory data. *Accident Analysis & Prevention*, 110, 1–8. doi: 10.1016/j.aap.2017.10.015.

Philipp, A. and Goehring, D. (2019). Analytic collision risk calculation for autonomous vehicle navigation. In *2019 International Conference on Robotics and Automation (ICRA)*, 1744–1750. IEEE, Montreal.

Pinnow, J., Masoud, M., Elhenawy, M., Glaser, S. (2021). A review of naturalistic driving study surrogates and surrogate indicator viability within the context of different road geometries. *Accident Analysis and Prevention*, 157, 106185. doi: 10.1016/j.aap.2021.106185.

Rivoirard, L., Wahl, M., Sondi, P., Berbineau, M., Gruyer, D. (2017). From multipoint relaying to chain-branch-leaf: Improving the clustering in OLSR for vehicular ad hoc networks. In *24th IEEE symposium on Communications and Vehicular Technology (SCVT 2017)*, University of Leuven, Leuven.

Rivoirard, L., Wahl, M., Sondi, P., Berbineau, M., Gruyer, D. (2018). A cooperative vehicle ego-localization application using V2V communications with CBL clustering. In *29th IEEE Intelligent Vehicles Symposium (IV 2018)*, 26–29 June, Changshu, Suzhou.

Rohani, M., Gingras, D., Gruyer, D. (2016). A novel approach for improved vehicular positioning using cooperative map matching and dynamic base station DGPS concept. *IEEE Transaction on Intelligent Transportation Systems*, 17(1), 230–239. doi: 10.1109/TITS.2015.2465141.

Shalev-Shwartz, S., Shammah, S., Shashua, A. (2017). On a formal model of safe and scalable self-driving cars. *arXiv:1708.06374*.

St-Aubin, P., Saunier, N., Miranda-Moreno, L.F. (2015). Comparison of various time-to-collision prediction and aggregation methods for surrogate safety analysis. In *Transportation Research Board 94th Annual Meeting*, 11–15 January. Transportation Research Board, Washington, DC.

Ward, J.R., Agamennoni, G., Worrall, S., Bender, A., Nebot, E. (2015). Extending time to collision for probabilistic reasoning in general traffic scenarios. *Transportation Research Part C: Emerging Technologies*, 51, 66–82. doi: 10.1016/j.trc.2014.11.002.

Westhofen, L., Neurohr, C., Koopmann, T., Butz, M., Schütt, B., Utesch, F., Neurohr, B., Gutenkunst, C., Böde. E. (2023). Criticality metrics for automated driving: A review and suitability analysis of the state of the art. *Archives of Computational Methods in Engineering*, 30, 1–35. doi: 10.1007/s11831-022-09788-7.

10

Resilience of Collective Perception in C-ITS – Deep Multi-Agent Reinforcement Learning

Imed GHNAYA[1], Hasnaâ ANISS[2], Marion BERBINEAU[3], Mohamed MOSBAH[1], and Toufik AHMED[1]

[1] *LaBRI, Bordeaux INP, France*

[2] *ERENA, COSYS, Gustave Eiffel University, Marne-la-Vallée, France*

[3] *COSYS-LEOST, Gustave Eiffel University, IFSTTAR, Villeneuve-d'Ascq, France*

Intelligent Transport Systems (ITS) encompass a wide range of technologies and systems designed to improve the efficiency, safety and sustainability of transport systems. Connected and autonomous vehicles (CAVs) constitute the most significant component of ITS. CAVs use on-board sensors, such as radars and cameras, to perceive the driving environment.

Through V2X communications, CAVs periodically share awareness and perception data with each other and with infrastructure such as traffic lights, signs and road-side units, to broaden their limited horizons. Multiple CAVs can simultaneously detect and exchange identical objects, resulting in a multitude of redundant data in the network. These unnecessary exchanges can cause an overloading of the communication network, thus reducing its reliability. The European Telecommunications Standards Institute (ETSI) has established a

Cooperative Intelligent Transport Systems,
coordinated by Léo MENDIBOURE.

collective perception message format and a set of generation techniques in order to find a compromise between network overload and awareness level. However, these approaches present many disadvantages when applied to scenarios with variable traffic densities.

In this chapter, we present a method of collective perception based on multi-agent deep reinforcement learning. This method enables each CAV, treated as an agent, to learn an independent interactive strategy that will be used to select and exchange detected objects in the network. The results of the simulations show that the proposed method, compared to the current works of the state of the art, improves network reliability without sacrificing the level of sensitivity within short ranges of under 100 m.

10.1. Introduction

10.1.1. *Background and issue*

Recent developments in Intelligent Transport Systems (ITS) have focused on autonomous driving as an emerging technology capable of reducing accidents by improving road safety and respecting driving rules. The majority of autonomous vehicles (AVs) rely on local sensors, such as cameras, RADAR and LiDAR to perceive their environment and interact with other road users. Despite significant advances in "sensor" performance, the detection range and field of view (FOV) are still limited by physical constraints. Especially in urban traffic environments where obstructing objects such as buildings, trees and other road users cause significant perception problems (Soto et al. 2022).

Through the use of V2X communication technologies (Chen et al. 2017), cooperative ITSs (C-ITS) enable connected AVs (CAV) to share information with each other, as well as with infrastructure and other connected road users. This sharing happens through vehicle-to-vehicle (V2V), vehicle-to-infrastructure (V2I) and vehicle-to-cloud (V2C) communication scenarios. In this regard, the European Telecommunications Standards Institute (ETSI) has standardized the ITS-G5 communication profile (ETSI 2020) based on the IEEE 802.11p standard, specifically for CAVs in order to exchange information on their current states (speed, position, direction, etc.) via cooperative awareness messages (CAM) (ETSI 2014).

V2X communication can overcome the limits of local perception by exchanging sensor data between CAVs. This process is called collective perception (CP) (ETSI 2019). As shown in Figure 10.1, CAVs or infrastructure perceive their environment locally through local sensors and then perform additional processing tasks, such as multi-sensor data merging, to improve the quality of local information. The merged information is then transmitted to the collective perception component. The latter:

1) merges this information with that received by V2X communication to better support autonomous driving functions;

2) selects the information to be shared with other connected road users.

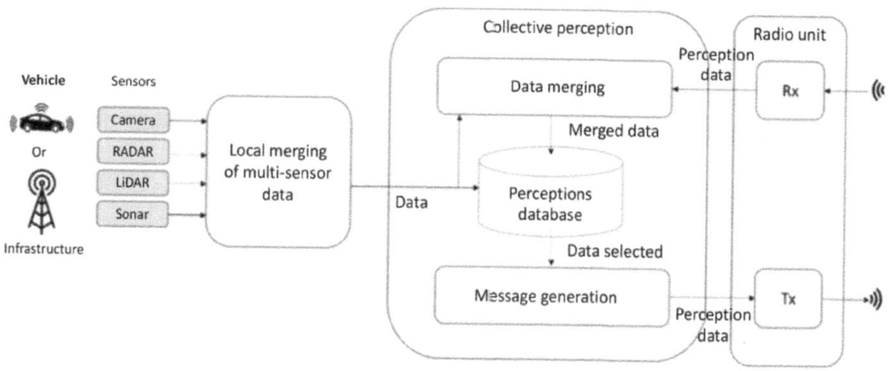

Figure 10.1. *Basic architecture of collective perception*

While CP offers a potential solution to the limits of local sensors, it is highly dependent on network resilience. In a cooperative driving environment, objects on the road can be detected simultaneously by several CAVs and, subsequently, their exchange by the CP can result in large amounts of redundant data being transmitted over the network. As a result, these unnecessary exchanges can reduce network reliability, which increases the loss rate and reduces the level of awareness.

10.1.2. *Motivation and contribution*

In order for perception to be beneficial, CAVs must continually perceive and share useful perception data with each other, both autonomously and in real time. To meet these constraints, a transmitting CAV must evaluate all potential combinations of detected objects in order to produce the best message to be broadcast in the network.

Recently, paradigms such as supervised, unsupervised and reinforcement learning (RL) have been extensively studied and have yielded better results on C-ITS-related problems (Haydari and Yılmaz 2022). In particular, RL represents a powerful paradigm that could be used as an automatic decision-making method in our context. Figure 10.2 shows the main components of RL. Its objective is to guide a CAV to make sequential decisions using a Markov decision process to maximize a reward value.

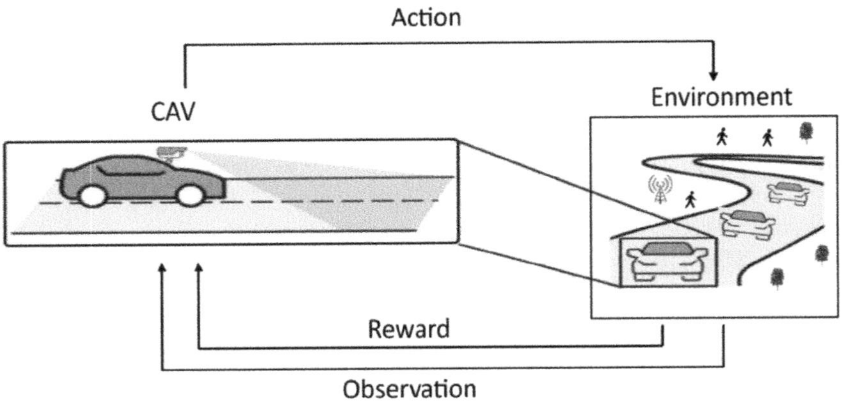

Figure 10.2. *Main components of reinforcement learning. For a color version of this figure, see www.iste.co.uk/mendiboure/transportsystems.zip*

Q-learning (Arulkumaran et al. 2017) is an RL-based method for a CAV to learn an optimal strategy for perception data selection that maximizes the utility of the receiving CAVs. However, the exponential increase in state-action spaces caused by the increase in the number of detected objects continues to represent a problem of significant complexity. In this context, the deep neural network (DNN) associated with RL is called deep RL (DRL) (Arulkumaran et al. 2017), which represents a promising strategy to improve learning capabilities in difficult challenges such as ours.

Recently, DRL-based algorithms, such as DQN (Arulkumaran et al. 2017), have aroused significant interest in the resolution of ITS problems, notably traffic management (Wu et al. 2018, 2019) and automatic decision-making in autonomous driving (Sallab et al. 2017). However, some traffic safety applications require a rapid system response, making centralized learning difficult to perform for challenges involving complex environments. Multi-agent RL (MARL) systems have demonstrated notable performance in learning decentralized strategies for a group of agents operating to achieve the same system objective (Oroojlooy and Hajinezhad 2022).

This chapter proposes a DRL-based multi-agent learning method in a multi-vehicle driving environment. The main objective of this method is to enable each CAV to learn an independent strategy for selecting perception data to be exchanged in the network. This strategy will be used to select and exchange data that maximizes the benefit of the receiving CAVs, thereby conserving network resources and preserving network reliability. The results obtained demonstrate that the proposed method effectively improves the reliability of the network by reducing the information redundancy compared to current state-of-the-art works.

The remainder of this chapter is organized as follows. Section 10.2 summarizes the standardization of PC and its methods of data exchange in the network. Section 10.3 formulates the utility of perception data in a cooperative driving environment. Section 10.4 presents the design and the learning algorithm for the proposed method. We present the results and evaluations in section 10.5. Finally, section 10.6 concludes this work.

10.2. State of the art

Several studies have been conducted by the C-ITS research community on the exchange of perception data between CAVs. Initially, this exchange was carried out by the direct transmission of sensor data over the V2X communication networks. Li and Nashashibi (2011) developed a first exchange technique between two connected vehicles in this field. This technology incorporated a "see-through" sensor that enables the road to be perceived as if no other vehicle or obstacle obstructed its view. In this technology, communication between connected automobiles and augmented reality provided an augmented view by removing the obstructive part of the road. Kumar et al. (2012) introduced CarSpeak technology, which enables CAVs to share sensor data in the form of 3D point clouds. This work encodes the raw data of the local sensors using an octree scheme. The CAVs then exchange 3D data from their regions via V2X communication networks.

However, these approaches often require significant network resources, such as throughput and bandwidth. Günther et al. (2016) proposed to perform CP by periodically exchanging environmental perception messages (EPMs) including descriptions of the tracked objects, such as their position, speed and size. These messages comprise data containers indicating the status of the transmitting CAV, the properties of its sensors and its detected objects.

10.2.1. *Standardization of collective perception by ETSI*

ETSI standardization of CP is described in ETSI (2019). On the basis of Günther et al. (2016) and a set of generation criteria, this standardization defined the specification of a collective perception message. CPMs are used to exchange collective perception messages via the network. Figure 10.3 describes the structure of a CPM, which is composed of the following information containers:

– a station container that defines the location of the transmitting station, its type (vehicle or infrastructure) and its direction;

– set of containers describing the identification, coverage and detection angle of each on-board sensor;

– set of containers containing information on each detected object, notably its distance from the perceiving vehicle, its speed, its length and its width;

– set of free-space addendum containers, storing the information required by the receivers to decode a received CPM, such as the relationship between the sensor IDs and the IDs of the available areas.

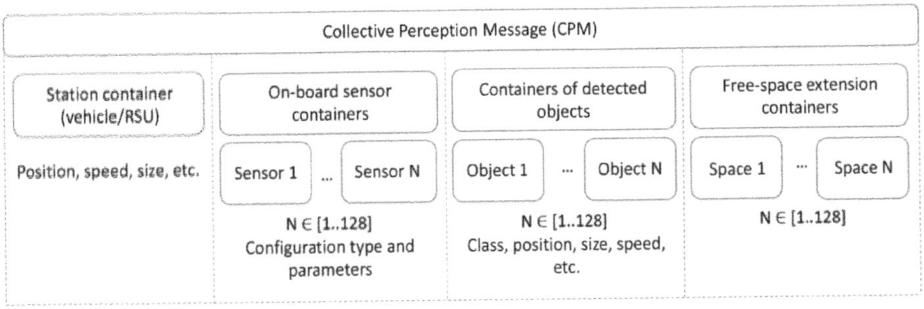

Figure 10.3. *Format of a CPM*

ETSI has established a set of generation rules based on the dynamics of detected objects to determine when a CAV needs to generate and send a CPM in light of the large amount of data provided by local sensors and the limitations of resources dedicated to V2X communication (Garlichs et al. 2019). A CAV generates and sends a CPM if one of the following conditions is met:

– it detects a new object;

– the difference in its position or velocity has changed by 4 m or 0.5 m/s, respectively, in relation to the latest information included in its CPM (Garlichs et al. 2019);

– the last inclusion of the detected object in its CPM is greater than or equal to one second (1 s);

– otherwise, none of these rules are satisfied, and the CAV produces a CPM with empty object containers every 1 s.

10.2.2. *Perception data selection and exchange techniques*

Huang et al. (2019) and Thandavarayan et al. (2019) demonstrated that the rules for generating a CPM result in the exchange of large amounts of redundant data in the network. Depending on their evaluation scenario, between 25 and 50 times per second, the CAVs receive identical updates on the same detected object.

Although redundancy is useful for improving information accuracy and quality, excessive use could reduce the reliability of the network and, consequently, degrade the quality of the CP. Recently, ETSI has established a set of techniques for mitigating perception data redundancy in the network (ETSI 2019) that can be applied when the channel load exceeds a predefined threshold. We discuss the techniques most used in the state of the art:

– Frequency-based technique: a CAV includes the objects detected by its on-board sensors in its current CPM if the number of its previous CPMs containing the same objects does not exceed a predefined redundancy threshold, *N_ Redundancy*.

– Distance-based technique: an object is removed by a CAV from its current CPM if it has been received more *than N_Redundancy* times during a predetermined time interval, *W_Redundancy*.

– Dynamic-based technique (Thandavarayan et al. 2020): the locally perceived object is omitted from the current CPM if none of the previous generation rules are fulfilled (Garlichs et al. 2019).

– Entropy-based technique: a CAV anticipates the prior knowledge of its neighbors regarding the current state of a locally perceived object based on an entropy computation. This object is omitted from the CPM if all neighbors within a predefined radius are supposed to perceive it and the relative entropy is less than a predetermined threshold (Higuchi et al. 2019).

However, we highlight three major disadvantages for the techniques proposed by ETSI. Each technique:

1) is based on static thresholds whose values may not be suitable in circumstances where traffic volumes are variable;

2) requires neighboring CAVs to use the same or a similar redundancy mitigation technique, which is not necessarily the case in mixed scenarios where one or more vehicles cannot exchange information;

3) broadcasts CPMs regardless of the state of the environment or the utility of the objects in the coverage of each receiver.

10.3. Mathematical modeling of the cooperative driving environment

The driving environment is characterized by a set of AVs of different sizes and classes. One or more of them are CAVs that require a CP when other vehicles or obstacles obstruct their fields of vision.

Figure 10.4. *The need for collective perception. For a color version of this figure, see www.iste.co.uk/mendiboure/transportsystems.zip*

Figure 10.4 describes a simple scenario in which CP is necessary in a driving context. In this scenario, CAVs A, B, D and E are equipped with cooperative perception capabilities such as LiDAR, RADAR, cameras and V2X communication. As *B* obstructs *A*'s line of sight, the latter is unable to perceive *C*; whereas *B* and *E* detect and broadcast *C* simultaneously via the network. Due to the limited resources of the network, inappropriate sharing of this information can lead to network congestion and, as a result, CAVs should only select and broadcast information that will be used by their neighbors.

In the remainder of this section, we present a mathematical description of data exchange and its utility in the driving environment.

10.3.1. *Awareness and perception data exchange*

Figure 10.5 presents the image of the driving environment with respect to an Ego CAV at an instant, t, in a 2D global coordinate system.

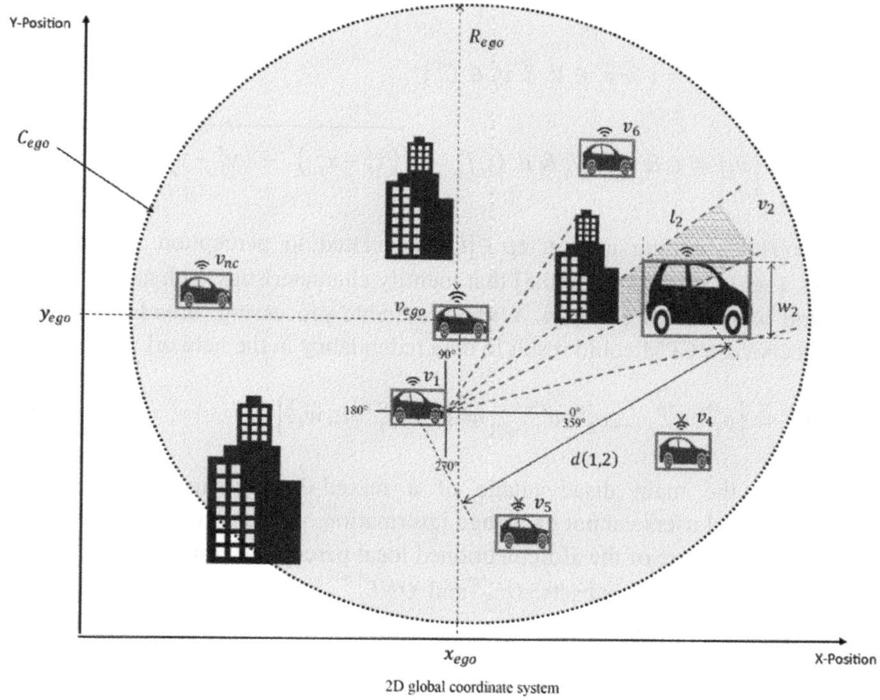

Figure 10.5. *Image of the driving environment at an instant, t. For a color version of this figure, see www.iste.co.uk/mendiboure/transportsystems.zip*

To begin with, we use the notation V [10.1] to represent the set of vehicles circulating in the environment at instant t. We consider the following hypotheses:

– Each vehicle $v_i \in V$ is represented by a rectangle of length l_i and width w_i.

– A subset, V_c, consists of $nc > 1$ CAV equipped with a wireless communication technology with a circular coverage of radius R and on-board 360°-vision sensors with a maximum detection range, m.

– Each CAV $v_i \in V_c$ transmits information on its current state, S_i^t [10.2] and receives all other states, SR_i^t [10.3] from the other CAVs in its communication coverage, C_i^t [10.4]. We define S_i^t as the set comprising the current position, x_i^t and y_i^t, in a 2D global coordinate system, the length, l_i, and the width, w_i.

$$V = \{v_i,\ v_i = (x_i, y_i, l_i, w_i), i \in [1, n]\} \quad [10.1]$$

$$S_i^t = \{x_i^t, y_i^t, l_i, w_i\} \quad [10.2]$$

$$SR_i^t = \{S_j^t; j \neq i \;\&\; v_j \in V_c \;\&\; v_j \in C_i^t\} \quad [10.3]$$

$$C_i^t = \left\{v_j j \neq i \;\&\; v_j \in V_c \;\&\; d^t(i,j) = \sqrt{\left(x_i^t - x_j^t\right)^2 + \left(y_i^t - y_j^t\right)^2} \leq R_i\right\} \quad [10.4]$$

We consider a time instant, $tp \in]t, 2t[$, related to perception. Each $v_i \in V_c$ perceives a set of objects O_i^{tp} [10.5] that identify characteristics such as the position, length and width of each object. Without suitable processing, broadcasting all of these objects via a CPM could result in data redundancy in the network:

$$O_i^{tp} = \left\{o_1^{tp}, o_2^{tp}, \ldots, o_k^{tp}; o_{i\in[1,k]}^{tp} = \left\{x_i^{tp}, y_i^{tp}, l_i, w_i\right\}\right\} \quad [10.5]$$

However, the main disadvantage of a mixed-driving environment is that unconnected road users cannot exchange information, which makes their perception more difficult because of the aforementioned local perception constraints. Therefore, we divide O_i^{tp} into two subsets, OC_i^{tp} and ONC_i^{tp}, to differentiate connected and unconnected objects, respectively.

10.3.2. *Utility of perception data in the driving environment*

We consider a $utility_i^{tp}$ function [10.6] that returns a utility value in [0,1] of a set of objects, cpm_i^{tp} [10.7] broadcast by $v_i \in V_c$ at time tp in its coverage, C_i^t:

$$utility_i^{tp}\left(cpm_i^{tp}, C_i^t\right) = utilitye_i^{tp}\left(cpmOC_i^{tp}, C_i^t\right),$$

$$cpm_i^{tp} = cpmOC_i^{tp} \cup cpmONC_i^{tp} \quad [10.6]$$

To this end, the objective of each $v_i \in V_c$ is to maximize its utility function as follows:

$$maximize_{cpmOC_i^{tp} \in PS\left(OC_i^{tp}\right)}\, utility_i^{tp}\left(cpmOC_i^{tp}, C_i^t\right) \quad [10.7]$$

with:

$$utility_i^{tp}(cpmOC_i^{tp}, C_i^t) = 1 - detection^{tp}(cpmOC_i^{tp}, C_i^t) \quad [10.8]$$

$$detection^{tp}(cpmOC_i^{tp}, C_i^t) = \frac{1}{|C_i^t|} \sum_{v_j \in C_i^t, j \neq i} detection^{tp}(cpmOC_i^{tp}, v_j) \quad [10.9]$$

$$detection^{tp}(cpmOC_i^{tp}, v_j) = \frac{1}{|cpmOC_i^{tp}|} \sum_{o \in cpmOC_i^{tp}} detection^{tp}(o, v_j) \quad [10.10]$$

$$detection^{tp}(o, v_j) = \begin{cases} 0 & if \quad d^{tp}(j,o) > m \\ (1 - \frac{d^{tp}(j,o)}{m})(1 - \frac{obstruction_{j\to o}^{tp}}{vision_{j\to o}^{tp}}) & otherwise \end{cases} \quad [10.11]$$

Equations [10.6] and [10.8] define a "utility" function that gives the probability of non-detection of the set of connected objects of a CPM, $cpmOC_i^{tp}$, of a transmitting CAV $v_i \in V_c$ in its coverage C_i^t.

– Equation [10.9] denotes this probability as an average of the detection probabilities calculated for all CAVs in C_i^t.

– Equation [10.10] defines the probability of detection of a $cpmOC_i^{tp}$ by an average of the probabilities computed for all of the objects it includes.

– Equation [10.11] determines the probability of detection of an object, o, by a CAV, $v_{j \neq i} \in V_c$. This probability is equal to 0 if the object appears outside its detection range. Otherwise, this value is defined by a product of two factors:

1) The first factor is related to the distance and is defined proportionally to the distance $d^{tp}(j,o)$.

2) The second factor is related to the obstructions and refers to the value by which o is visible (directly in the line of sight) by v_j.

10.4. Multi-agent learning with DRL for selection and exchange of perception data

This section describes the design of the system and the learning algorithm.

10.4.1. *System design*

CAVs only have at their disposal partial information on the environment. For this, we employ decentralized Markov decision processes with partial observability (Oliehoek and Amato 2016), where each CAV represents an instance.

10.4.1.1. *State and observation*

We define the state of the environment of $v_i \in V_c$ at time t by:

$$s_i^t = \{d^t(j,k)\,, vision_{j\to k}^t, l_j, w_j; j \neq i\,\&\, v_j \in V_c \,\&\, v_j \in C_i^t\}, \qquad [10.12]$$

with $d^t(j,k)$ and $vision_{j\to k}^t$ representing the distance and the viewing angle of v_i towards another $v_j \in V_c$ in its communication coverage, respectively; l_j and w_j being the dimensions of v_j. As the environment is partially observable, we define an observation, $z_i^t \subseteq s_i^t$, by a partial set of s_i^t.

10.4.1.2. *Action*

Given the subset of connected objects, OC_i^t, perceived by $v_i \in V_c$ at time t, we defined an action, a_i^t, by selecting a subset of OC_i^k, which maximizes the "utility" function in C_i^t. Due to the dynamic nature of the driving environment, we divide the FOV of each CAV into p paths and s sectors, as shown in Figure 10.6.

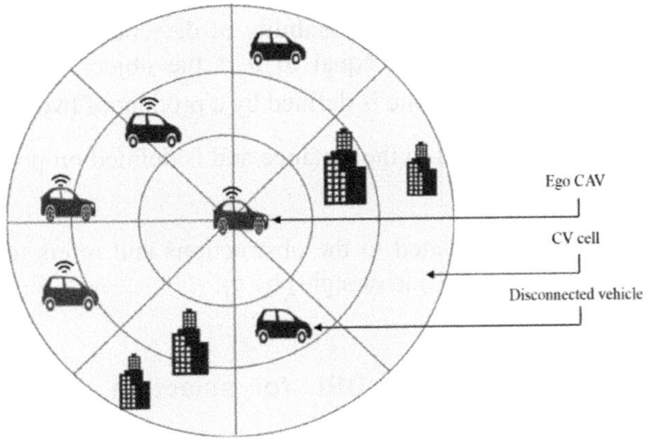

Figure 10.6. *Circular field of view of an Ego CAV*

We represent the FOV of $v_i \in V_c$ mathematically by a set of cells of size $p * s$:

$$FOV_i^t = \{P_{1,1}^0(t), \dots, P_{1,s}^1(t), P_{1,0}^2(t), \dots, P_{,p,s}^{p*s-1}(t)\}, \qquad [10.13]$$

where $P^j_{p',s'}(t)$ is the cell indexed by path p' and sector s'. To this end, a_i^t becomes the selection of cells whose objects are useful in C_i^t.

10.4.1.3. *Reward*

The reward, r_i^t, of a value $v_i \in V_c$ is the value provided by [10.6], which represents the utility of its generated connected objects resulting from the chosen action.

10.4.2. *Learning algorithm*

The objective of using RL as an autonomous decision-making tool in a cooperative driving environment is to enable each CAV to learn a data selection strategy that maximizes the utility of its neighbors. Considering discrete time instances, t, each $v_i \in V_c$:

1) observes the state of its environment, s_i^t;

2) executes an action, a_i^t , in accordance with a certain strategy, $\pi_i(s_i^t, a_i^t)$. This strategy denotes the probability of selecting an action, a_i^t, conditioned by the current state of the environment, s_i^t;

3) obtains a reward, r_i^t.

The environment then passes from the current state, s_i^t, to the next state, s_i^{t+1}. This process forms a learning experience at time t described by $e_i^t = (s_i^t, a_i^t, r_i^t, s_i^{t+1})$. The cumulative reward can be defined by $R_i^t = \sum_{\tau=0}^{t} \gamma^{t-\tau} s_i^{\tau}$, where $\gamma \in [0,1]$ is a reduction factor given to previous rewards to decrease their influence on the current value. In this context, Q-learning (Arulkumaran et al. 2017) is one of the RL-based algorithms that can be applied. This algorithm leads a CAV, $v_i \in V_c$, to learn a value function, $Q_i^{\pi}(s_i^t, a_i^t) = R_i^t$, when it performs a_i^t in the state s_i^t by following its data selection strategy, π_i. Q-learning uses a table that stores the "state-action" value functions of all actions in the action space. However, this algorithm is no longer suitable for problems involving large state-action spaces due to the high time and memory consumption needed to examine each entry in the table of values in order to establish the optimal strategy. Recently, deep-Q-network (DQN) (Arulkumaran et al. 2017) has been used to overcome this limitation. A deep neural network parameterized by a vector of weight θ is used to learn an optimal strategy without going through all of the possible states of a system. This method allows a $v_i \in V_c$ to approximate its "state-action" value function, $Q_i(s_i^t, a_i^t, \theta) \approx Q_i^*(s_i^t, a_i^t)$, where $Q_i^*(s_i^t, a_i^t)$ gives the function that returns the ideal value of the selected action.

Algorithm 10.1 describes the learning process in a cooperative driving environment.

Inputs:
Number of $N_{episodes}$ episodes, learning steps per episode N_{steps}, ϵ-greedy $\epsilon \in]0,1[$
Output:
An optimal strategy for perception data selection for each CAV
Begin:
Initialize two deep neural networks, θ_i, $\theta_i^{'} \leftarrow \theta_i$, and a buffer memory, D_i, for each CAV i
$episodes \leftarrow 0$
While $episodes < N_{episodes}$ **do**
    $steps \leftarrow 0$
    if $steps < N_{steps}$ **do**
        Each CAV i:
            Builds s_i^{steps} as a function of [10.12]
            Samples a number, c, from the uniform (0,1)
            if $c \leq \epsilon$ **then**
              Predicts a_i^{steps}, a possible combination of cells in its FOV
              Defined by [10.13]
            else
              $a_i^{steps} = arg\max_{a_i^{steps}} Q_i(s_i^{steps}, a_i^{steps}, \theta)$
            End if
            Obtains a_i^{steps} and observes $s_i^{steps+1}$
            Stores $e_i^{steps} \leftarrow (s_i^{steps}, a_i^{steps}, a_i^{steps}, s_i^{steps+1})$ in D_i
        $steps \leftarrow steps + 1$
    End while
    Each CAV i:
        Samples a minibatch from D_i
        Updates θ_i as a function of [10.14]
    $episodes \leftarrow episodes + 1$
    After each number of N_{init} episodes, each CAV i resets $\theta_i^{'} \leftarrow \theta_i$
End while
End

Algorithm 10.1. *Learning algorithm in a cooperative driving environment*

In order to carry out the learning process, each $v_i \in V_c$ uses two DQN networks: a learning DQN and a target DQN, parameterized by θ_i' and θ_i, respectively. The learning network is updated at the end of each episode, which consists of a fixed number of discrete instants. This update is performed by minimizing the learning loss computed in a minibuffer uniformly sampled from the set of experiments $D_i = [e_i(0), e_i(1), e_i(2), \ldots]$ of $v_i \in V_c$, as indicated below:

$$loss(\theta_i) = E_{(s,a,r,s')\sim U(D_i)}\left[y_i^{DQN} - Q_i(s, a, \theta)\right], \quad [10.14]$$

where $y_i^{DQN} = r_i' + \gamma \max_{a_i'} Q_i(s_i', a_i', \theta_i')$, and s_i', a_i' and r_i' represent the state, action and reward of the next time instant.

10.5. Simulations, results and evaluations

Based on simulations, this section examines the performance of the proposed method.

10.5.1. *Simulation tools, scenarios and parameters*

The simulations are based on the Artery (Riebl et al. 2015) and SUMO (Simulation of Urban MObility) tools (Krajzewicz et al. 2012), in which we generate at random vehicles of different classes and sizes and equip them with the same local perception capabilities. For each vehicle, we configure sensors with a maximum detection range of 100 m and an FOV of 360 degrees.

Table 10.1 summarizes the communication parameters used in this simulation.

Parameter	Value
Technology and channel of communication	ETSI ITS-G5/CCH
Communication coverage	500 m
Transmission power	23 dBm
Data rate	6 Mbps
CAM message generation	Every 100 ms
CPM message generation	Every 150 ms

Table 10.1. *Communication parameters*

10.5.2. *Results and evaluations*

Figure 10.7 shows the number of redundant objects generated by the CPM generation rules as a function of the distance between the detected object and the CAV that receives it. On average, CAVs receive about 11 updates of the same object over short distances. This high level of redundancy is explained by the fact that CAVs are able to simultaneously perceive and exchange objects regardless of the state of the environment and what their neighbors can also perceive. However, this metric decreases as the distance increases, making perception more difficult as objects move further away from the perceiver.

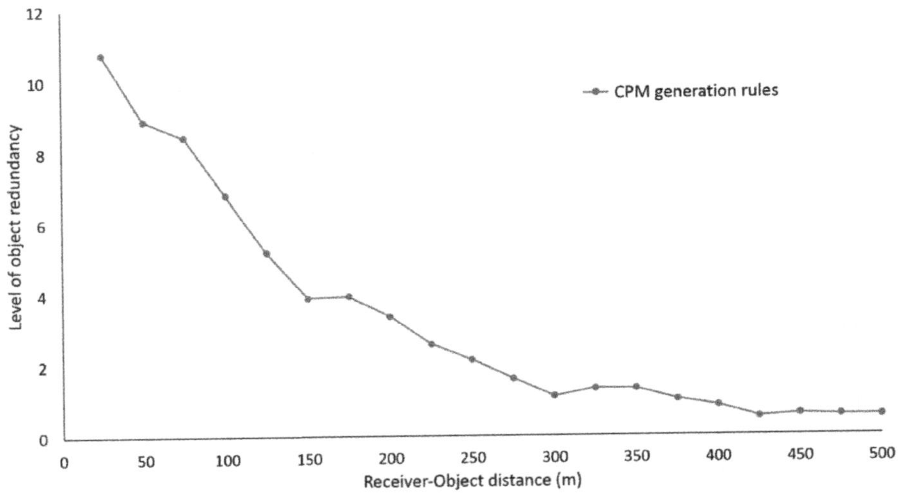

Figure 10.7. *Redundancy level of objects in the network: performance of the basic ETSI method. For a color version of this figure, see www.iste.co.uk/mendiboure/transportsystems.zip*

Unnecessary sharing of perception data reduces the reliability of ITS-G5 technology and, consequently, decreases its reliability. The proposed method overcomes this limitation. To do this, we build deep neural networks in each CAV. The learning phase is composed of 6,000 episodes, each episode consisting of 10 simulation instants. This phase employs the RMSprop optimizer (Ruder 2016) with a learning rate of 0.001, a minibatch 64 in size, a reduction factor of 0.99 and a buffer 10^6 in size. For reasons of complexity, we divide the CV of each CAV into three paths and three sectors, resulting in 512 possible unique cell combinations.

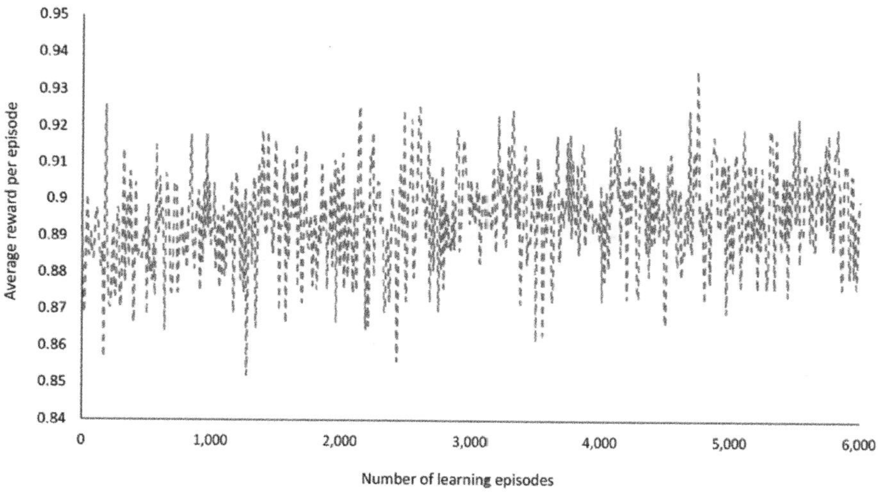

Figure 10.8. *Variation in average CAV reward as a function of the learning episodes. For a color version of this figure, see www.iste.co.uk/mendiboure/transportsystems.zip*

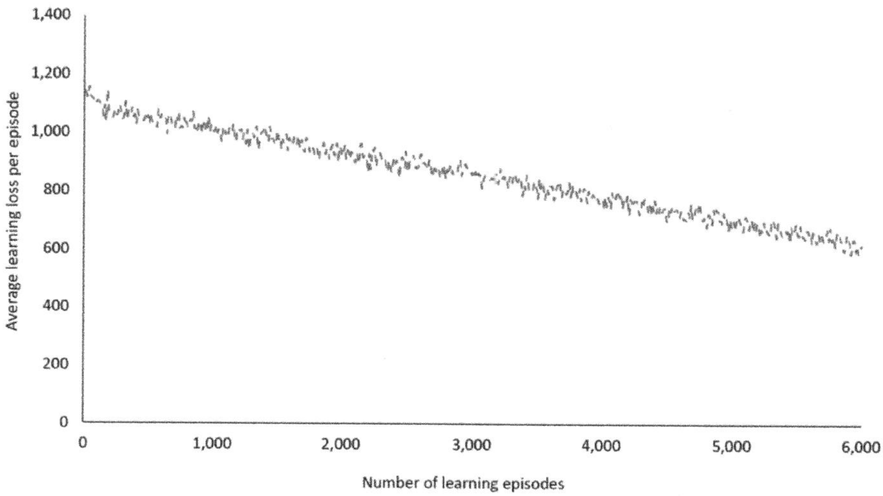

Figure 10.9. *Variation in average loss as a function of the learning episodes. For a color version of this figure, see www.iste.co.uk/mendiboure/transportsystems.zip*

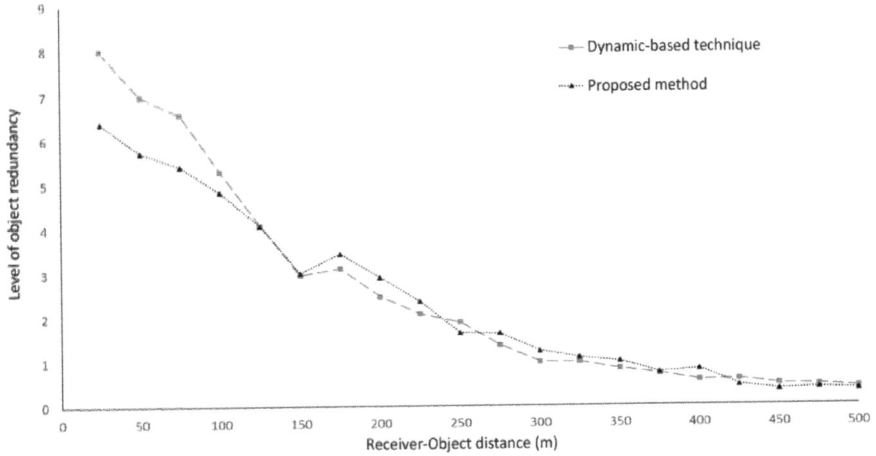

Figure 10.10. *Comparison of the performance of the proposed method with the dynamic-based technique for the reduction of redundant perception data in the network. For a color version of this figure, see www.iste.co.uk/mendiboure/transportsystems.zip*

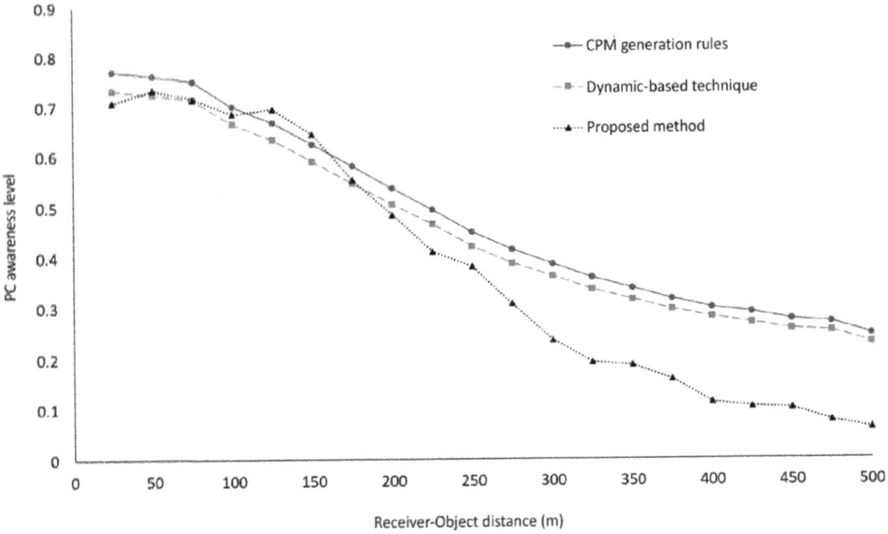

Figure 10.11. *Level of consciousness generated by the CP in the network as a function of the distance between the perceived object and the receiving CAV. For a color version of this figure, see www.iste.co.uk/mendiboure/transportsystems.zip*

Figure 10.8 shows the variation in the average reward of CAVs, representing the average utility of their CPMs (equation [10.6]) as a function of the learning episodes. We can observe that the average reward in the majority of episodes fluctuates considerably. This is because the environment is constantly changing and the CAVs move quickly, enabling them to explore new states and take different actions. This has a negative impact on the efficiency of data to be exchanged in the network. Figure 10.9 shows that the average learning loss gradually converges to zero with an increase in the number of episodes. This means that the learning process improves over time.

Based on the level of redundancy of objects in the network, we compare the performance of the proposed method with the dynamic-based technique (Thandavarayan et al. 2020). Figure 10.10 shows that the proposed method offers better performance over distances of less than 100 m. These improvements are obtained without sacrificing the level of perception in the network. This is demonstrated in Figure 10.11, which compares the level of perception obtained with the CPM generation rules, the dynamic-based technique and the method proposed in this chapter. This level is obtained by dividing the number of unique objects perceived by a CAV by the total number of objects in its coverage.

However, such a redundancy mitigation technique must also maintain a level of perception close to the rules for generating CPMs, which broadcast all of the perceived objects and necessarily lead to better results. Figure 10.11 shows that the dynamic-based technique and the proposed method reach levels of perception close to the rules for generating CPMs at distances of less than 150 m. In contrast, the latter gives less satisfactory results for this metric over distances greater than 175 m. This is because CAVs, using their selection policies, only examine a subset of the combinations representing the cells closest to their action spaces.

10.6. Conclusion

This work proposes a new learning method for connected and autonomous vehicles in order to optimize the selection and exchange of its detected objects with other connected road users. This method is based on a deep multi-agent reinforcement learning approach. Compared to state-of-the-art approaches, the proposed method minimizes the number of redundant objects in the network while maintaining an adequate level of consciousness at short distances of less than 150 m.

However, the simulations also reveal that the proposed method is unable to achieve a sufficient level of awareness in the network at distances greater than 175 m. This is due to the huge fluctuations in the average reward of connected and autonomous vehicles, which result from the great complexity of the state-action space and the constant variety of the driving environment; the main perspective of this work will be to have a flexible method that optimizes learning.

10.7. References

Arulkumaran, K., Deisenroth, M.P., Brundage, M., Bharath, A.A. (2017). Deep reinforcement learning: A brief survey. *IEEE Signal Processing Magazine*, 34, 26–38.

Chen, S., Hu, J., Shi, Y., Peng, Y., Fang, J., Zhao, R., Zhao, L. (2017). Vehicle-to-everything (v2x) services supported by LTE-based systems and 5G. *IEEE Communication Standards Magazine*, 1, 70–76.

ETSI (2014). EN 302 637-2 V1.3.2; Intelligent Transport Systems (ITS); Vehicular communications; Basic set of applications; Part 2: Specification of cooperative awareness basic service. ETSI Standard.

ETSI (2019). TR 103 562 V2.1.1; Intelligent Transport System (ITS); Vehicular communications. Basic set of applications; Analysis of the Collective Perception Service (CPS); Release 2. ETSI Technical Report.

ETSI (2020). EN 302 663 V1.3.11; Intelligent Transport Systems (ITS); ITS-G5 Access layer specification for Intelligent Transport Systems operating in the 5 GHz frequency bands. ETSI Standard.

Garlichs, K., Günther, H., Wolf, L. (2019). Generation rules for the collective perception service. In *Proceedings of the IEEE Vehicular Networking Conference (VNC)*, Los Angeles, 1–8.

Günther, H.J., Mennenga, B., Trauer, O., Riebl, R., Wolf, L. (2016a). Realizing collective perception in a vehicle. In *Proceedings of the 2016 IEEE Vehicular Networking Conference (VNC)*, Columbus.

Günther, H., Riebl, R., Wolf, L., Facchi, C. (2016b). Collective perception and decentralized congestion control in vehicular ad-hoc networks. In *Proceedings of the IEEE Vehicular Networking Conference (VNC)*, Columbus.

Günther, H., Riebl, R., Wolf, L., Facchi, C. (2018). The effect of decentralized congestion control on collective perception in dense traffic scenarios. *Comput. Commun.*, 122, 76–83.

Haydari, A. and Yılmaz, Y. (2022). Deep reinforcement learning for intelligent transportation systems: A survey. *IEEE Transactions on Intelligent Transportation Systems*, 23, 11–32.

Higuchi, T., Giordani, M., Zanella, A., Zorzi, M., Altintas, O. (2019). Value-Anticipating V2V communications for cooperative perception. In *Proc. the 30th IEEE Intelligent Vehicles Symposium*, Paris, 1690–1695.

Hobert, L., Festag, A., Llatser, I., Altomare, L., Visintainer, F., Kovacs, A. (2015). Enhancements of V2X communication in support of cooperative autonomous driving. *IEEE Communication Magazine*, 53, 64–70.

Huang, H., Fang, W., Li, H. (2019). Performance modelling of V2V based collective perceptions in connected and autonomous vehicles. In *Proceedings of the IEEE 44th Conference on Local Computer Networks (LCN)*, 356–363.

Krajzewicz, D., Erdmann, J., Behrisch, M., Bieker, L. (2012). Recent development and applications of SUMO – Simulation of Urban MObility. *International Journal on Advances in Systems and Measusrements*, 5(3&4), 128–138.

Kumar, S., Shi, L., Ahmed, N., Gil, S., Katabi, D., Rus, D. (2012). Carspeak: A content-centric network for autonomous driving. In *Proc. of ACM SIGCOMM*, 259–270. Association for Computing Machinery, New York.

Li, H. and Nashashibi, F. (2011). Multi-vehicle cooperative perception and augmented reality for driver assistance: A possibility to 'See' through front vehicle. *Proceedings of the IEEE Intelligent Transportation Systems Conference (ITSC)*, Washington, DC, 242–247.

Oliehoek, F.A. and Amato, C. (2016). *A Concise Introduction to Decentralized POMDPs*. Springer, Cham.

Oroojlooy, A. and Hajinezhad, D. (2022). A review of cooperative multi-agent deep reinforcement learning. *Applied Intelligence*, 53(11), 13677–13722.

Riebl, R., Günther, H.C.F., Wolf, W. (2015). Artery: Extending veins for VANET applications. In *International Conference on Models and Technologies for Intelligent Transportation Systems (MT-ITS)*, 450–456. IEEE, Budapest.

Ruder, S. (2016). An overview of gradient descent optimization algorithms. *arXiv:1609.04747*.

Sallab, A.E., Abdou, M., Perot, E., Yogamani, S. (2017). Deep reinforcement learning framework for autonomous driving. *Electronic Imaging*, 19, 70–76.

Soto, I., Calderón, M., Amador, O., Urueña, U. (2022). A survey on road safety and traffic efficiency vehicular applications based on C-V2X technologies. *Vehicular Communications*, 33, 100428.

Thandavarayan, G., Sepulcre, M., Gozalvez, J. (2019). Analysis of message generation rules for collective perception in connected and automated driving. In *Proceedings of the IEEE Intelligent Vehicles Symposium (IV)*.

Thandavarayan, G., Sepulcre, M., Gozalvez, J. (2020). Redundancy mitigation in cooperative perception for connected and automated vehicles. In *Proceedings of the 2020 IEEE 91st Vehicular Technology Conference (VTC2020-Spring)*, Paris, 134–139.

Wu, Y., Tan, H., Ran, B. (2018). Differential variable speed limits control for freeway recurrent bottlenecks via deep reinforcement learning. *arXiv:1810.10952*.

Wu, Y., Tan, H., Peng, J., Zhang, H., He, H. (2019). Deep reinforcement learning of energy management with continuous control strategy and traffic information for a series-parallel plug-in hybrid electric bus. *Applied Energy*, 247, 454–466.

PART 4

Securing Cooperative Intelligent Transport Systems

11
Distance-Bounding Protocols

David GÉRAULT[1], Pascal LAFOURCADE[2] and Léo ROBERT[3]

[1]*Technology Innovation Institute, Abu Dhabi, UAE*
[2]*LIMOS, Université Clermont-Auvergne, Clermont-Ferrand, France*
[3]*MIS, Université de Picardie Jules Verne, Amiens, France*

Security of communications has become indispensable. It ensures the confidentiality of exchanges, the authenticity of correspondents, the integrity of sent messages and other properties, depending on the application case.

Autonomous vehicles must guarantee a sufficient level of security for the good of the actors involved in this context, such as:

– Users: their physical safety must be guaranteed, and also their private lives.

– Manufacturers: no attack should be able to leak information to competitors or tarnish the brand image by exhibiting malfunctions of their vehicles.

– Companies offering services: they need assurances as to the identity of their clients to ensure that services are addressed to the right people.

The various current technological components communicate using *protocols*. If these protocols are not built correctly, then security flaws will allow attackers to jeopardize the guarantees mentioned above.

In this chapter, we will focus on a specific class of protocol (*Distance-Bounding protocol* or DB). In wireless communications, they make it possible to ensure that entities have valid behavior in a defined geographical area.

Let us consider the following example to illustrate the importance of DB protocols in the case of autonomous vehicles. Suppose Alice is in her vehicle and arrives at a toll. Between her and the barrier is the vehicle of Eve, an ill-intentioned person who does not want to pay the toll. In the absence of a DB protocol, Eve could mount an attack which targets Alice and makes her pay in her place. Indeed, when Alice arrives, Eve can simply act as a relay and retransmit the exchanges between the barrier and Alice. Therefore, Alice's car thinks it is exchanging with the toll gate, while in reality, it is Eve who is retransmitting the barrier's messages. From the barrier's perspective, it is Eve's vehicle that is in order and it therefore lets her pass. Eve therefore impersonated Alice to the toll barrier by making the latter believe that she was Alice. A DB protocol prevents this attack by verifying that the paying vehicle is indeed the vehicle situated in front of the barrier.

The purpose of this chapter is to formally define these DB protocols and to compare existing constructions.

NFC and RFID technologies are increasingly present in our lives. These technologies allow a transponder (also called a tag) to communicate without contact with a reader. In wireless communication, an intruder can still listen and transmit a signal, allowing them to mount what is called a *wormhole attack*. In recent decades, several DB protocols have been introduced to avoid such attacks. In this context, there are several threat models: terrorist fraud, mafia fraud, distance fraud, etc.

We first show the links between existing threat models. Next, we explain how to improve attacks on some of the existing protocols.

11.1. Introduction

In the present day, radio-frequency identification (RFID) and near-field communication (NFC) technologies, and more generally wireless technologies, are being increasingly developed. They are commonly used in payments, access-control applications and even in many electronic passports (Finkenzeller 2003). Modern vehicles are no exception to this phenomenon, and technologies such as the passive keyless entry and start (PKES) system, which opens the doors of a car by detecting an approaching smart key, are one example.

The main objective of these technologies is to allow a reader (or verifier) to communicate wirelessly with tags (or provers) implanted in objects. In this context, an intruder can mount relay attacks by simply transmitting a signal and then deceive the reader by using the signal from a tag that can be very far away. Such attacks have been successfully applied against the PKES system, allowing the theft of vehicles (Francillon et al. 2011; Yang et al. 2012b).

To avoid these attacks, DB (*Distance Bounding*) were introduced by Brands and Chaum (1993). These constitute a countermeasure against relay attacks, since they measure round-trip times during a rapid challenge/response phase. The DB protocols verify that the provers are close to the verifier, i.e. in an area of trust. In the scientific literature, there exist several threat models, depending on the power and objective of the intruder:

– *Distance Fraud* (Brands and Chaum 1993): a faraway malicious prover tries to convince the verifier that they are close, while the prover is not in the area of trust. A practical example is house arrest, where a prisoner wearing an electronic bracelet cannot leave a given area. If they can mount a distance fraud against the electronic surveillance device, then they can deceive the device over their presence within the allowed area.

– *Mafia Fraud* (MF) (Desmedt 1988): an adversary, placed between a faraway honest prover and a verifier, tries to take advantage of their position to authenticate the prover close to the verifier. The adversary can simply relay the message, but they can also modify the messages involved or create new messages (for the prover or the verifier). To illustrate this attack, imagine a queue in which an attacker would relay the signal between the payment card of a client who is at the end of the queue and the payment terminal. This allows the attacker to make someone else pay in their place. It is also the attack described involving vehicles and the toll barrier.

– *Terrorist Fraud* (TF) (Desmedt 1988): a faraway malicious prover, aided by an adversary, tries to convince the verifier that they are close. In fact, the adversary is close to the verifier and the prover gives information to the adversary, but the adversary cannot impersonate the prover during further running of the protocol. One example is a person who wants to let a friend open their locker once but does not want to allow them to do it later. In other words, they are only willing to provide help if that help cannot be used by the friend to authenticate again in the future.

– *Impersonation Fraud* (IF) (Avoine and Tchamkerten 2009): an adversary tries to impersonate the prover to the verifier. The purpose of such an attack would be to have someone else accused of their own ill actions.

– *Distance Hijacking* (Cremers et al. 2012): a faraway prover takes advantage of some honest, active provers (one of which is close) to have the verifier grant privileges to the distant prover. This can be used to forge an alibi.

Chapter contents: we first explain the relations between the different threats by distinguishing between threats where the prover is dishonest, on the one hand, and threats where the prover is honest, on the other hand. Next, we present an overview of DB protocols, as well as 24 attacks.

DB protocol history: Distance Bounding was introduced by Brands and Chaum (1993) to combat relay attacks. They also introduced the attack that we call "distance fraud": their protocol detects response bits sent in advance. Before the existence of distance-bounding protocols, in 1988, Desmedt identified terrorist fraud and mafia fraud (Desmedt 1988). Then, Avoine and Tchamkerten (2009) studied impersonation fraud. In 2012, Cremers et al. (2012) proposed an attack that they called "distance hijacking" (DH). At the same time, some of them laid the foundations for formal modeling of DB protocols (Avoine et al. 2009; Dürholz et al. 2011). Avoine et al. (2009) proposed a formal framework for cryptanalyzing the DB protocols. In particular, they defined the adversary strategies for mafia and terrorist fraud, which they called *no-ask, pre-ask* and *post-ask* strategy. Two years later, Dürholz et al. (2011) proposed the first computational formal framework for providing properties of DB protocols that are based on shared symmetric keys. They give rigorous cryptographic security models for mafia, terrorism and distance fraud. The BMV model proposed by Boureanu et al. (2013a) generalizes the previous fraud into three groups of threats: distance fraud, man-in-the-middle and collusion fraud. By their definitions, distance fraud includes DH and previous distance fraud, man-in-the-middle contains mafia fraud and impersonation fraud, and collusion fraud extends the concept of terrorist fraud. However, they do not establish relations between these three different threat models. Some articles (Boureanu et al. 2013b, 2013c; Baghernejad and Nasour Bagheri 2014; Boureanu and Vaudenay 2014, 2015; Gambs et al. 2014a) compare 4 to 14 protocols to their success probabilities for distance fraud, mafia fraud, terrorist fraud and/or identity fraud attacks. Distance bounding has been studied in the context of RFID and also in ad hoc networks, as in the study proposed by Meghdadi et al. (2011).

11.2. Relations between threats for DB protocols

Boureanu et al. (2013a) proposed a framework, called the BMV model, which generalizes the definitions of the previously enumerated common frauds into three possible threats: distance fraud, man-in-the-middle fraud and collusion fraud.

We present the formal definitions given in Boureanu et al. (2013a), and then we show how these definitions cover the usual threat models and prove some relations between several of these concepts.

11.2.1. *Threat models*

In what follows, provers are denoted by P, verifiers by V, the adversary by $\mathcal{A}$ and P^* denotes dishonest verifiers. Provers do not have output, and verifiers have an output of one bit Out_V, where $\mathsf{Out}_V = 1$ means acceptance and $\mathsf{Out}_V = 0$ means rejection.

DEFINITION 11.1 (Distance-Bounding protocol (Boureanu et al. 2013a)).– *A* DB protocol *is defined by a tuple* $(Gen, P, V, \mathbb{B})$, *where:*

1) Gen *is a randomized key generation algorithm such that* $Gen(1^s; r_k) \mapsto (x, y)$, *where* r_k *designates random coins of* Gen *and* s *a security parameter.*

2) $P(x; r_P)$ *e is a probabilistic polynomial-time interactive Turing machine (PPT ITM) running the prover algorithm with an input* x *and a random input* r_P.

3) $V(y; r_V)$ *is a PPT ITM running the verifier algorithm with an input* y *and a random input* r_V.

4) $\mathbb{B}$ *is a (*distance-bound*).*

They must be such that the following two properties are respected, where we denote by $d(loc_V, loc_P)$ the distance between the positioning of V and P:

- *Termination:* $(\forall s)(\forall \mathbf{R})(\forall r_k, r_V)$ *if* $(., y) \rightarrow Gen(1^s; r_k)$ *and* $\mathbf{R} \leftrightarrow V(y; r_V)$ *model the running, then* V *stops in* $Poly(s)$ *computational steps, where* $\mathbf{R}$ *is any set of (unbounded) algorithms.*
- *p-completeness:*$(\forall s)(\forall loc_V, loc_P$ *such that* $d(loc_V, loc_P) \leq \mathbb{B})$, *we have:*

$$\Pr_{r_k, r_P, r_V}\left[Out_V = 1 : \begin{array}{c}(x, y) \leftarrow Gen(1^s; r_k)\\ P(x, r_P) \leftrightarrow V(y; r_V)\end{array}\right] \geq p$$

The notation "Pr_r[event: experience]*" denotes the probability that an event will occur after the experiment, taken on the set of random coins* r *underlying the experiment. The random variable associated with the event is defined via the experiment leading to the description of a random variable.*

This model implicitly assumes *concurrency* involving participants who do not share the secret inputs with each other. In the remainder of the chapter, $\alpha, \beta, \gamma, \gamma' \in [0; 1]$ and view of a participant on an experiment is the collection of all their initial inputs (including coins) and incoming messages.

11.2.1.1. *Distance fraud (DF)*

This type of fraud corresponds to the conventional concept, but concurrent runs with many participants are also taken into account, i.e. they include other possible provers (with other secrets) and verifiers. Consequently, this generalized distance fraud also includes DH.

DEFINITION 11.2 (α-resistance to DF).– *A protocol is* α*-resistant to DF if:* $\forall s, \forall P^*, \forall loc_V$ *such that* $d(loc_V, loc_{P^*}) > \mathbb{B}$ *and* $\forall r_k$, *we have:*

$$\Pr_{r_V}\left[Out_V = 1 : \begin{array}{c}(x, y) \leftarrow Gen(1^s; r_k)\\ P^*(x) \leftrightarrow V(y; r_V)\end{array}\right] \leq \alpha$$

where P^* is an (unbounded) dishonest prover. In a concurrent setting, a bounded (polynomial) number of honest $P(x')$ and $V(y')$ close to $V(y)$ with independent (x', y') are implicitly allowed.

In other words, a protocol is α-resistant to DF if a faraway prover cannot be authenticated by a verifier with a probability greater than α.

11.2.1.2. *Man-in-the-middle (MiM) fraud*

This formalization considers an adversary that works in two phases. During a *learning phase*, this adversary interacts with many honest provers and verifiers. Then, the *attack phase* involves a faraway honest prover of a given ID and possibly many other honest provers and verifiers. The adversary's aim is to make the verifier accept in a session with ID. Clearly, this generalizes mafia fraud and includes impersonation fraud.

DEFINITION 11.3 (β-resistance to man-in-the-middle (MiM)).– *A protocol is β-resistant to MiM attacks if: $\forall s, \forall m, l, z$ bounded by a polynomial, $\forall \mathcal{A}_1, \mathcal{A}_2$ bounded by a polynomial, for all locations such that $d(loc_{P_j}, loc_V) > \mathbb{B}$, with $j \in \{m+1, ..., l\}$ we have:*

$$Pr\left[Out_V = 1 : \begin{array}{c} (x, y) \leftarrow Gen(1^s) \\ P_1(x), ..., P_m(x) \leftrightarrow \mathcal{A}_1 \leftrightarrow V_1(y), ..., V_z(y) \\ P_{m+1}(x), ..., P_l(x) \leftrightarrow \mathcal{A}_2(View_{\mathcal{A}_1}) \leftrightarrow V(y) \end{array} \right] \leq \beta$$

over all random coins, where $View_{\mathcal{A}_1}$ is the final view of $\mathcal{A}_1$. In a simultaneous setting, a polynomial number of $P(x')$, $P^*(x')$ and $V(y')$ with independent (x', y') is implicitly allowed everywhere.

A protocol is β-resistant to MiM if the probability that an adversary authenticates a faraway prover to a verifier is β at most, even though the adversary has access to information of a first session run between provers close to the verifiers. Definition 11.3 separates a learning phase (with adversarial behavior $\mathcal{A}_1$) from an attack phase (with adversarial behavior $\mathcal{A}_2$). This definition models a practical framework in which an attacker would have cloned several tags (provers) and would have made them interact with several readers (verifiers) with which they are registered. Thanks to this multi-party communication, the attacker can potentially obtain more benefits, in a shorter period of time. To increase their gain, the attacker can put in place the learning phase as they wish (otherwise, the learning phase is not obligatory). Therefore, the attacker can place the prover tags close to the verifiers/readers, even though being an active adversary between two neighboring P and V is technically more difficult than interfering between two faraway parties.

11.2.1.3. *Collusion fraud (CF)*

This fraud considers a faraway verifier holding a secret x that assists an adversary in making the verifier accept. This can be done in the presence of many other honest participants. However, there should not be a man-in-the-middle (MiM) attack based on this malicious prover, i.e. the adversary should not take advantage of this prover to (later) run a MiM attack.

DEFINITION 11.4 ((γ, γ^t)-resistance to CF).– *A protocol is* (γ, γ^t)*-resistant to CF if:* $\forall s, \forall P^*, \forall loc_{V_0}$ *such that* $d(loc_{V_0}, loc_{P^*}) > \mathbb{B})$ *and* $\forall \mathcal{A}^{CF}$*ppt. such that:*

$$Pr\left[Out_{V_0} = 1 : \begin{array}{l}(x, y) \leftarrow Gen(1^s) \\ P^*(x) \leftrightarrow \mathcal{A}^{CF} \leftrightarrow V_0(y)\end{array}\right] \geq \gamma$$

over all random coins, there exists a MiM attack $m, l, \mathcal{A}_1, \mathcal{A}_2, P_i, P_j, V_{i'}$ *using* P *and* P^* *in the learning phase, such that:*

$$Pr\left[Out_V = 1 : \begin{array}{l}(x, y) \leftarrow Gen(1^s) \\ P_1^{(*)}(x), ..., P_m^{(*)} \leftrightarrow \mathcal{A}_1 \leftrightarrow V_1(y), ..., V_z(y) \\ P_{m+1}(x), ..., P_l(x) \leftrightarrow \mathcal{A}_2(View_{\mathcal{A}_1}) \leftrightarrow V(y)\end{array}\right] \geq \gamma^t$$

where P^* *is a(ny) number of dishonest provers and* $P^{(*)}$ *runs either* P *or* P^**. According to the MiM requirements,* $d(loc_{P_j}, loc_V) > \mathbb{B}$*, for all* $j \in \{m+1, ..., l\}$*. In a concurrent setting, a polynomially bounded number of* $P(x')$*,* $P^*(x')$ *and* $V(y')$ *with independent* (x', y') *values is implicitly allowed, but no participant is close to* V_0*.*

In other words, a protocol is (γ, γ^t)-resistant to CF if, when an adversary succeeds in authenticating a remote prover to a verifier with a probability of at least γ, then there exists another MiM attack, where an adversary succeeds in authenticating a remote prover to the verifier with a probability of at least γ^t.

11.2.2. *Relation between different threat models*

We prove certain relations present between these properties. For a given protocol:

$$X \rightarrow Y$$

denotes that if the property X is satisfied, then Y is also satisfied. This amounts to saying that if there exists an attack for Y, then there also exists an attack for X.

For a given protocol, we also denote by:

$$X \dashrightarrow Y$$

the fact that if there exists an attack on property Y without sending the secret x, then there exists an attack on property X. For conventional threat models, we explain how they can be defined using this formal model.

THEOREM 11.1 (DF $\rightarrow$ DH).– *If a protocol is α-resistant to DF, then it is also α-resistant to DH.*

PROOF 11.1.– *Distance hijacking is included in the definition of resistance of Definition 11.2. An experiment in which a dishonest faraway prover P^* can use several provers to get authenticated as a single honest prover P that is close to the verifier is clearly included in the concurrent parameters:*

$$(x, y) \leftarrow Gen(1^s)$$
$$P^*(x) \leftrightarrow P_1(x'), .., P_n(x') \leftrightarrow V(y)$$

THEOREM 11.2 (MiM $\rightarrow$ MF and MiM $\rightarrow$ IF).– *If a protocol is β-resistant to MiM, then it is β-resistant to MF (Mafia Fraud) and β-resistant to IF (Impersonation Fraud).*

PROOF 11.2.– *In Definition 11.3, the conventional concept of impersonation corresponds to $m = z = 0$ and $l = 1$. The experiment performing this attack can be described as follows:*

$$(x, y) \leftarrow Gen(1^s)$$
$$P(x) \leftrightarrow \mathcal{A} \leftrightarrow V(y)$$

In Definition 11.3, the conventional concept of impersonation corresponds to $l = m$, i.e. there is no prover in the attack phase. The corresponding experiment is:

$$(x, y) \leftarrow Gen(1^s)$$
$$P_1(x), ..., P_m(x) \leftrightarrow \mathcal{A}_1 \leftrightarrow V_1(y), ..., V_z(y)$$
$$\mathcal{A}_2(View_{\mathcal{A}_1}) \leftrightarrow V(y)$$

THEOREM 11.3 (CF $\rightarrow$ TF).– *If a protocol is (γ, γ^t)-resistant to CF, then it is (γ, γ^t)-resistant to TF (Terrorist Fraud).*

PROOF 11.3.– *The concept of terrorist fraud is linked to Definition 11.4: It is sufficient to take $m = z = 1$, $l = 2$ and $\mathcal{A}_1$ simply runs $\mathcal{A}^{CF}$ in the learning phase, i.e. $\mathcal{A}^{CF}$ obtains information to directly impersonate the prover. We can model this fraud by:*

$$(x, y) \leftarrow Gen(1^s)$$
$$P^*(x) \leftrightarrow \mathcal{A}_1 \leftrightarrow V(y)$$
$$P(x) \leftrightarrow \mathcal{A}_2(View_{\mathcal{A}_1}) \leftrightarrow V(y)$$

THEOREM 11.4 (TF $\dashrightarrow$ DF).– *If a protocol is not α-resistant to DF, then there exists a TF attack succeeding with a probability of at least α.*

PROOF 11.4.– *Let us assume that there exists a DF attack A such that:* $Pr[A\ succeeds] = \alpha \Leftrightarrow Pr\left[Out_V = 1 : \begin{array}{c}(x,y) \leftarrow Gen(1^s; r_k)\\ P^*(x) \leftrightarrow V(y; r_V)\end{array}\right] = \alpha$, *where* α *is not negligible. Then, it is possible to find a TF attack, only if* P^* *does not transmit their secret in clear, i.e.:*

$$Pr\left[Out_{V_0} = 1 : \begin{array}{c}(x,y) \leftarrow Gen(1^s)\\ P^*(x) \leftrightarrow \mathcal{A}^{TF} \leftrightarrow V_0(y)\end{array}\right] \geq \alpha$$

with $\mathcal{A}^{TF}$ *simply relaying the messages and* P^* *playing the same role as before. However, if* P^* *and* $\mathcal{A}^{TF}$ *cooperate, the probability of success may be improved.*

In Figure 11.1, we summarize all these relations. On the left are the attacks where the prover is faraway and dishonest, and on the right are attacks where the prover is close to the verifier. The arrow from TF (or CF) to DH is obtained by transitivity, as is the arrow from CF to DF. Once an attack against one property is discovered, it is easy to extend it to other properties using these results.

Figure 11.1. *Relations between different threat models*

11.3. Overview of existing protocols

We list as many protocols as possible to understand their particularities. We do not consider the DH threat model in our study, since few articles study this property (Cremers et al. 2012) and mounting such attacks is difficult. Each DB protocol generally follows the following form: an initialization phase where the participants generally share secret data (often denoted x) or public information often denoted by N_V and N_P for the verifier and the prover, respectively. An encryption, a hash function or a *pseudo-random function* (PRF) is also often used. Then, a fast challenge response phase where, generally, a sequence of n challenges c_i is sent by the verifier to the prover, who responds with responses denoted r_i. Finally, the last verification or authentication phase generally consists of opening a commitment or verifying a common shared data item, called transcription. The general structure of a DB protocol is presented in Figure 11.2. In the rest of this section, we explain the new attacks discovered on certain protocols, and then the improvements made to protocols that use the PRF inappropriately.

The majority of protocols are available at http://www.avoine.net/rfid/.

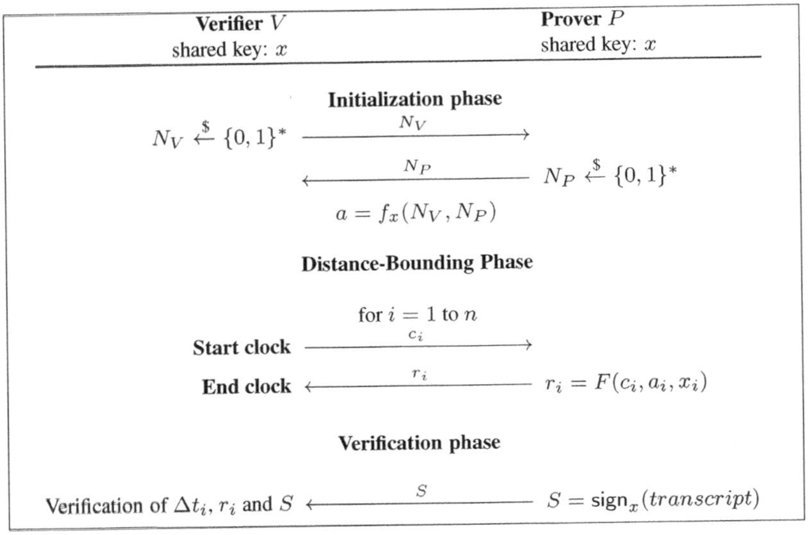

Figure 11.2. *General structure of DB protocols*

11.3.1. *Improvement of attacks*

For Avoine and Tchamkerten (2009), Tippenhauer and Capkun (2009), Benfarah et al. (2010), Trujillo-Rasua et al. (2010), Yum et al. (2011), Kardaş et al. (2012), Lee et al. (2012), Hoda Jannati (2012), Mitrokotsa et al. (2012), Entezari et al. (2014) and Trujillo-Rasua et al. (2014), there exist TF attacks. These attacks all follow the same scheme: the adversary is close to the verifier, while the prover is faraway. During the initialization phase, the adversary simply relays the messages from the prover to the verifier and the messages from the verifier to the prover. After this phase, the prover computes responses, with the help of their key (the adversary does not have access to this key and the prover does not want the latter to obtain it). Then, the prover sends their responses to their partner (the adversary), so that the adversary can respond to the challenges of the verifier in the fast challenge/response phase. The prover sends all the results of these computations because they know that the adversary cannot recover the shared secret key, even though they have access to all the results (except for the protocols (Gürel et al. 2011; Hoda Jannati 2012) where the adversary receives $n - v$ bits of the secret key). For the last phase, the adversary sends everything that they receive from the verifier to the prover, and thus the prover can close the session by sending their signature of the transcript. Using the result on the relation between threat models, we immediately deduce attacks for CF.

Many protocols use PRFs (Hancke and Kuhn 2005; Reid et al. 2007; Kapoor et al. 2008; Kim et al. 2008; Munilla and Peinado 2008a; Nikov and Vauclair 2008; Avoine and Tchamkerten 2009; Kim and Avoine 2009; Peris-Lopez et al. 2009; Benfarah et al. 2010; Trujillo-Rasua et al. 2010, 2014; Avoine et al. 2011; Gürel et al. 2011; Yum et al.

2011; Hoda Jannati 2012; Kardaş et al. 2012; Lee et al. 2012; Mitrokotsa et al. 2012; Yang et al. 2012a; Fischlin and Onete 2013; Boureanu and Vaudenay 2014; Entezari et al. 2014; Falahati and Jannati 2015; Bultel et al. 2016; Boureanu et al. 2017; Pagnin et al. 2018). It is possible to mount certain attacks if the PRF used follows a certain form. This type of attack was first introduced in Boureanu et al. (2012). Starting from their idea, it is possible to improve certain DF attacks on Munilla and Peinado (2008a), Kim and Avoine (2009), Benfarah et al. (2010), Trujillo-Rasua et al. (2010), (2014), Kardaş et al. (2012), Lee et al. (2012), Baghernejad (2014), Entezari et al. (2014) and Falahati and Jannati (2015) and a MiM attack on Yum et al. (2011).

We detail only one attack on DF and the one on MiM; however, for the other protocols, the PRF building to mount a successful attack follows the same structure.

To succeed in a DF attack, it is necessary for the prover to send their response before receiving the challenge from the verifier during the fast challenge/response phase. However, when the output of the PRF is divided into several parts to precompute responses using a special PRF, the dishonest prover is able to make the response independent of the challenge received. We show this type of attack in Figure 11.3. The PRF f is based on the other PRF g as follows, where z is a special value known by P, also called a *trapdoor*):

$$f_x(N_V, N_P) = \begin{cases} a||a \text{ if } N_P = z \\ g_x(N_V, N_P) \text{ otherwise} \end{cases}$$

This attack therefore succeeds with a probability of 1.

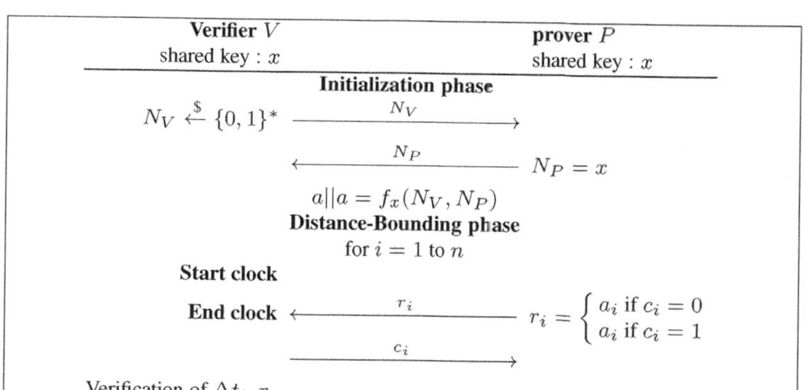

Figure 11.3. *Distance fraud attack against DB protocols using a PRF*

These functions are easy to compute in one direction but difficult in the other direction, if we do not know a secret associated with the function. For example, if f

is a trapdoor function with a secret s, then it is easy to compute (i.e. computable in polynomial time in input size) the value $y = f(x, s)$ and it is difficult to compute (i.e. no polynomial algorithm exists) the value x, knowing y but without knowing s.

These functions are widely used in cryptography, notably for public-key schemes (where encrypting a message is easy but decrypting it is difficult without knowing the associated secret key).

Munilla and Peinado (2008a) – DF: Let g be a PRF. Let us consider the PRF hash built such that:

$$\mathsf{hash}(K, N_a, N_b) = \begin{cases} P||v||v & \text{if } N_b = z \\ g(K, N_a, N_b) & \text{otherwise} \end{cases}$$

Let us take an instance of the Munilla and Peinado protocol where hash is used. In this instance, a faraway malicious prover P^* could easily perform a distance fraud attack. By choosing N_b equal to z, they can send the same response r_i independently of the challenge c_i sent by the verifier. Then, according to the protocol (Munilla and Peinado 2008a) (no challenge/response if $P_i = 1$), if $P_i = 0$ for all challenges c_i, the response is the ith bit of v. If $P_i = 1$, P^* awaits a delay before sending the next response r_{i+1}. Therefore, if the malicious prover applies this strategy, they can defeat the distance-bound.

Yum et al. (2011) – MiM: Let us consider an instance of the protocol of Yum et al., where $\tau = 1$ and K represents the shared keys. As $\tau = 1$, we have $D = f_\tau(K, N_V, N_P) = 0^n$. Therefore, for each round, the prover awaits a challenge from the verifier. Let g be a PRF used to compute $v = g(K, N_V, N_P)$. Let PRF be a PRF and h be the PRF, based on PRF as follows:

$$Z = h(K, N_V, N_P, C, R) = \begin{cases} K & \text{if } C = 0^{\frac{n}{2}}||v^{\frac{n}{2}} \\ \mathsf{PRF}(K, N_V, N_P, C, R) & \text{otherwise} \end{cases}$$

$0^{\frac{n}{2}}$ is a number composed of $\frac{n}{2}$ bits equal to 0 and $v^{\frac{n}{2}} = \mathbf{msb}_{\frac{n}{2}}(f(K, N_V, N_P))$, where $\mathbf{msb}(x)$ is the most significant bit of x. The attacker impersonates the verifier to the prover, sends an N_V to the prover and receives the prover's nonce N_P. The prover computes v.

Then, the rapid challenge/response phase begins: for the $\frac{n}{2}$ first rounds, the attacker sends $c_i = 0$, to receive the $\frac{n}{2}$ most significant bits ($\mathbf{msb}$) of v. For the remaining half of rounds, the attacker sends the received bits one by one. Then, the prover creates Z and $C = c_1||..||c_n$ such that $C = 0^{\frac{n}{2}}||v^{\frac{n}{2}}$. Therefore, by sending Z, the prover shares with the attacker their own secret key, K. The attacker is then able to impersonate the prover during future running of the protocol.

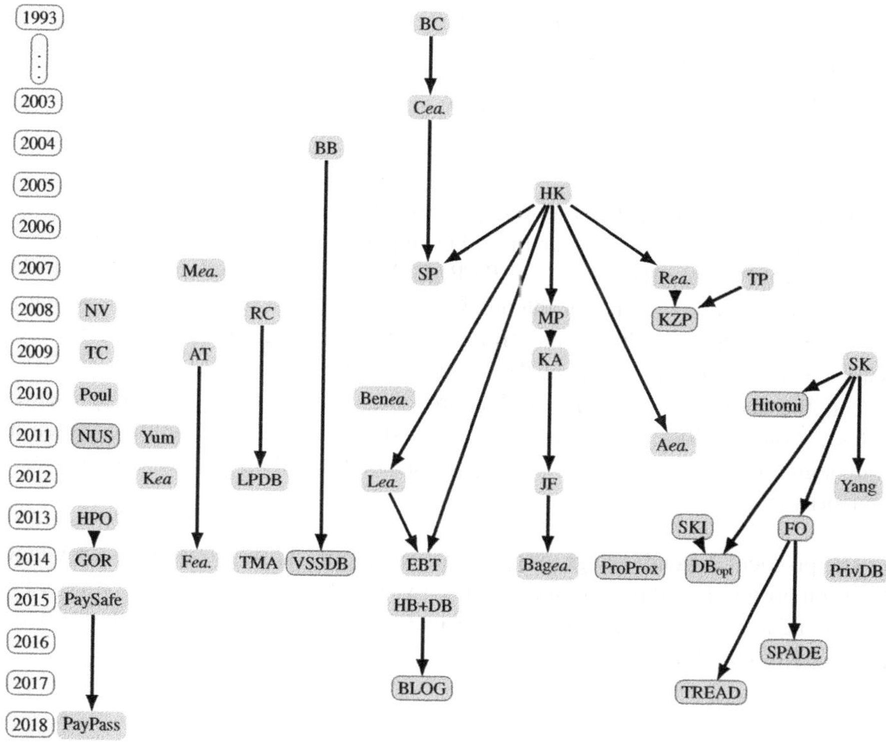

Figure 11.4. *Protocol-dependency graph. For a color version of this figure, see www.iste.co.uk/mendiboure/transportsystems.zip*

11.3.2. Comparison of DB protocols

To conclude this chapter, we give a comparison of DB protocols in order to highlight certain points. First, very few protocols are strong against all types of fraud; only 12 protocols assure security against all types of attacks. These are the protocols cited in red in Figure 11.4. These protocols are the following:

– KZP (2008) (Kapoor et al. 2008);

– Hitomi (2010) (Peris-Lopez et al. 2009);

– NUS (2011) (Gürel et al. 2011);

– SKI_{pro} (2013) (Boureanu et al. 2013a);

– FO (2013) (Fischlin and Onete 2013);

– DB1 (2014) (Boureanu and Vaudenay 2014);

– DB2 (2014) (Boureanu and Vaudenay 2014);

– ProProx (2014) (Vaudenay 2014);

– VSSDB (2014) (Gambs et al. 2014a);

– SPADE (2016) (Bultel et al. 2016);

– TREAD (2016) (Avoine et al. 2017);

– BLOG (2017) (Boureanu et al. 2017).

These protocols offer different compromises between security, performance and privacy, and the choice of one or the other depends on the desired properties. For example, the PayPass-RRP protocol-RRP (EMVCo 2018), implemented by EMV for contactless payments, is not intended to protect against terrorist fraud, which is considered insignificant due to the properties of payment cards. This simplification enables a more efficient protocol. While the TREAD version (Avoine et al. 2017) protects the anonymity of users and has provable security against attacks presented in this chapter, it does, however, use very costly cryptographic primitives, such as group signatures and has less resistance to distance fraud than protocols not resistant to terrorist fraud.

The protocol dependency graph, presented in Figure 11.4, shows the descendants of certain protocols and reveals six families, two of which are large (one composed of the descendants of Brands & Chaum and Hancke & Kuhn, the other of the descendants of Swiss-Knife and $\mathrm{SKI}_{\mathrm{pro}}$).

11.4. References

Abyaneh, M.R.S. (2011). Security analysis of two distance-bounding protocols. *arXiv: 1107.3047.*

Aumasson, J., Mitrokotsa, A., Peris-Lopez, P. (2011). A note on a privacy-preserving distance-bounding protocol. *ICICS 2011*, 78–92. doi: 10.1007/978-3-642-25243-3_7.

Avoine, G. and Tchamkerten, A. (2009). An efficient distance bounding RFID authentication protocol: Balancing false-acceptance rate and memory requirement. *Information Security, 12th International Conference, ISC 2009*, 7–9 September, Pisa, 250–261. doi: 10.1007/978-3-642-04474-8_21.

Avoine, G., Bingöl, M.A., Kardas, S., Lauradoux, C., Martin, B. (2009). A formal framework for cryptanalyzing RFID distance bounding protocols. *IACR Cryptology ePrint Archive*, 543 [Online]. Available at: http://eprint.iacr.org/2009/543.

Avoine, G., Lauradoux, C., Martin, B. (2011). How secret-sharing can defeat terrorist fraud. In *Proceedings of the Fourth ACM Conference on Wireless Network Security, WiSec '11*, New York, 145–156. doi: 10.1145/1998412.1998437.

Avoine, G., Bultel, X., Gambs, S., Gérault, D., Lafourcade, P., Onete, C., Robert, J.-M. (2017). A terrorist-fraud resistant and extractor-free anonymous distance-bounding protocol. In *AsiaCCS 2017*, 800–814. doi: 10.1145/3052973.3053000.

Baghernejad, F. and Nasour Bagheri, M.S. (2014). Security analysis of the distance bounding protocol proposed by jannati and falahati. *Electrical and Computer Engineering Innovations*, 2(2), 85–92.

Bay, A., Boureanu, I., Mitrokotsa, A., Spulber, I., Vaudenay, S. (2012). The Bussard-Bagga and other distance bounding protocols under man-in-the-middle attacks. In *Proceedings of Inscrypt'2012, 8th China International Conference on Information Security and Cryptology*. Springer, Beijing.

Benfarah, A., Miscopein, B., Gorce, J., Lauradoux, C., Roux, B. (2010). Distance bounding protocols on TH-UWB radios. In *Proceedings of the Global Communications Conference, 2010, GLOBECOM 2010*, 6–10 December, Miami, 1–6. doi: 10.1109/GLOCOM.2010.5684000.

Boureanu, I. and Vaudenay, S. (2014). Optimal proximity proofs. In *Information Security and Cryptology, 10th International Conference, Inscrypt 2014*, 13–15 December, Beijing, 170–190 [Online]. doi: 10.1007/978-3-319-16745-9_10.

Boureanu, I. and Vaudenay, S. (2015). Challenges in distance bounding. *IEEE Security & Privacy*, 13(1), 41–48. doi: 10.1109/MSP.2015.2.

Boureanu, I., Mitrokotsa, A., Vaudenay, S. (2012). On the pseudorandom function assumption in (secure) distance-bounding protocols – PRF-ness alone does not stop the frauds! *LATINCRYPT*, 100–120.

Boureanu, I., Mitrokotsa, A., Vaudenay, S. (2013a). Practical & provably secure distance-bounding. *IACR Cryptology ePrint Archive*, 465 [Online]. Available at: http://eprint.iacr.org/2013/465.

Boureanu, I., Mitrokotsa, A., Vaudenay, S. (2013b). Secure and lightweight distance-bounding. In *Lightweight Cryptography for Security and Privacy - Second International Workshop (LightSec 2013)*, Gebze, 97–113.

Boureanu, I., Mitrokotsa, A., Vaudenay, S. (2013c). Towards secure distance bounding. In *Fast Software Encryption - 20th International Workshop, FSE 2013*, Singapore, 55–67.

Bourenau, I., Gérault, D., Lafourcade, P., Onete, C. (2017). Breaking and fixing the HB+DB protocol. In *WiSec 2017*, 241–246. doi: 10.1145/3098243.3098263.

Brands, S. and Chaum, D. (1993). Distance-bounding protocols (extended abstract). In *EUROCRYPT'93*, 344–359. Springer-Verlag.

Bultel, X., Gambs, S., Gérault, D., Lafourcade, P., Onete, C., Robert, J-M. (2016). A prover-anonymous and terrorist-fraud resistant distance-bounding protocol. In *WISEC 2016*, 121–133 [Online]. doi: 10.1145/2939918.2939919.

Bussard, L. and Bagga, W. (2005). Distance-bounding proof of knowledge to avoid real-time attacks. In *IFIP SEC 2005, 20th IFIP International Information Security Conference*, 30 May–1 June, Chiba [Online]. Available at: http://www.eurecom.fr/publication/1607.

Capkun, S., Buttyán, L., Hubaux, J.-P. (2003). Sector: Secure tracking of node encounters in multi-hop wireless networks. In *ACM Workshop on Security of Ad Hoc and Sensor Networks (SASN)*, 21–32.

Cremers, C., Rasmussen, K.B., Schmidt, B., Capkun, S. (2012). Distance hijacking attacks on distance bounding protocols. *IEEE Symposium on Security and Privacy*.

Desmedt, Y. (1988). Major security problems with the "unforgeable" (feige-) fiat-shamir proofs of identity and how to overcome them. *Congress on Computer and Communication Security and Protection, Securicom'88*, 147–159. SEDEP, Paris.

Dürholz, U., Fischlin, M., Kasper, M., Onete, C. (2011), A formal approach to distance-bounding RFID protocols. In *Information Security, 14th International Conference, ISC 2011*, 26–29 October, Xi'an, 47–62 [Online]. doi: 10.1007/978-3-642-24861-0_4.

EMVCo (2018). Book C-2 Kernel 2 Specification v2.7. EMV Contactless Specifications for Payment System [Online]. Available at: www.emvco.com/wp-content/plugins/pmpro-customizations/oy-getfile.php?u=/wp-content/uploads/documents/C-7_Kernel_7_V_2_7_Final.pdf.

Entezari, R., Bahramgiri, H., Tajamolian, M. (2014). A mafia and distance fraud high-resistance RFID distance bounding protocol. In *ISCISC*, 67–72.

Falahati, A. and Jannati, H. (2015). All-or-nothing approach to protect a distance bounding protocol against terrorist fraud attack for low-cost devices. *Electronic Commerce Research*, 15(1), 75–95 [Online]. doi: 10.1007/s10660-014-9167-y.

Finkenzeller, K. (2003). *RFID Handbook: Fundamentals and Applications in Contactless Smart Cards and Identification*, 2nd edition. John Wiley & Sons, New York.

Fischlin, M. and Onete, C. (2012). Provably secure distance-bounding: An analysis of prominent protocols. *IACR Cryptology EPrint Archive*, 128 [Online]. Available at: http://eprint.iacr.org/2012/128.

Fischlin, M. and Onete, C. (2013). Terrorism in distance bounding: Modeling terrorist-fraud resistance. In *Proceedings of the 11th International Conference on Applied Cryptography and Network Security, ACNS'13*, 414–431, Springer-Verlag, Berlin, Heidelberg. doi: 10.1007/978-3-642-38980-1_26.

Francillon, A., Danev, B., Capkun, S. (2011). Relay attacks on passive keyless entry and start systems in modern cars. In *Proceedings of the Network and Distributed System Security Symposium, NDSS 2011*, 6 February–9 February, San Diego, CA [Online]. Available at: https://www.ndss-symposium.org/ndss2011/relay-attacks-on-passive-keyless-entry-and-start-systems-in-modern-cars.

Gambs, S., Killijian, M.-O., Lauradoux, C., Onete, C., Roy, M., Traoré, M. (2014a). VSSDB: A verifiable secret-sharing and distance-bounding protocol. In *International Conference on Cryptography and Information security (BalkanCryptSec'14)*, Istanbul [Online]. Available at: https://hal.inria.fr/hal-01090056.

Gambs, S., Onete, C., Robert, J. (2014b). Prover anonymous and deniable distance-bounding authentication. *IACR Cryptology ePrint Archive*, 114 [Online]. Available at: http://eprint.iacr.org/2014/114.

Gürel, A.O., Arslan, A., Akgün, M. (2011). Non-uniform stepping approach to RFID distance bounding problem. In *Proceedings of the 5th International Workshop on Data Privacy Management and 3rd International Conference on Autonomous Spontaneous Security, DPM'10/SETOP'10*, 64–78, Springer-Verlag, Berlin, Heidelberg [Online]. Available at: http://dl.acm.org/citation.cfm?id=1964419.1964427.

Hancke, G.P. and Kuhn, M.G. (2005). An RFID distance bounding protocol. In *Proceedings of the First International Conference on Security and Privacy for Emerging Areas in Communications Networks, SECURECOMM '05*, 67–73. IEEE Computer Society, Washington, DC. doi: 10.1109/SECURECOMM.2005.56.

Hermans, J., Peeters, R., Onete, C. (2013). Efficient, secure, private distance bounding without key updates. In *Sixth ACM Conference on Security and Privacy in Wireless and Mobile Networks, WISEC'13*, 17–19 April, Budapest, 207–218. doi: 10.1145/2462096.2462129.

Hoda Jannati, A.F. (2012). Mutual implementation of predefined and random challenges over RFID distance bounding protocol. In *Information Security and Cryptology (ISCISC)*, 43–47.

Ju-Tu, Y. and Piramuthu, S. (2007). RFID distance bounding protocols. In *First International EURASIP Workshop on RFID Technology*.

Kapoor, G., Zhou, W., Piramuthu, S. (2008), Distance bounding protocol for multiple RFID tag authentication. In *2008 IEEE/IPIP International Conference on Embedded and Ubiquitous Computing (EUC 2008)*, 17–20 December, Shanghai, 115–120. doi: 10.1109/EUC.2008.48.

Kardaş, S., Kiraz, M.S., Bingöl, M.A., Demirci, H. (2012), A novel RFID distance bounding protocol based on physically unclonable functions. In *Proceedings of the 7th International Conference on RFID Security and Privacy', RFIDSec'11*, 78–93. Springer-Verlag, Berlin, Heidelberg. doi: 10.1007/978-3-642-25286-0_6.

Kim, C.H. (2011). Security analysis of YKHL distance bounding protocol with adjustable false acceptance rate. *IEEE Communications Letters*, 15(10), 1078–1080. doi: 10.1109/LCOMM.2011.080811.111299.

Kim, C.H. and Avoine, G. (2009). RFID distance bounding protocol with mixed challenges to prevent relay attacks. In *Proceedings of the 8th International Conference on Cryptology and Network Security, CANS '09*, 119–133. Springer-Verlag, Berlin, Heidelberg. doi: 10.1007/978-3-642-10433-6_9.

Kim, C.H., Avoine, G., Koeune, F., Standaert, F., Pereira, O. (2008). The swiss-knife RFID distance bounding protocol. In *Information Security and Cryptology - ICISC 2008, 11th International Conference*, 3–5 December, Seoul, 98–115. doi: 10.1007/978-3-642-00730-9_7.

Lee, S., Kim, J.S., Hong, S.J., Kim, J. (2012). Distance bounding with delayed responses. *IEEE Communications Letters*, 16(9), 1478–1481. doi: 10.1109/LCOMM.2012.073112.121134.

Meadows, C., Poovendran, R., Pavlovic, D., Chang, L., Syverson, P.F. (2007). Distance bounding protocols: Authentication logic analysis and collusion attacks. In *Secure Localization and Time Synchronization for Wireless Sensor and Ad Hoc Networks*, 279–298. Springer, Boston, MA. doi: 10.1007/978-0-387-46276-9_12.

Meghdadi, M., Ozdemir, S., Güler, I. (2011). A survey of wormhole-based attacks and their countermeasures in wireless sensor networks. *IETE Technical Review*, 28(2), 89–102 [Online]. Available at: http://www.tandfonline.com/doi/abs/10.4103/0256-4602.78089.

Mitrokotsa, A., Onete, C., Vaudenay, S. (2012). Mafia fraud attack against the RČ distance-bounding protocol. In *2012 IEEE International Conference on RFID-Technologies and Applications, RFID-TA 2012*, 5–7 November, Nice, 74–79. doi: 10.1109/RFID-TA.2012.6404571.

Munilla, J. and Peinado, A. (2008a). Distance bounding protocols for RFID enhanced by using void-challenges and analysis in noisy channels. *Wireless Communications and Mobile Computing*, 8(9), 1227–1232. doi: 10.1002/wcm.v8:9.

Munilla, J. and Peinado, A. (2008b). Security analysis of Tu and Piramuthu's protocol. In *NTMS 2008, 2nd International Conference on New Technologies, Mobility and Security*, 5–7 November, Tangier, 1–5. doi: 10.1109/NTMS.2008.ECP.88.

Nikov, V. and Vauclair, M. (2008). Yet another secure distance-bounding protocol. In *SECRYPT 2008, Proceedings of the International Conference on Security and Cryptography*, 26–29 July, Porto, 218–221.

Pagnin, E., Yang, A., Hu, Q., Hancke, G., Mitrokotsa, A. (2018). HB+DB: Distance bounding meets human based authentication. *Future Generation Computer Systems*, 627–639. doi: 10.1016/j.future.2016.05.031.

Peris-Lopez, P., Castro, J.C.H., Estévez-Tapiador, J.M., van der Lubbe, J.C.A. (2009). Shedding some light on RFID distance bounding protocols and terrorist attacks. *arXiv:0906.4618*.

Rasmussen, K.B. and Capkun, S. (2008). Location privacy of distance bounding protocols. In *Proceedings of the 2008 ACM Conference on Computer and Communications Security, CCS 2008*, 27–31 October, Alexandria, VA, 149–160. doi: 10.1145/1455770.1455791.

Reid, J., Nieto, J.M.G., Tang, T., Senadji, B. (2007), Detecting relay attacks with timing-based protocols. In *Proceedings of the 2nd ACM Symposium on Information, Computer and Communications Security, ASIACCS '07*, New York, 204–213. doi: 10.1145/1229285.1229314.

Singelée, D. and Preneel, B. (2007). Distance bounding in noisy environments. In *Security and Privacy in Ad-hoc and Sensor Networks*, 101–115. Springer, Berlin, Heidelberg.

Tippenhauer, N.O. and Capkun, S. (2009). ID-based secure distance bounding and localization. In *Computer Security – ESORICS 2009, 14th European Symposium on Research in Computer Security*, 21–23 September, Saint-Malo, 621–636. doi: 10.1007/978-3-642-04444-1_38.

Trujillo-Rasua, R., Martin, B., Avoine, G. (2010). The poulidor distance-bounding protocol. In *Proceedings of the 6th International Conference on Radio Frequency Identification: Security and Privacy Issues, RFIDSec'10*, 239–257. Springer-Verlag, Berlin, Heidelberg [Online]. Available at: http://dl.acm.org/citation.cfm?id=1926325.1926352.

Trujillo-Rasua, R., Martin, B., Avoine, G. (2014). Distance-bounding facing both mafia and distance frauds: Technical report. *arXiv:1405:5704*.

Vaudenay, S. (2013). On modeling terrorist fraud. In *Provable Security - 7th International Conference, ProvSec 2013*, 23–25 October 2013, Melaka, 1–20.

Vaudenay, S. (2014). Proof of proximity of knowledge. *IACR Cryptology ePrint Archive*, 695 [Online]. Available at: http://eprint.iacr.org/2014/695.

Vaudenay, S. (2015). Private and secure public-key distance bounding – Application to NFC payment. In *Financial Cryptography and Data Security 2015*, Puerto Rico.

Yang, A., Zhuang, Y., Wong, D.S. (2012a). An efficient single-slow-phase mutually authenticated RFID distance bounding protocol with tag privacy. In *Proceedings of the 14th International Conference on Information and Communications Security, ICICS'12*, 285–292. Springer-Verlag, Berlin, Heidelberg. doi: 10.1007/978-3-642-34129-8_25.

Yang, T., Kong, L., Xin, W., Hu, J., Chen, Z. (2012b). Resisting relay attacks on vehicular passive keyless entry and start systems. In *2012 9th International Conference on Fuzzy Systems and Knowledge Discovery*, 2232–2236.

Yum, D.H., Kim, J.S., Hong, S.J., Lee, P.J. (2011). Distance bounding protocol with adjustable false acceptance rate. *IEEE Communications Letters*, 15(4), 434–436. doi: 10.1109/LCOMM.2011.030911.110066.

12

Context-Aware Security and Privacy as a Service for the Connected and Autonomous Vehicle

Tidiane SYLLA[1], **Mohamed Aymen CHALOUF**[2], **Léo MENDIBOURE**[3], **Francine KRIEF**[4]

[1] *LaMIA, Université des Sciences, des Techniques et des Technologies de Bamako (USTT-B), Mali*

[2] *IRISA, Université de Rennes, Lannion, France*

[3] *ERENA, COSYS, Université Gustave Eiffel, Marne-la-Vallée, France*

[4] *LaBRI, Bordeaux INP, France*

12.1. Introduction

On-road communication needs are set to increase significantly with the advent of the connected and autonomous vehicle, which will allow the deployment of a wide variety of applications. While these new applications will primarily aim to improve road safety and the comfort of users during travel, new services will emerge, such as platooning, which will allow a vehicle with a driver to guide autonomous vehicles assembled in convoy on the urban network. The Internet of Things will further expand the scope of services offered. These increasingly numerous applications will have very varied security requirements, which will need to be met.

The solutions proposed in many works have focused on the implementation of mechanisms to ensure one or more security services, targeting road safety

Cooperative Intelligent Transport Systems,
coordinated by Léo MENDIBOURE.

applications in particular. However, it is necessary to have security services supporting all the applications in the connected and autonomous vehicle domain. Moreover, the vast majority of proposed security solutions do not take into account the characteristics of the connected and autonomous vehicle. These characteristics include the following:

– Intelligence: this is generally implemented by a combination of algorithms, advanced data processing techniques (automatic learning and reasoning) and computational capacity. This intelligence could be used to choose the level of security and privacy that best meets application needs and user preferences.

– Sensitive data: personal data collected by the vehicle, such as geolocation, is sensitive data. From this data, it is possible to track a person. This sensitive data requires protection so that it does not fall into the wrong hands. A thief could, for example, take advantage of knowing that the user is away from home in order to break into their house.

– Context awareness: to make the right decisions, the vehicle must continuously collect data in order to be informed of changes taking place within its environment. This data, such as the types of access networks available, may be used when deciding on the security mechanisms to be implemented.

– Heterogeneity of devices and means of communication: the connected and autonomous vehicle is led to dialog with a wide variety of devices, such as neighboring vehicles, pedestrians' smartphones, connected objects (road signs, intersection lights, rail barriers, etc.) and road-side units (RSU). It also communicates with heterogeneous information systems and allows access to various networks, such as the Internet. It needs to coexist with the surrounding computer systems and infrastructure. This heterogeneity increases the attack surface and will therefore have an impact on the security mechanisms to be applied.

– Diversity of applications: for simplicity, these applications can be divided into two broad categories: road safety and traffic management applications (safety, navigation, diagnostics and remote operations) and commercial applications (insurance, vehicle sharing, entertainment, etc.). Performance, security and privacy needs vary, depending on the type of application. Therefore, road safety applications need to process the data collected in real time to detect collisions or pedestrians on the road. In addition, they must be able to rely on an environment of trust. Entertainment applications, for their part, do not have the same high requirements in terms of priorities or security level.

– Processing, storage and communication capacity: the vehicle has significant enough processing, storage and communication capabilities that they can be made available to other vehicles (or a group of vehicles). It must be able to do so with complete confidence.

– Diversity of users: the connected and autonomous vehicle can assist the driver with different levels of automation (Society of Automotive Engineers). These levels range from level 0, which presents no advanced driver assist, to level 5, which no longer requires the presence of the driver. The number of applications offered to users (driver and passengers) inside the vehicle increases with the level of automation, as each user may have different security and privacy requirements. This is even more true in the case of car sharing, where some users may wish to not have their travel habits disclosed (journeys, locations frequented, services used, etc.). The vehicle will need to consider these disparities when implementing security and privacy mechanisms.

Given these characteristics, using context-aware security as a tool in implementing security and privacy seems appropriate.

In section 12.2, we present a taxonomy of the applications that will become available in the domain of connected and autonomous vehicles, as well as the services required to ensure security in this environment. Next, in section 12.3, we describe the context-aware security and privacy architecture that we propose. Section 12.4 then presents a use case. In section 12.5, we introduce the main challenges. Lastly, section 12.6 concludes this chapter.

12.2. Security, privacy and trust of connected and autonomous vehicle applications

This section presents the main vehicular applications and these applications' needs in terms of security, privacy and trust.

12.2.1. *Main applications of the connected and autonomous vehicle*

Thanks to the different types of V2X communication (V2I, V2V, V2S, V2R and V2P), many applications are envisaged.

Figure 12.1. *Five types of vehicular communication. For a color version of this figure, see www.iste.co.uk/mendiboure/transportsystems.zip*

Vehicle applications are generally divided into three main categories of services: road safety, traffic flow and passenger comfort and entertainment. However, a finer breakdown is possible:

– Road safety: the primary goal of the connected and autonomous vehicle is to reduce the risk of road accidents. This can be achieved by means of systems warning the driver of external risks and by putting in place automatic reactions following the detection of these risks. It involves, for example, detection of obstacles on the road, coordination of intersections and cooperative collision prevention.

– Traffic flow: the primary goal of these services is to manage the movements of road users and to optimize them according to different factors: pollution, duration, cost, etc. This involves, for example, managing traffic in real time, reserving parking spaces or creating maps cooperatively.

– Vehicle maintenance: the primary goal of these services is to monitor the state of the vehicle in real time in order to optimize maintenance costs and simplify its control. They include, for example, remote maintenance (reconfigurable vehicles, and data analysis) and a personal virtual assistant for simplified appointment scheduling (parts reservation, roadworthiness test, etc.). Virtualization applied to the vehicle will facilitate dynamic reconfigurations (Kambiré et al. 2020).

– Passenger entertainment: the primary purpose of these services is to enable road users to entertain and educate themselves, and to provide visibility to advertisers. Examples include multimedia applications (music, video, social networks, etc.), augmented reality and points of interest (stores, promotions, etc.).

– Driver assistance: the primary goal of these services is to simplify driving of the vehicle and to limit the tasks assigned to the driver. This involves, for example, advanced driver-assist systems (ADAS), remote driving of the vehicle or total automation of the vehicle.

– Monitoring of state of health: the primary purpose of these services is to ensure the well-being and health of the driver and their passengers. This involves, for example, cooperative detection of fatigue, stress or alcohol levels, adaptation of the environment to passengers' mood or medical assistance.

– Car sharing: the primary goal is to improve the use of resources while reducing transport costs. This can be achieved by traveling in a group (carpooling). New applications of this type will emerge, such as cloud-based car-sharing applications. These will allocate requests to the car owner based on a certain number of criteria, such as local address, work place, timing, gender, age and employment position held. The registration of car owners and requesters of automotive services will have to be verified by the service provider.

– Insurance: this domain is set to evolve with the connected and autonomous vehicle. These applications will aim to reduce the cost of insurance, thus optimizing the cost of ownership of the vehicle. New applications include insurance based on driving statistics: this is an insurance system for vehicles that automatically computes insurance fees using the driving statistics information. This information will include the place of usage, the number of kilometers driven daily/monthly/yearly, daily/monthly/annual traffic violations, or behavior behind the wheel.

– Other: the cloud enables the creation of new services by either forming an autonomous cloud for a group of vehicles, or by connecting the vehicles to a traditional cloud. In both cases, the resources of connected vehicles are available to be used as a cloud service. Vehicles can use smart cloud services. The cloud makes it possible to overcome vehicles' computation and storage limitations. Virtualization further expands the scope of possible applications by offering dynamicity and flexibility (dynamic addition/withdrawal of services, etc.), facilitating scalability and future developments (software developments and new services).

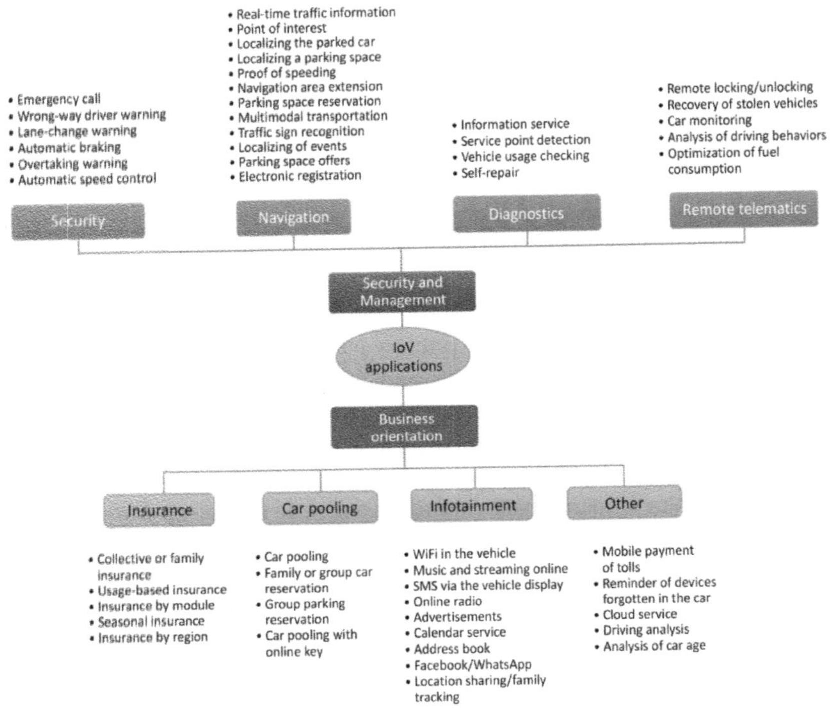

Figure 12.2. *Taxonomy of Internet of Vehicles (IoV) client applications. For a color version of this figure, see www.iste.co.uk/mendiboure/transportsystems.zip*

12.2.2. Security, privacy and trust services for the connected and autonomous vehicle

Security and privacy are among the most significant challenges faced by the connected and autonomous vehicle, which, like any connected system, can be subject to cyberattacks. Indeed, these systems are often collaborative and communicate with each other, with devices, or with heterogeneous information systems, and allow access to various networks, such as the Internet. The interconnection of these networks increases the attack surface and can give rise to the possibility of being made the target of intrusions and computer attacks. The damage from these attacks can be dramatic. It is therefore essential to take into account the security requirements of all applications in the connected and autonomous vehicle domain. It is also indispensable to consider user privacy.

In this section, we present the services required to ensure the security of applications in the connected and autonomous vehicle domain; these are: authentication, access control, integrity, confidentiality, availability and non-repudiation. We also describe privacy protection and trust management, which are complementary to security (Contreras-Castillo et al. 2017a, 2017b).

12.2.2.1. *Security services*

– Authentication: this service consists of verifying the identity of an entity in order to authorize this entity to exchange information and/or access protected resources. Several authentication methods are to be differentiated according to the security level requested, using different elements of identification and information, called authentication factors. Authentication mechanisms must allow for the preservation of privacy during the authentication process by protecting the identities of authenticating entities. In the vehicular context, authentication concerns the following entities: users, applications, vehicles, connected objects and exchanged messages. Applications must authenticate with the vehicles in order to access the data collected by them (e.g. insurance applications) or to control them (e.g. unlocking the vehicle remotely). Vehicles must also be able to authenticate with each other (e.g. autonomous cloud of a group of vehicles). The purpose of authentication here is to ensure that the dialog takes place between recognized entities. Messages exchanged between vehicles or between a vehicle and an object, or between vehicles and the application must also be authenticated.

– Access control: this mechanism ensures that only entities possessing the necessary authorizations can access a protected resource. In other words, it is a mechanism that protects access to sensitive resources/data/information by only allowing access for entities holding the necessary rights; for example, allowing an insurance application to access a vehicle's data. This access control can contribute, inter alia, to protecting user privacy. In addition, it limits the actions that an entity can undertake on the vehicle, by granting it only the necessary authorizations. This service is assured by a server that receives access requests in order to obtain authorization or denial of access. In the connected and autonomous vehicle domain, an entity having authorization creates the possibility for that entity to access the information collected by the vehicle, to control opening a door, etc. To do this, several access control techniques exist: access control based on an access control list (ACL), which is access control based on roles, capabilities, attributes, etc.

– Data confidentiality: this must be ensured from collection, during transmission and up until storage. It ensures that only the entities involved in a communication have access to the data exchanged. Encryption is the primary mechanism used to implement this security service. The data is encrypted using a cryptographic algorithm. Therefore, when the user, the application or the vehicle needs to communicate certain sensitive data, an encrypted channel is then established for this transmission.

– Data integrity: this ensures that the data has not undergone any alteration, deliberate or accidental, during its transmission, processing or storage. It is essential for the security of transmissions, and in particular those relating to road safety applications, as it ensures that the information received by one or more destinations is indeed that transmitted by the source. The mechanism used to control data integrity is the hash, which consists of generating a unique fingerprint from the data to be transmitted.

– Authentication of the origin of a message: this ensures that during a communication, between RSU, for example, the data received does indeed come from one of the legitimate interlocutors. The message authentication code (MAC) and the electronic signature are the most commonly used mechanisms to both authenticate the origin of a message and to verify its integrity.

– Non-repudiation: one of the principles of computer security, consisting of ensuring that a sender cannot deny having sent a message and that a recipient cannot deny having received a message. Generally, electronic signature is the means used to ensure non-repudiation.

– Availability: this security service consists of ensuring that the services and resources offered by the system are available at all times, and accessible by authenticated, authorized users. The mechanisms used to implement availability are denial of service (DoS) attack protection and distributed denial of service (DDoS) attack protection. Therefore, an RSU or a vehicle making its resources available to other vehicles could be the target of DoS/DDoS attacks.

12.2.2.2. *Privacy*

Privacy means a person having control over their personal information. The user may, for example, agree to use certain geolocation applications to improve their journeys, but does not want to be geolocated permanently and in real time. Privacy also means guaranteeing to a user that only authorized entities, i.e. those holding the necessary authorizations, will be able to access their personal data. In addition, laws protecting users' privacy and data emphasize greater involvement of users in the protection of their data; this is by allowing them to exercise more control over the collecting of their data.

12.2.2.3. *Trust management*

With trust, it can be ensured that the connected and autonomous vehicle evolves in a safe environment and that its interlocutors are the legitimate ones (e.g. autonomous cloud of a group of vehicles). Establishing and managing trust are tasks essential and complementary to security. As the context relating to autonomous vehicles can change frequently, trust management must be conducted dynamically. Generally, reputation-based mechanisms are used to manage trust.

12.3. Security and privacy architecture

Security and privacy concerns are among the barriers to the adoption of connected and autonomous vehicles. Indeed, it is to be noted that connected and autonomous vehicles are exposed to several security threats, such as impersonation, denial of service and network eavesdropping. User privacy is also exposed to risks of disclosure, route tracing, etc. These risks relate to the absence of appropriate protection mechanisms.

12.3.1. *Context-aware security and privacy*

Here, we introduce the concepts of context-aware security in the connected and autonomous vehicle domain. This context-aware security will enable us to implement optimal security and privacy that is self-adaptive to different situations and contexts. Firstly, we present the limitations of conventional security methods.

12.3.1.1. *Limitations of conventional security methods*

The connected and autonomous vehicle must consider the frequent change in situation (access networks available, entities with which to dialog, etc.), certain users preferring a high level of security while others are less concerned. With users being led to change (car sharing and carpooling), we consider the user profile as contextual information that the vehicle will have to take into account when it makes decisions.

Moreover, intelligence and context awareness are also not considered by conventional security mechanisms. Intelligence and context awareness will enable security and privacy risks to be dynamically and intelligently assessed, and appropriate security and privacy measures to be implemented in real time.

For example, in the presence of different access networks, the choice will be one that will offer the most appropriate level of security to meet the needs of the application and the user's preferences (e.g. prioritizing security over QoS). In the end, traditional approaches to security and privacy show their limits in terms of effectiveness and adaptability.

12.3.1.2. *Context-aware security*

Context-aware security has recently emerged to overcome the new security problems brought about by the increased presence of ubiquitous, heterogeneous, mobile and pervasive computing systems (Sylla et al. 2021a, 2022). We define context-aware security in the connected and autonomous vehicle domain as follows: "Security is context-aware if the choice and implementation of security mechanisms are context-based and are conducted without explicit user intervention". This definition applies well to the vehicular environment, subject to frequent changes of context (available networks, entities with which to dialog, user preferences, etc.).

Security risks can vary, depending on the context. For example, the vehicle receiving information that a security service is compromised will seek to raise the level of security. Therefore, context-aware security by taking contextual parameters into account makes it possible not only to render the security service dynamic but also to reinforce it when the context requires it, while limiting the user's intervention. This allows the user to be permanently protected, by dynamically deploying security and privacy mechanisms that respond to the threat model characterizing the user's current context without their intervention (Sylla et al. 2021b).

Communication security that takes account of the context can help identify communications that need to be secured in a given situation (underlying networks, nature of endpoints, types of data, etc.) and define the best means of securing them. Therefore, when the user, the application or the vehicle needs to communicate certain sensitive data, an encrypted channel can be automatically established for this transmission.

Table 12.1 summarizes the advantages of context-aware security.

Security and privacy services	Context-aware security	Advantages
Authentication	– Strengthening of authentication according to context	– Adaptive authentication – Impersonation very difficult
Access control	– Dynamic allocation or revocation of authorization tokens according to context – Contextual verification of roles and allocations	– Dynamic and adaptive management of access control: addition and revocation of authorizations at any time and on the fly – Dynamic protection against unauthorized access
Privacy	– Context-adaptive anonymization/pseudonymization – Controlled and automatic disclosure of data according to preferences	– Better user control over their data – Contextual and dynamic access to data – Sensitive data protected in all contexts
Communication security	– Selection of the level of security of communications based on context and risk	– Dynamic and adaptive security of communications regardless of context and underlying networks

Table 12.1. *Advantages of context-aware security*

12.3.2. *Gaps in existing solutions*

The vast majority of the proposed solutions focus on a particular security service, or on privacy, independently of the security services, without supporting the characteristics of the connected and autonomous vehicle identified in the introduction. However, the implementation of security services and privacy are complementary.

Several significant issues need to be resolved:

– A specification of the context-aware security and privacy approach in the connected and autonomous vehicle domain: this approach allows a better supporting of user characteristics in the implementation of security and privacy. This will, at the same time, promote the adoption of applications by connected and autonomous vehicle users.

– Complete and global context-aware security and privacy architecture: context-aware security in the vehicular environment will have to consider all security and privacy services. It should not be vertical, i.e. specific to an application, as proposed in the existing works. It will have to be horizontal; in other words, allow a variety of applications to be supported. Therefore, an architecture capable of providing context-aware security that is complete and horizontal needs to be proposed.

– Integration into new network architectures: the design of a context-aware security and privacy architecture that integrates new architectures, particularly edge computing and 5G, is necessary in the vehicular environment. Indeed, this integration will allow the constraints of real-time or near-real-time applications to be supported and scaled up.

12.3.3. *Proposed solution*

As we have mentioned, existing solutions for the security and privacy of users in the vehicular environment present several limitations: given the absence of mechanisms capable of assuring all security and privacy services, the proposed solutions are specific to one type of application and do not support the full range of characteristics of the connected and autonomous vehicle. To address these issues, we propose a context-aware security and privacy architecture as a service. This architecture implements context-aware security and considers the characteristics of the connected and autonomous vehicle.

12.3.3.1. *Context-aware security as a service*

We propose a new context-aware security architecture based on the "as a service" paradigm. "As a service" computing is a more flexible variant of service-oriented information based on service-oriented architecture, whose scope is at the application level. Indeed, in these architectures, the design of systems and applications is based on reusable specialized components, also called services or micro-services. Loosely coupled, reusable and specialized components work independently and collaborate to accomplish application tasks.

These properties facilitate the implementation of increasingly distributed applications, which were previously designed using the monolithic approach. In addition, computing as a service enables advantage to be taken of new computing paradigms, such as containerization and service composition, native cloud, etc. The service-oriented approach addresses several challenges in terms of securing vehicular applications, notably dynamicity, flexibility, heterogeneity, mobility and customization (Sylla et al. 2019). In addition, this approach allows the scalability of the proposed architecture. Indeed, it allows instantiation on-demand of the necessary services. As a result, car manufacturers, road infrastructure operators and service providers can better benefit from the advantages offered by cloud computing and new network architectures.

12.3.3.2. *Context-aware security and privacy architecture*

In this section, we present context-aware security and privacy in the form of "as a service" architecture. This architecture supports context-aware security and privacy, dynamicity, flexibility, mobility, customization of security and privacy services, and support for different types of vehicular applications. In addition, it can be seen as an extension to the reference architecture.

The reference communication architecture for Cooperative Intelligent Transportation Systems (C-ITS) is composed of three main planes:

– The management plane (MP): this allows global management of the C-ITS communication architecture. Among the features of this plane are the prioritization of communications according to the importance of applications and congestion control management.

– The data and control plane (DCP): this plane is also sometimes referred to as the "communication protocol plane". It is composed of three main layers. The "Access" layer provides C-ITS applications with seamless communication via different radio access technologies. The "Networking and Transport" layer provides the set of transport and network protocols needed to establish communications between C-ITS devices. Lastly, the "Facilities" layer corresponds to the services needed to operate the different C-ITS applications, such as interactive map support.

– The security and privacy plane (SPP): this plane contains the functionalities needed to secure C-ITS applications, C-ITS devices and C-ITS communication protocols. It also ensures the privacy of road users, such as with route tracing. The functionalities of this plane include: the implementation of firewalls and intrusion management mechanisms, identity management and introduction of mechanisms guaranteeing anonymity (or pseudoanonymity), and the definition of modules having to guarantee the physical security of devices (HSM, Hardware Security Module).

The operation of C-ITS applications (road safety, traffic flow, entertainment, etc.), presented as a horizontal layer in the reference architecture, is based on the interaction between these three planes.

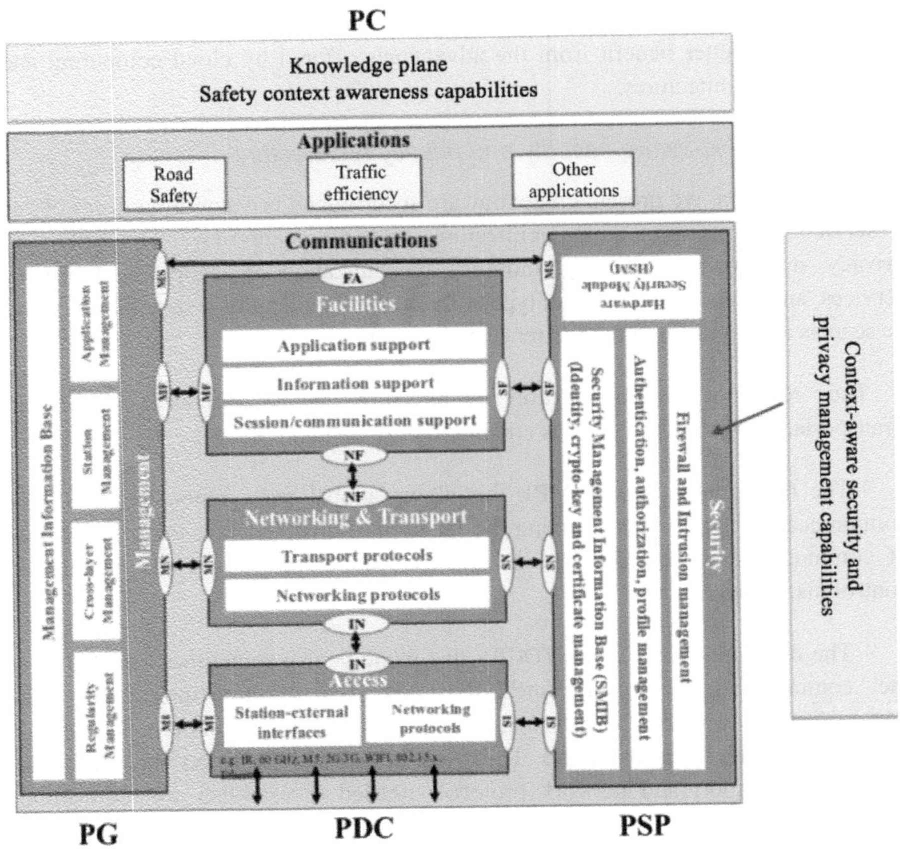

Figure 12.3. *Integration of our proposal into the C-ITS reference architecture (ISO 2020). For a color version of this figure, see www.iste.co.uk/mendiboure/transportsystems.zip*

The architecture that we propose aims to improve the SPP plane in order to provide context-aware security as a service for the connected and autonomous vehicle. It also aims to add a knowledge plane (KP) in the C-ITS reference architecture that will offer information processing and analysis services, as well as decision-making support, to all other planes. The benefits that could result from defining a knowledge plane for C-ITS architecture are numerous (Mendiboure et al. 2021). Examples include improved usage of resources through information sharing between the different planes and layers of the C-ITS architecture. We propose integrating security context-awareness capabilities into this plane. Therefore, our proposal will allow the C-ITS reference architecture to support context-aware security and privacy, while being compatible with the integration of next-generation networks.

Figure 12.3 shows the integration of the proposed architecture into the C-ITS reference architecture (ISO 2020).

The separation of intelligence and enforcement of security decisions allows for greater modularity and more flexibility. Therefore, we propose integrating security context-awareness capabilities into the PC plane that we add to the C-ITS architecture, as well as enhancing the SPP plane by adding context-aware security and privacy management capabilities.

Figure 12.4 shows the interactions between modules of the proposed architecture.

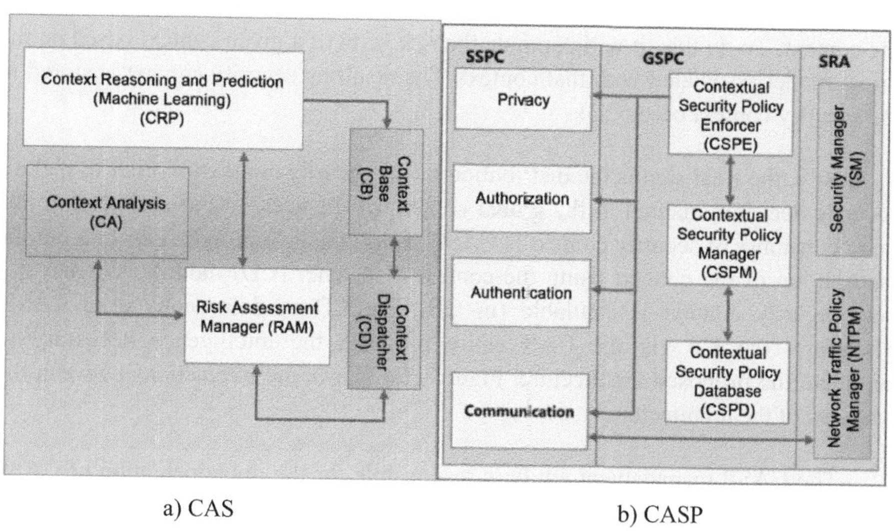

Figure 12.4. *Interactions between modules of the proposed architecture. For a color version of this figure, see www.iste.co.uk/mendiboure/transportsystems.zip*

Each entity of the architecture plays a key role:

– The primary objective of the context-aware security (CAS) entity of the knowledge plane will be to provide specific and relevant contexts for the context-aware security and privacy (CASP) management entity. On the basis of the context provided, the CASP entity will implement the security and privacy mechanisms most appropriate to the situation encountered. Therefore, the CAS entity is responsible for modeling, determining and distributing the security context to the CASP entity. This will allow the CASP entity to dynamically compose the appropriate services/micro-services according to the context.

The CAS entity (Figure 12.4(a)) comprises several modules. Contextual information is available in the KP knowledge base, such as the vehicle's position or the available access networks. The first step in determining the security context is the analysis of this context, which is performed by the context analysis (CA) module. The next step is context reasoning. Contextual reasoning is the process of deducing the security context from several items of contextual information. Therefore, the CA output is used by the context reasoning and prediction (CRP) module to determine the security context. Indeed, it infers on the contextual information provided by CA using a reasoning technique (e.g. rules, automatic learning and fuzzy logic) to determine the security context. It also relies on the risk assessment manager (RAM) module, which assesses the level of risk associated with the context. To do this, it will compute the risk level of a given context based on the threat model associated with that context. The resulting security context is stored in the security context base (CB).

Lastly, the final step is the distribution of the security context in order to make a security decision adapted to the global context of the user. In our architecture, the main consumer of security context is CASP. To do this, we propose carrying out the distribution of the context using the context dispatcher (CD) module. Whenever a new security context is available (in CB), the CD module sends it to CASP. Therefore, the KP via the CAS entity provides the intelligence necessary to automate the proposed architecture. Figure 12.4 shows the interactions between the modules of the architecture.

– The CASP management entity is responsible for the implementation of CASP mechanisms, such as authentication, access control, security of communications and privacy. Indeed, depending on the context provided by CAS, CASP will compose the services and/or micro-services necessary for the implementation of mechanisms required for this context.

The CASP entity (Figure 12.4(b)) is divided into two functional components: CSPS (Contextual Security and Privacy Services) and CSP Mgmt. (Contextual Security and Privacy Management). The objective of the CSP Mgmt. component is to make security and privacy decisions based on the context that will be provided by CAS. The CSPS component includes the modules responsible for applying CASP decisions made by the CSP Mgmt. component.

In addition to the functional components, the CASP entity includes a non-functional component, network and architecture security (NAS). Its objective is to protect architecture from attacks targeting its proper functioning, and to manage the policies of network traffic emanating from the CSP Mgmt. component.

This functional division is used to process high-level contexts, provided by the CAS entity, before enforcing the CASP policy decision. Security policies enable the definition of the security mechanisms to be deployed, based on a clearly defined context.

The contextual security policy manager (CSPM) is in charge of selecting the contextual security policy corresponding to the context and risk level provided by the CAS entity. To do this, it obtains the corresponding policy from the contextual security policy database (CSPD) and sends it to the contextual security policy enforcer (CSPE). The latter will need to orchestrate the composition of services corresponding to the appropriate modules of the CSPS component in accordance with the policy provided.

Once the contextual security policy decision has been processed, the selected policy must be enforced by CASP mechanisms. Therefore, at the CSPS level, we offer privacy, authentication, authorization (access control) and communication modules for the implementation of CASP mechanisms.

The NAS comprises two modules: the architecture security manager and the network traffic policy manager. The architecture security manager implements deep packet inspection and filtering mechanisms aimed at the architecture modules. In the event of an attack, it instructs the network traffic policy manager (NTPM) on which network traffic policy to apply. The NTPM acts as an SDN (Software-Defined Networking) application (Shin et al. 2012). It is responsible for passing the paths to be used by the data plane when an attack occurs to the SDN controller. The NTPM also transmits the flow rules to the connected and autonomous car so that it can better secure the traffic in the case of implementing a contextual security policy. The car thus acts as an element of the SDN data plane.

In section 12.4, we present an example of implementing our architecture in the case of self-adaptive selection of network access technologies.

12.4. Self-adaptive selection of network access technologies

The connected and autonomous vehicle must be able, in real time, to autonomously and securely select the radio access technology (or technologies) best meeting the needs of each type of communication flow it has to process (V2I, V2V, V2S, V2R or V2P) and for defined use cases (see section 12.2.1). To do so, it must be able to take into account various constraints, such as quality of service, security, energy saving in case of low battery, or user preferences such as costs or a concern for limiting the environmental impact. In this case study, we will focus on security and privacy. Therefore, we will consider, in particular, the security needs of the application, the level of security offered by the different access networks available and the user preferences in terms of privacy. Indeed, the different passengers in a vehicle may each have different security and privacy requirements: the first may want to launch a geolocation application but without being geolocated; the second may wish to stream a video using the infotainment service offered on-board the vehicle without needing to worry about security; the third may ask the vehicle, via a voice assistant, to reserve a parking space with complete peace of mind, and therefore in a reliable and secure manner, etc. In addition, the vehicle itself may want to communicate with nearby vehicles, in the event of an accident, for example, in order to advise them or update their traffic navigation information – and do so in a reliable and secure manner. More broadly, in the case of the smart city, the vehicle will have to interact with numerous devices, both fixed (traffic lights, road signs, toll station, etc.) and mobile (bus, train, drone, bicycle, scooter, motorcycle, etc.) in order to best manage urban mobility.

In Kaiwartya et al. (2016), the authors ranked wireless communications (vehicular, mobile and wireless communications) according to the type of vehicular application (see section 12.2.1). They also ranked wireless communications, notably taking account of security support. This order of preference of wireless access technologies based on the application type and the level of security offered can be used to choose the most appropriate access technology for each vehicular application.

The autonomous vehicle must also be capable of predicting the likely evolution of the level of service in order to enable proactive action to ensure continuity of service in mobility. To do so, it must have a cognitive loop, as presented in Figure 12.5.

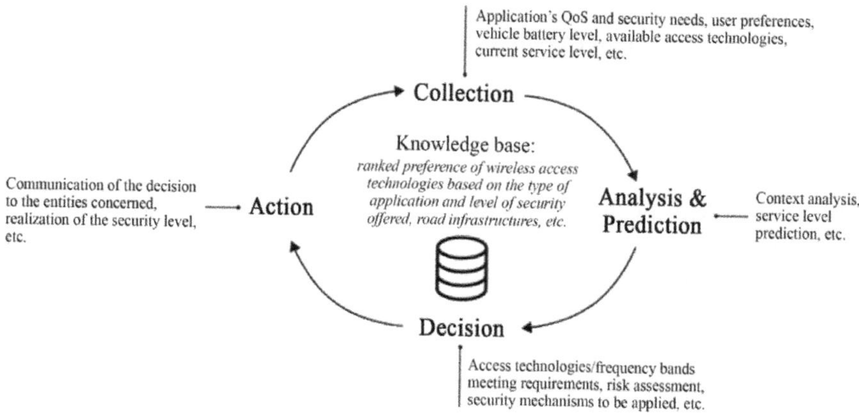

Figure 12.5. *Cognitive loop*

Collection, analysis and prediction and decision-making (choice of access network) will be carried out in the knowledge plane. The decision-making on security mechanisms to be put in place and the realization of the security level (Action) will be carried out in the CASP entity.

12.4.1. *Infrastructure edge computing*

The CASP architecture that we propose (section 12.3.3.2) is designed using the "as a service" approach. This enables its implementation in new network architectures, such as in an edge computing (EC) infrastructure.

The application of EC technology in vehicular networks presents numerous advantages (Mendiboure et al. 2021):

– Flexible deployment: with the EC approach, it becomes possible to locally deploy resources that meet the needs of a given application in a particular location. Therefore, low latency and high availability can be guaranteed.

– Scalability: using network equipment resources (fog computing or MEC), or deploying additional servers (Cloudlet), increases the amount of processing and storage resources available. With this approach, exchanges to the cloud and the amount of bandwidth used are also reduced. Therefore, EC ensures significant scalability. With EC, the increase in the number of connected vehicles and vehicular applications could therefore be managed efficiently.

– Efficient data processing: certain C-ITS applications require short response times (road safety and traffic flow). With EC, the services required for these applications would be deployed at the network edge. Therefore, processing times could be reduced for these applications (Ha et al. 2013) and data processing could also be more efficient (Mendiboure et al. 2019).

In the absence of EC infrastructure, if a vehicle has sufficient processing, storage and communication capabilities, it can make them available to other vehicles (or a group of vehicles). It thus forms an autonomous cloud for a group of vehicles, as shown in Figure 12.6. This approach enables the placement of new services like the one we propose, even in the absence of EC infrastructure, while benefiting from the advantages cited above.

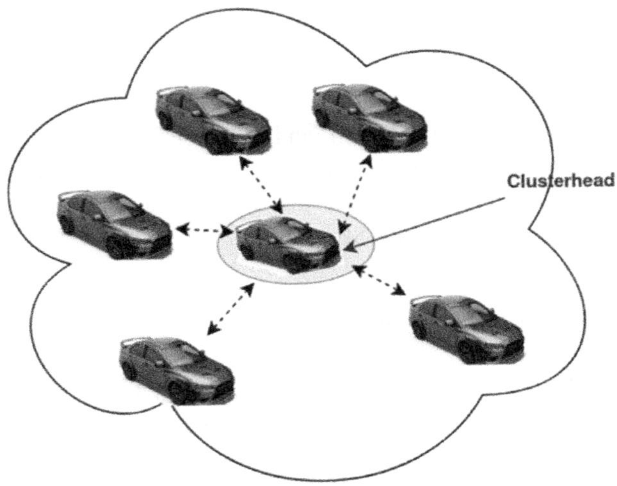

Figure 12.6. *Autonomous cloud. For a color version of this figure, see www.iste.co.uk/mendiboure/transportsystems.zip*

The design of security and privacy services following the "as a service" approach allows the design of distributed services in modular, flexible and independent form, which can be placed in different ways in the edge, or indeed on an autonomous cloud, as presented in the previous section. The components of this service can be placed as a single service on a single edge/vehicle node or distributed over several edge/vehicle nodes according to the constraints of placement, quality of service and/or the user's quality of experience (Petri et al. 2019).

Our architecture has been designed to be placed either as a single block on a single node or in several blocks on several nodes, depending on the resources available. Figure 12.7 presents an overview of the architecture considered. RSU constitute the EC infrastructure, where they are present. In their absence, a vehicle with sufficient resources available will make them available to the surrounding vehicles, with this group of vehicles constituting an autonomous cloud. The RSUs are interconnected by SDN (Software-Defined Networking) switches. SDN switches are controlled by an SDN controller, the set of physical devices that constitute the logical SDN controller. The edge orchestrator, deployed near the SDN controller, will orchestrate and manage the placement and dynamic migration of security and privacy service components in the edge.

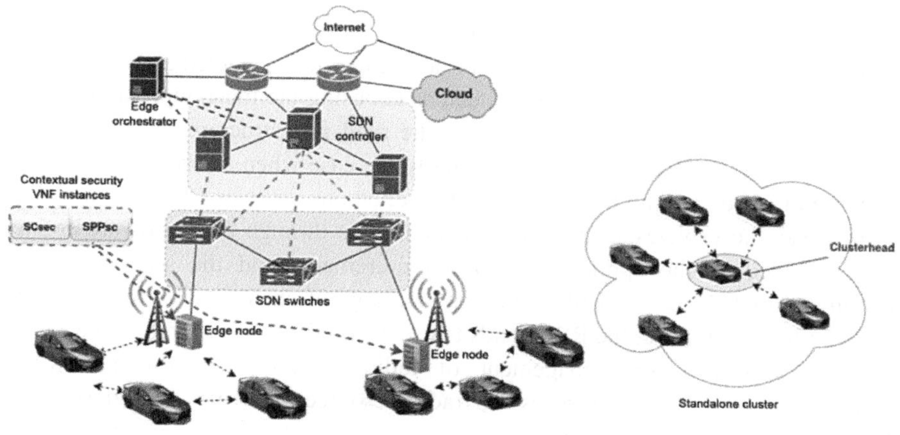

Figure 12.7. *General view of the retained architecture. For a color version of this figure, see www.iste.co.uk/mendiboure/transportsystems.zip*

12.4.2. *Orchestration and placement of services*

Orchestration is a process consisting of planning, managing and executing the placement and/or migration of virtualized computing services and other computing workloads (such as machine learning tasks) in a computing infrastructure; here, it is the EC infrastructure.

Service placement consists of seeking favorable execution locations to instantiate the execution of a service. To do this, the orchestrator must assess (online or offline) several parameters in order to place the service. Among these parameters, we find the QoS constraints to place the service.

Service migration consists of moving the service from one node to another node, while maintaining continuity of service.

Placement and migration of services are the most difficult tasks of orchestration in an EC infrastructure. A sound orchestration strategy should be based on algorithms that use optimization methods to minimize service placement failures. These optimization methods can be combined with artificial intelligence techniques for tangible placement results, supporting the characteristics of edge computing (Salaht et al. 2020).

12.4.2.1. *Example of placement of security and privacy services in EC infrastructure*

The CAS entity of the knowledge plane determines the security context. It is located in the knowledge plane (PC), itself located in the connected and autonomous vehicle to allow it to make decisions, such as choosing the most appropriate access network. It may be of interest to involve the road infrastructure in this selection process when it is present. This is particularly the case when smart radio is used to meet the on-road communication needs of autonomous vehicles (Aniss et al. 2021). In this case, the vehicles send the detected channels through their software-defined radio, as well as the application's QoS requirements and the security context determined by CAS to the RSU, which is in charge of channel selection. Once the selection has been made, the RSU sends the selection decisions to the vehicles. This approach allows better management of radio resources. It also allows the management of conflicts in access to radio resources between secondary users. RSUs also communicate with each other to share information on the detected channels, the selection processes in progress and the radio channels used in their area. They may thus need to reassess the level of risk if they have received information that, for example, a radio channel is compromised. In order to make all of their decisions, it would be useful if each RSU could also have a knowledge plane. This knowledge plane would also integrate the CAS entity in order to facilitate security monitoring of their respective area.

The CASP management entity composes the services and/or micro-services necessary to implement the security and privacy mechanisms required by the security context provided by CAS. It is located in the SPP, which is located in the connected and autonomous vehicle to allow it to evolve in a secure environment.

When RSUs are involved in the decision-making process (e.g. use of smart radio to meet the on-road communication needs of autonomous vehicles), they must also have at their disposal the CASP entity in order to implement the security mechanisms required by the security context provided by their knowledge plane, via the CAS entity. In this case, the connected and autonomous vehicle simply applies the security decisions corresponding to the security context provided by the RSU. Here, we can note the separation of intelligence from the application of security decisions.

12.4.2.2. *Example of placement of security and privacy services in the autonomous cloud*

When smart radio is used to meet the on-road communication needs of autonomous vehicles (Aniss et al. 2021), it may again be of interest here to make use of the autonomous cloud in the selection process, in the absence of road infrastructure. The vehicle making its processing, storage and communication capabilities available to the other vehicles can then be seen as a clusterhead. Its role is similar to that played by the RSU, presented in section 12.4.2.1. Therefore, based on the information received from the vehicles of its cluster, it is this vehicle that will take charge of selecting the channels and sending the selection decisions to the vehicles of its cluster. This approach, again, offers better management of radio resources and makes it possible to manage conflicts of access to radio resources between secondary users. Clusterheads also communicate with each other in order to share their information and may be led to reassess the risk level, in particular, using the CAS entity.

In this example, the CAS and CASP entities, integrated into the clusterhead knowledge plane and security plane, respectively, are made available to the vehicles of the cluster. Here again, we can note the separation of intelligence from the enforcement of security decisions on the side of the connected and autonomous vehicle, the latter applying security decisions corresponding to the security context provided by the clusterhead.

12.5. Main research works to be conducted

The CASP architecture, in the form of a service that we have presented, would allow users to have trust in the connected and autonomous vehicle, thus favoring its adoption. However, to address all the issues raised by CASP, adapted to the characteristics of the connected and autonomous vehicle, several works remain to be conducted, in particular:

– The implementation and assessment of this architecture as a service. The service to be implemented can be broken down into several services representing different parts of the architecture.

– The definition and implementation of the risk assessment management module (RAM module). This aims to assess the levels of security and privacy violation risks associated with a context based on the threats that are associated with that context.

– The definition and implementation of a service block placement strategy in an edge computing infrastructure. The proposed solution would support the dynamic evolution of resources in an edge computing infrastructure and ensure a low placement failure rate and good resource utilization, when the infrastructure is saturated.

– The assessment of the architecture proposed and implemented as a service under real conditions, for example, in a 5G network. This would make it possible to determine the performance of the architecture under real conditions, in particular, in terms of security and adaptive privacy in real time.

– The definition, implementation and assessment of a service placement strategy in multi-domain edge computing infrastructures. Indeed, the various edge computing infrastructures used by the connected and autonomous vehicle throughout its journey can be managed by different operators.

– The assessment of authentication and trust management solutions scaling-up for the implementation of SD-IoV (Software-Defined – Internet of Vehicles) applications within the framework of autonomous vehicles. Indeed, the SD-IoV paradigm was introduced to solve problems related to mobility and managing QoS in IoV networks with the implementation of SDN technology in these environments. In this context, several works have been produced to ensure the security of SD-IoV applications (Mendiboure et al. 2018). However, these solutions are not suited to the CASP architecture for autonomous vehicles that is proposed in this chapter. Therefore, research needs to be conducted in order to bridge this gap in these domains and to provide comprehensive security, and supporting secure communications between secure SDN entities (applications and vehicles).

12.6. Conclusion

Vehicular applications will continue to develop, especially with the development of the connected and autonomous vehicle. Their implementation implies several security and privacy issues. Several works have been carried out and have led to the development of security solutions.

In this chapter, we focused on context-aware security adapted to the characteristics of the connected and autonomous vehicle. We started by presenting the security and privacy needs in vehicular applications. This led us to propose a new architecture suitable for C-ITS and new network architectures. The solution presented is based on the "as a service" approach to support a variety of vehicular applications. It aims to enhance the SPP of the C-ITS reference architecture to provide context-aware security as a service for the connected and autonomous vehicle. It also aims to add a knowledge plane in the C-ITS reference architecture, which will offer information processing and analysis services, as well as decision-making support, to all other planes.

We have presented an example of the implementation of our architecture in the case of self-adaptive selection of network access technologies. This example notably allowed us to highlight the modularity and flexibility of the proposed solution. Lastly, we mentioned several research works remaining to be carried out.

12.7. References

Aniss, H., Krief, F., Berbineau, M., Le Page, K. (2021). Cognitive radio contribution to meeting vehicular communication needs of autonomous vehicles. In *Intelligent Network Management and Control: Intelligent Security, Multi-criteria Optimization, Cloud Computing, Internet of Vehicles, Intelligent Radio*, Benmammar, B. (ed.). ISTE Ltd, London, and John Wiley & Sons, New York.

Contreras-Castillo, J., Zeadally, S., Guerrero-Ibañez, J.A. (2017a). Internet of vehicles: Architecture, protocols, and security. *IEEE Internet of Things Journal*, 5(5), 3701–3709.

Contreras-Castillo, J., Zeadally, S., Guerrero Ibáñez, J.A. (2017b). A seven-layered model architecture for internet of vehicles. *Journal of Information and Telecommunication*, 1(1), 4–22.

Ha, K., Pillai, P., Lewis, G.A., Simanta, S., Clinch, S., Davies, N., Satyanarayanan, M. (2013). The impact of mobile multimedia applications on data center consolidation. In *2013 IEEE International Conference on Cloud Engineering, IC2E 2013*, San Francisco, 25–27 March, 166–176.

ISO (2020). ISO 21217:2020; Systèmes de transport intelligents – Architecture du station et du communication, disponible en ligne. ISO Standard [Online]. Available at: https://www.iso.org/fr/standard/80257.html.

Kaiwartya, O., Abdullah, A.H., Cao, Y., Altameem, A., Prasad, M., Lin, C.-T., Liu, X. (2016). Internet of vehicles: Motivation, layered architecture, network model, challenges, and future aspects. *IEEE Access, Special Section on Future Networks: Architectures, Protocols, and Applications*, 4.

Kambiré, S.J., Aniss, H., Krief, F., Maaloul, S., Berbineau, M. (2020). Radio access technologies selection in vehicular networks: State-of-the-art and perspectives for autonomous connected vehicles. *Communication Technologies for Vehicles*, 99–112.

Mendiboure, L., Chalouf, M.A., Krief, F. (2018). Towards a blockchain-based SD-IoV for applications authentication and trust management. In *Internet of Vehicles. Technologies and Services Towards Smart City*, Andrzej, M.J., Skulimowski, Z., Sheng, S., Khemiri-Kallel, C.C., Ching-Hsien, H. (eds). Springer International Publishing, Cham, 265–277.

Mendiboure, L., Chalouf, M.-A., Krief, F. (2019). Edge computing based applications in vehicular environments: Comparative study and main issues. *Journal of Computer Science and Technology*, 34(4), 869–886.

Mendiboure, L., Chalouf, M.-A., Krief, F. (2021). Toward new intelligent architectures for the internet of vehicles. In *Intelligent Network Management and Control: Intelligent Security, Multi-criteria Optimization, Cloud Computing, Internet of Vehicles, Intelligent Radio*, Benmammar, B. (ed.). ISTE Ltd, London, and John Wiley & Sons, New York.

Myung-Ki, S., Nam, K.-H. Kim, H.-J. (2012). Software-defined networking (SDN): A reference architecture and open APIs. In *2012 International Conference on ICT Convergence (ICTC)*. IEEE.

Petri, I., Rana, O., Zamani, A.R., Rezgui, Y. (2019). Edge-cloud orchestration: Strategies for service placement and enactment. In *2019 IEEE International Conference on Cloud Engineering (IC2E)*, Prague, 67–75.

Salaht, F.A., Desprez, F., Lebre, A. (2020). An overview of service placement problem in fog and edge computing. *ACM Computing Surveys*, 53(3), 1–35.

Shin, S. and Guofei, G. (2012). CloudWatcher: Network security monitoring using OpenFlow in dynamic cloud networks (or: How to provide security monitoring as a service in clouds?). In *2012 20th IEEE International Conference on Network Protocols (ICNP)*, 1–6.

Sylla, T., Chalouf, M.-A., Krief, F. (2019). Towards a context-aware security and privacy as a service in the internet of things. In *IFIP International Conference on Information Security Theory and Practice (WISTP)*, 240–252. Springer, Cham.

Sylla, T., Chalouf, M.-A., Krief, F. (2021a). Context-aware security in the internet of things: A survey. *International Journal of Autonomous and Adaptive Communication Systems*, 14(3), 1.

Sylla, T., Mendiboure, L., Chalouf, M.-A., Krief, F. (2021b). Blockchain-based context-aware authorization management as a service in IoT. *Sensors*, 21(22), 7656. doi: 10.3390/s21227656.

Sylla, T., Chalouf, M.-A., Krief, F. (2022). Adapting the security level of IoT applications. In *Intelligent Security Management and Control in the IoT*, Chalouf, M.-A. (ed.). ISTE Ltd, London, and John Wiley & Sons, New York.

13

Vehicular Wireless Communications: Risks and Detection of Attacks

Jonathan VILLAIN[1], Virginie DENIAU[2] and Christophe GRANSART[2]

[1] *COSYS-LEOST, IMT Nord Europe, Lille, France*
[2] *COSYS-LEOST, Gustave Eiffel University, Villeneuve-d'Ascq, France*

13.1. Introduction

Wireless communications present new opportunities in numerous domains, notably in road domains where they enable interactions between vehicles and the transfer of information between road infrastructures and vehicles. However, the conditions that can affect the performance of wireless communicating solutions are much less manageable than in the case of wired communications. Another example is radio coverage, i.e. the strength of communication signals according to the distances between connected objects and infrastructure radio frequency terminals. We can also count the propagation conditions of communication signals that can be affected by relief, obstacles and surrounding materials. Also worth mentioning is possible electromagnetic interference, which can degrade the reception capacities of communication signal receivers. Lastly, wireless communications can also fall victim to specific attacks that exploit the fact that communication signals are transmitted in the air and are not confined to the cables protecting them.

However, the functions envisaged are varied and may concern optimization and comfort as well as user safety. Therefore, despite all of these potential impacting factors, it is essential to guarantee a minimum performance, indispensable for the deployment of safety-related functions.

In this chapter, we assume the position that vehicles, with each other or with infrastructures, necessarily employ wireless communication solutions. We are therefore interested in vulnerability to attacks that precisely exploit the "wireless" nature of these communications. Signals do not benefit from any physical protection as they do in a cable, and certain attacks exploit this lack of signal containment.

Therefore, protection and barriers to attacks require solutions for detecting these attacks at the level of the wireless links. Detection solutions that aim to detect attacks targeting wireless protocols generally rely on what are known as layers 1 and 2 of the OSI model, namely the physical (PHY) and data link (DL) layers. In this chapter, we will present the different methods proposed on these layers.

To begin with, we will describe the main characteristics of wireless communications in the domain of connected vehicles, differentiating in particular between direct vehicle-to-vehicle communications and vehicle-to-infrastructure communications. We will then discuss the main communication protocols valid for vehicles connected for short- or long-range links. Next, we will detail the attacks that target the lowest layers of communication protocols, namely the physical and data link layers, and discuss the risks incurred in vehicular applications. Lastly, the chapter concludes with the methods of detecting these attacks and in particular those based on artificial intelligence (AI).

13.2. General characteristics of wireless communications for connected vehicles

In this section, after returning to the main issues represented by connected vehicles, we explain the definitions of vehicle-to-vehicle (V2V) and vehicle-to-infrastructure (V2I) communications.

13.2.1. *Challenges related to the connected vehicle*

A connected car is a car with in-vehicle communication systems that allow wireless communications with the environment. With the advent of current information and communication technologies, it is increasingly easy to embed these tools that transmit different information gathered by sensors into vehicle technologies. The most commonly used in cars is GPS (Global Positioning System), which retrieves the position of the vehicle. Thanks to this satellite connection, vehicles can embed applications offering navigation assistance and sometimes travel time optimization, if the application allows the collection of information on traffic conditions. Depending on the communicating solutions available within the vehicle, the traffic-condition information can be obtained from GPS traces of users or can be signaled by the users themselves.

Another example is the fact that all new vehicles are equipped with cellular communication solutions to enable an automatic emergency call to be established in the event of an accident. This system is called "eCall", a contraction of "emergency call". eCall can be generated either manually by the vehicle's passengers or automatically by activating in-vehicle sensors in the event of a serious road accident. When activated, the in-vehicle eCall system establishes a 112-voice connection directly with the relevant public safety answering point. Then, if no passenger is able to speak, for example, due to being injured, a "minimum set of data" is sent to the answering point. It contains the exact location of the accident site, the triggering mode (automatic or manual), the vehicle identification number and a time stamp. In this way, information valuable to emergency responders can reach them as quickly as possible.

Other applications are dedicated to "infotainment", a contraction of the terms "information" and "entertainment". While they may be viewed as more recreational, they could present significant commercial interest in the future. Other applications may, for example, make it possible to put in place driving optimization functions with a view to better energy efficiency or virtual toll applications, which facilitate the daily life of users and optimize the usage of the road network.

Lastly, beyond services to users, the data generated and exchanged by connected vehicles, such as their positions, journeys, speeds, etc. constitute commercial challenges. Analyses of this data can lead to much more elaborate information on habits, lifestyles or driving behaviors. All of this data opens the door to numerous applications that can be considered at the service or without the knowledge of the user.

In the future, new applications will undoubtedly have a security vocation. These may be applications that operate on data exchanges between vehicles, or via relay terminals located along the road network (RSU: Road-Side Unit), or via a cellular network. Of course, for these applications, the performance of communicating solutions will have to be scrutinized in detail as well as the risks introduced by any attacks.

13.2.2. *V2V communications*

Vehicle-to-vehicle, or V2V, communications are those established directly between vehicles. They are made possible if the vehicles are equipped with terminals operating with appropriate protocols that do not require passage through a ground-based communication network. We also discuss ad hoc networks, which are networks constituted solely by terminals on-board vehicles and reconfigurable over time.

The requirements on these V2V communications are not necessarily the same depending on the information that need to be transmitted between vehicles. Indeed, when speaking of infotainment, which provides a comfort or leisure service, the requirement is not the same as when speaking of informing the user that automatic emergency braking has just been activated on a vehicle located on the vehicle's trajectory. Therefore, in order to enable such security functions, well-adapted protocols are required, allowing very low latency and very frequent V2V communications so that the information can be rapidly updated. Performance must also be stable, regardless of the density of the vehicles, the environment and any interference present. Therefore, it is also necessary to consider evaluating the effect of possible attacks on this performance, which is essential to the implementation of security functions. A successful attack is not only an attack that prevents communication, but can be an attack that degrades the performance of V2V communications resulting in latency incompatible with applications such as assistance entering a fast lane, warning of heavy braking or of an object or person on the road, etc.

13.2.3. *V2I communications*

Vehicle-to-infrastructure (V2I) communications are those established between vehicles and connected road equipment and communications between vehicles and access points or base stations of a ground-based communication network. These include cellular communications in 2G, 3G, 4G or 5G between an embedded terminal in a vehicle and a base station on the ground, as well as Wi-Fi communication[1] between a device or sensor built into the vehicle and a Wi-Fi access point located in a car park or along a road. These communications are essential in some communication protocols even when information must pass between two vehicles in proximity to each other. Indeed, for example in 2G, 3G or indeed 4G, the protocol requires passing by the ground-based communication infrastructure. Indeed, in these protocols, the allocation of the resource (channel, instant) to any terminal is managed by the infrastructure. Therefore, in the case of V2I communications, attackers can seek to affect the reception of in-vehicle terminals, or indeed that of infrastructure terminals.

1. The protocol dedicated to connected vehicles, which originates from the Wi-Fi family or protocol 802.11, is referred to as ITS-G5 and will be described further on in this chapter.

13.3. Characteristics of wireless communications

In this section, we summarize the main communication protocols dedicated to connected vehicles for long- and short-range links.

13.3.1. *Principle of wireless communications*

The ancestor of current wireless technologies dates back to the 1970s. The ALOHA protocol (Abramson 1970) was invented at the University of Hawaii. This protocol was invented to enable hotel room bookings between the different islands of the archipelago.

Its principle is very simple: if a computer needs to communicate, it sends its message (in the air). This protocol was not very effective: numerous collisions occurred. However, it is the ancestor of all of the versions of wireless protocols that we know today.

In the IEEE 802.3 (wired Ethernet) protocol, the carrier can sense multiple access/collision detection (CSMA/CD) protocol was defined. This allows the detection of a collision during a transfer on the medium when two machines communicate simultaneously.

This system of transmission and simultaneous listening cannot function in the context of wireless networks: if a terminal transmits, it cannot listen simultaneously if a collision occurs because it is dazzled by the transmission that is in progress.

This is what led to the development of a protocol such as RTS/CTS for IEEE 802.11. Each terminal that wishes to communicate firstly requests to reserve the resource via a request-to-send (RTS) message and the access point sends a clear-to-send (CTS) message, which serves to tell each terminal in the coverage area that the resource will be busy during the sending time.

In the context of cellular networks, management of the radio resources in terms of time and frequency is managed entirely by the cell in which the terminals are located. It is the cell that allocates communication slots to each terminal.

13.3.2. *Long-range communications*

Among the different 802.11 standards, 802.11p is dedicated to vehicular communications. The 802.11p standard defines a frequency band around 5.9 GHz for V2V and V2I communications. In Europe, ETSI is in charge of standardizing V2V/V2I communications for cooperative ITS (Intelligent Transport Systems). ETSI has published the ITS-G5 standard, based on 802.11p, notably supporting

GeoNetworking, which defines the area of validity of messages transmitted by a vehicle and the possibility of relaying messages by other vehicles to reach vehicles that are not within range of the vehicle transmitting the message.

ITS-G5 is based on units or access points deployed on the roadside by the road manager. ITS-G5 can be hybridized with existing cellular solutions (3G, 4G and 5G), notably for certain non-latency-critical services.

An alternative technology derived from 4G cellular networks is LTE-V2X. This has been standardized by 3GPP in its Release 14. It provides for V2V and V2I communications to be in the same 5.9 GHz band. However, V2I communications, which are not critical, will take place on the bands of traditional 4G networks.

As regards 5G, according to Release 16 of 3GPP TR 21.916, it should also allow for both V2V and V2I communications.

13.3.3. *Short-range communications*

This section illustrates some of the short-range communications that are already deployed on-board current vehicles.

The Bluetooth protocol allows very short-range dialog between various devices. Even if BLE (Bluetooth Low Energy) allows dialog at a distance of 100 m, the main usages remain in the passenger compartment of the vehicle between the smartphone and the dashboard.

Among other short-range systems, the tire pressure monitoring system (TPMS) informs the driver when one of the car's tires is under-inflated. This enables the tire pressure to be checked permanently to avoid a potential accident (tire blowout). This device operates in the frequency band of 433 MHz in Europe and 315 MHz elsewhere in the world. In addition to the tire pressure, an identifier unique to the sensor is sent to avoid interpreting any data from another vehicle that is in close proximity.

Electronic toll collection is based on DSRC (Dedicated Short-Range Communication) technology. This technology operates in the 5.9 GHz range and allows highway tolls to be passed through at a speed of up to 30 kph.

13.3.4. *Advent of 5G*

5G cellular communication technology is designed to allow much more flexible and variable usage of time/frequency resources than 4G. Indeed, 4G employs channels of at least 1.4 MHz, which makes it a well-designed technology of interest for providing minimum throughputs. With 5G allowing either wider or much narrower communication channels than 4G, it should also meet the needs of the Internet of Things. Given that connected objects do not require high throughputs, 5G can allow a large number of objects to be connected by limiting the resource each of them uses. It can therefore be configured in a variable manner within the network either for high-throughput applications or for low-throughput applications. This is network virtualization, one of the technologies used by 5G that divides a network into several virtual slices. Each of these slices is configurable according to the usages that they support, which allows the communication signal to be configured for a throughput, density or latency adapted to each usage.

Therefore, one of the aspects of 5G that presents a significant evolution for the vehicular domain compared to 4G technology lies in the possibility of supporting applications that require low latency. This low latency is enabled by what are known as "stand-alone" network architectures[2], and thanks to the distributed deployment of application servers closer to access (edge computing). It is still unclear whether this solution will be deployed for vehicular applications and whether it will respond to short-range applications.

Currently, stand-alone formats are mainly deployed within the context of private networks, while 5G access available to the general public continues to rely on 4G network cores.

13.4. Susceptibility of communications and risks incurred

In this section, we will consider the modes of attacks that target communication system layers 1 and 2, according to the OSI model. These are the physical and data link layers. In our description, denial of service is not considered as an attack but a consequence. Indeed, several modes of attack (or indeed the majority of attacks) on the PHY layer lead to denial of service. However, one of the differences can be the duration of this denial of service and its geographical extent. Therefore, depending on the applications affected, the scope and duration of the denial of service, the consequences for vehicles and the risks for occupants can vary in how critical they are.

13.4.1. *Principle of attacks targeting layers 1 and 2 of communication systems*

Communication systems are conventionally represented by the OSI model, which comprises seven layers ranging from layer 1, which corresponds to what is known

2. In 5G, it is important to make the distinction between stand-alone 5G and non-stand-alone 5G, which uses a 4G network core and therefore does not reduce latency times.

as the "physical" layer, i.e. the signals exchanged between two terminals, to layer 7, which corresponds to the application layer. In this chapter, we focus on attacks that disturb the two lowest layers, namely the "physical" layer and the "data link" layer.

Layer-1 attacks degrade the reception mechanisms of in-vehicle receivers. They can either degrade signal-to-interference ratios and prevent messages from being read correctly, or make the resource appear to be constantly busy. Indeed, where the communication protocol incorporates algorithms for verifying the availability of the resource, the attack makes the resource appear permanently occupied. All attacks lead to denial of service. For this, the attack signals do not need to respect the waveforms of the communication signals: it is sufficient for them to cover the frequencies of the communication channels of the victim communication system.

Regarding attacks that target layer 2, they are attack signals that respect the format defined by the protocol, at layers 1 and 2. This is essential for these signals to pass layer 1. It either involves sending too many messages, thus saturating the receivers, or preventing messages from being transmitted to the recipient, retrieving data or ordering the network to detach. The consequences can be denial of service, malfunction or information leakage.

13.4.2. *Sybil attack*

This type of attack can also be referred to as an "identity deception". The Sybil attack aims to degrade the stability of a peer-to-peer network by creating a large quantity of identities within that network and using those identities to influence the network. The Sybil attack is an attack in which a legal node is converted into a Sybil node, which is a replica of a node with a different personality but using a similar identifier. The Sybil node can adopt several identities by changing its MAC and IP addresses or other identity information. The presence of such Sybil nodes mainly influences routing protocols and operations such as voting, data aggregation and reputation. Therefore, if such operations are influenced, this can lead, at a minimum, to an increase in the time required for the transmission of certain data.

13.4.3. *Deauthentication frame attack*

The deauthentication frame is a packet type defined in the IEEE 802.11 Wi-Fi standard. It is part of the management frames.

This frame is used to terminate a connection between a station and a Wi-Fi access point. It can be sent by the access point or station to let the other party know that the connection is interrupted.

This deauthentication frame can be sent for various reasons. The station may send a deauthentication frame to the access point because it wishes to detach, in order to switch to another Wi-Fi network or pass through another access point. The access point may also send a deauthentication frame to the station because the router needs to restart. Therefore, deauthentication can be used both ways, and it is not a request but rather a notification, which cannot be refused.

The deauthentication frames have relatively simple structures. However, the MAC address of the sender or receiver must be accessed to spoof a deauthentication packet. This can be obtained by searching for nearby Wi-Fi devices using software such as WireShark.

The target receiving the deauthentication command can reconnect immediately, perhaps without the user realizing that the connection had been interrupted. However, if these deauthentication packets are sent continuously, this results in a denial of service, and network access is blocked for the duration of the attack. In some cases, this form of attack can be used in combination with the creation of a fake access point. This is so that a station detaches from a lawful access point and attaches to an illicit access point that can carry out a man-in-the-middle attack.

13.4.4. *Black-hole attack*

The use of sensors in vehicles is essential for mobility and modern lifestyles. Wireless sensor networks have brought about great changes in the communication process of sensor networks as well as great possibilities for transport applications. However, the more communicating sensors multiply, the more it is necessary for them to be self-coordinated and for communicating solutions to guarantee constant supervision. The vehicular ad hoc network (VANET) is a specific wireless network among mobile ad hoc networks (MANET). It includes vehicles that communicate wirelessly with each other and form networks that evolve during road travel. Wireless transmission allows vehicles to communicate with each other, without passing through a telecommunications network infrastructure, to the lane comprising base stations or access points. VANET shall allow for rapid and efficient exchange of data for the benefit of driver comfort and safety and can be used to develop intelligent traffic information systems. VANETs are organized by nodes of sensors dispersed in space and unified. Sensor nodes detect the situation and use their communication modules to transmit data detected over wireless channels to other nodes via a selected node. The head node gathers the data transmitted to it in order to redistribute this data to other vehicles or to other networks.

Therefore, these networks can be vulnerable to routing attacks. One of the routing attacks is the black-hole attack targeting node heads. A malicious node may be chosen as the node head and absorb all data received from the members associated with it. Therefore, the assailant absorbs any message received and can prevent its retransmission. Selective transmission is a specific type of black-hole attack. Instead of discarding all received packets, the adversarial node randomly or maliciously selects which packets will be transmitted or which packets will be discarded.

13.4.5. *Jamming attack*

Jamming can occur whenever several systems transmit on the same frequencies. The principle is that a receiver receiving a communication signal within its bandwidth simultaneously receives another signal that it is not supposed to receive. Depending on the strength ratios between the expected signal and the unwanted signal, a jamming situation may occur. The receiver may be unable to detect or demodulate the expected signal due to the presence of the unwanted signal at its input.

Therefore, there exist intentional jammers capable of transmitting in multiple frequency bands and thus jamming the reception of various communication terminals. To be more precise, the jamming may nevertheless occur in a slightly different manner depending on the communicating protocols attacked and depending on its strength compared to the expected communication signals. It can either overlap with sufficient strength ratios the signals of the communication system, thus making it impossible to detect and recognize the preamble signals used for synchronization, equalization, channel estimation, etc. or prevent data demodulation. Sometimes, in the case of protocols for which the terminals check the resource availability, with lower jamming strengths, the presence of the jamming signal can deceive the availability analysis algorithms, giving the effect of a permanently used resource.

Most intentional jamming transmits permanently, generally frequency-scanning jamming signals. They scan repeatedly, with periods of several μs, the frequency bands allocated to wireless communication systems.

There may also exist slightly more targeted jamming signals depending on the communication protocols against which they are designed. Therefore, different terms are found in the literature to denote the different jamming solutions (Pirayesh and Zeng 2022). We speak of constant jamming attacks to denote those that transmit permanently, and reactive jamming attacks for those that rely on listening and transmit only when they detect the transmission of a communication signal. Lastly, the transmission instants can also vary according to the targeted jamming strategies, or even be random, to simply reduce the jammer consumption. The transmission periods and instants can be adjusted to cover only the preambles of the communication signals or the cyclic prefixes of the symbols, or even the periods for controlling the availability of time–frequency resources.

13.4.6. *Flooding attack*

A flooding attack consists of sending more packets than the receiver is able to process. This leads to a denial of service for the receiver. This attack is the basis of distributed denial of service (DDoS) against Internet servers (Afanasyev et al. 2013).

In the context of vehicular communications, this attack can be carried out towards an RSU that will no longer be able to communicate with the vehicles surrounding it but can also lead to flooding the control center if the packets are propagated by the RSU.

This attack on a vehicle will make the vehicle no longer capable of processing the information that reaches it and could lead to a poor decision being made, leading to an off-road accident or a collision with another vehicle.

A variant of this attack consists of giving the impression that the network is completely saturated by continuously sending CTS frames serving to inform users of a radio resource reservation request (Sakiz and Sen 2017).

13.4.7. *Risks and performance indicators*

The attacks we have discussed can lead to different consequences. Attacks can cause performance degradation, for example, on throughputs, error rates or latency times. They can also produce total breakdowns of communication that can be long or short in duration, with varying levels of difficulty in restoring service. This thus results in the non-transmission of certain information. Lastly, some attacks can lead to information leaks.

In the case of connected vehicles, the risks introduced by these different attacks and their consequences are greatly dependent on applications carried by wireless connections. In particular, a degradation of performance with an increase in latency can sometimes be more damaging than a total breakdown of communication, especially in platooning or obstacle/incident detection applications. Indeed, the communication breakdown is more easily identified while performance can vary regularly depending on different external parameters and not only in the presence of attacks. Therefore, for certain security applications, the standards evolve to take account of this danger.

We may mention, for example, Version 2 of Release 16 of the ETSI TR 121 916 standard published in July 2022 (ETSI 2022). This includes a section dedicated to V2X applications and outlines the requirements for vehicle quality of service. This section specifies that a V2X application must be informed in a timely manner of a planned or estimated change in quality of service (QoS). Indeed, if communication error rates risk increasing or decreasing, certain V2X applications need to adjust their parameters. If we take the example of the platooning application, it needs to increase or decrease the inter-vehicle distance.

This necessitates V2X application servers that may present specific quality-of-service requirements. They may also be informed whether the requested quality of service requirements can or cannot be met.

With this information, V2X applications need to adapt their behavior according to the quality of service that can be provided by the 3GPP system.

13.5. Attack detection

In this section, we will focus on detection methods and their usage for detecting attacks against connected vehicles.

13.5.1. *Need for a detection system*

The connected vehicles are connected to devices via wireless communication networks. The evolution of the IoT has seen the data transmitted by these vehicles become increasingly large in number and diversified. With the increase in transmission possibilities, vehicles no longer only share positioning or connection data with a telephone, but also use these technologies to transmit data related to vehicle safety data, and to communication between vehicles.

For example, the automotive IoT enables high-throughput vehicle-to-vehicle and vehicle-to-infrastructure communications that today provide:

– adaptive cruise control;

– automatic route planning based on real-time conditions;

– diversion of traffic away from congestion;

– automatic updating of road signage, to signal road and traffic conditions;

– communications to drivers, informing them of accident sites or drivers in front heading in the wrong direction;

– automatic braking of vehicles to avoid collisions (undergoing testing);

– control of autonomous and semi-autonomous vehicles (undergoing testing).

With the constant quest to make roads safer and the real interest surrounding autonomous cars, the usage of these networks will continue to grow, along with the critical nature of these communications.

It is then easy to imagine a system in which vehicles would communicate with each other in order to share information regarding their position, their speed, the presence of obstacles and what they will do next (e.g. stop). In this context, we can conceive of what an attack on these networks could cause.

If we consider the case of a Sybil attack, it can simulate a traffic jam situation, by creating fake vehicles on a road. This could lead to this road not being used by real vehicles, which would be diverted to another route.

It is therefore essential to protect and be able to estimate the quality and reliability of the data retrieved from these networks. Thus, it is necessary to detect the presence of attacks against these networks in order to deploy an effective countermeasure.

13.5.2. *Detection method*

There exist several ways of using mathematics to highlight data information collected by sensors. Depending on the data collected, as well as the experts' prior knowledge regarding this data, the usage of mathematics can have different objectives (see Figure 13.1). It is possible to wish to highlight the characteristics of this data through descriptive or inferential statistical analyses, but it is also possible to wish to put in place models to characterize the data through learning. In the latter case, we generally seek to differentiate groups, or to model a phenomenon.

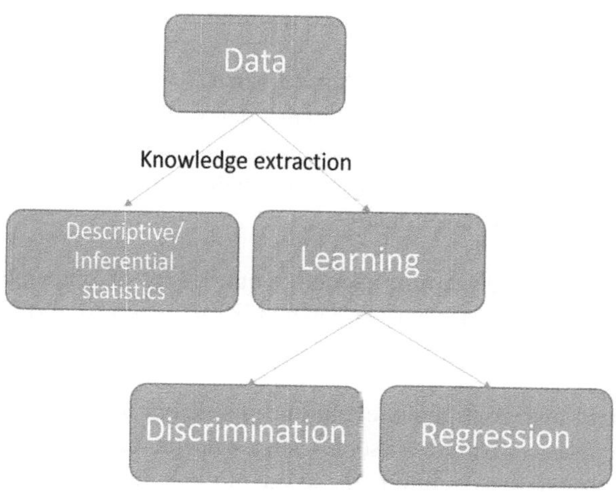

Figure 13.1. *Mathematics and data*

In the domain of mathematics, detection models aim to detect, through learning, a known or unknown profile using collected or simulated information. In their usages, the primary objective of these algorithms is to recognize the presence or absence of a form or combination of values within the data and thus to detect them.

These algorithms are generally based on discrimination approaches (classification, or anomaly detection). These approaches can be broken down into several methods (see Figure 13.2).

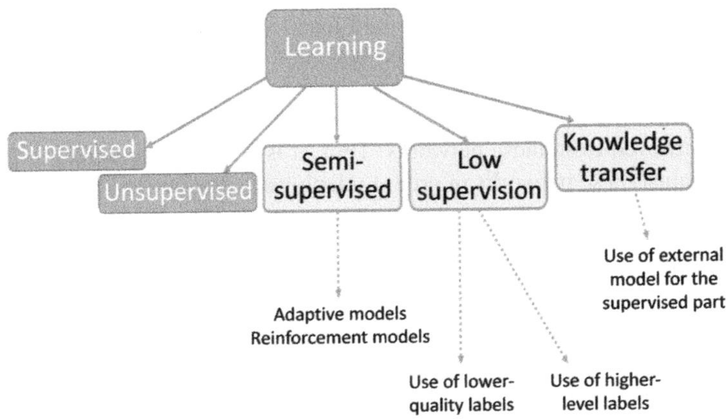

Figure 13.2. *Learning methods*

Among the classification algorithms used for detection, two families are generally differentiated: supervised and unsupervised learning.

The learning methods known as "unsupervised" are not trained by an expert providing knowledge on the collected data. They depend on learning methods that identify shapes or profiles by fine-combing unlabeled datasets, studying the correlations between this data. Models trained with this method are not directed to find a result or identify particular data.

Conversely, supervised learning models require a lot of preparatory work from experts. The data collected needs to be labeled by an expert to allow models to best add the boundaries enabling a decision to be made.

With the rapid and continuous evolution of sensors as well as of storage and communication space, we find ourselves with ever-increasing amounts of data to process. In response to this evolution, the usage of more flexible methods has become necessary. Three new approaches were then developed: weak learning methods, semi-supervised models and knowledge transfer methods.

Weak learning methods are the most intuitive methods. The concept is to reduce the complexity of learning by reducing the complexity of labels. For example, in the case of connected communication, there exist different types of communication. If we wish to characterize the source of the communication in a V2V/V2I system, we may want to have a label corresponding to each of the elements of the network by its identifier. The number of labels would be far too large for the models to be efficient. Weak learning thus consists of reducing the number of labels in order to obtain a good compromise between performance and need. In this case, it could be decided to label the source of the signal as "vehicle", "infrastructure" or "other".

Semi-supervised learning methods combine labeled data (as in the supervised case) and unlabeled data (as in the unsupervised case). These algorithms are a mix of supervised and unsupervised approaches. Algorithms of this type feed on certain information by means of labeled categories, suggestions and examples. Then, they create their own labels by exploring the data itself, following a rudimentary scheme or the indications of experts. One example of the usage of these approaches is the classification of attacks. Indeed, with attacks constantly evolving, these algorithms make it possible to classify known attacks as well as to evolve with the emergence of new attacks.

Lastly, knowledge transfer methods are approaches that use existing concepts or models to guide learning or serve as a basis for future learning. In this family of models, we find reinforcement learning that is based on reward and punishment systems. The algorithm is assigned an objective and seeks to obtain a maximum reward. It builds on prior information and learns from its previous actions. These approaches generally depend on previously transferred basic models that vary in complexity. For example, we could learn a simple communication model of IEEE 802.11n, 802.11ax, 802.11p, etc. exchange based on simulation and use reinforcement on real communications.

Numerous classification algorithms are presented in Clarke et al. (2009), James et al. (2013), Charu (2018) and Murphy (2022) and will allow us to deepen our knowledge in this domain.

While classification is primarily intended to label information using its similarities, detection also brings profile recognition in a larger set. Let us take the example of communication. Where a classification model allows us to determine what type of signal is present on a communication channel, the detection will be able to recognize the presence of these signals on different communication channels. In the following, we will present several AI use cases for the detection of attacks on vehicular communications.

13.5.3. *AI for detection*

Like any other communication and data processing system, vehicular networks are exposed to different types of threats and attacks. Because of the critical nature of the usage of these communications, research in the domain of detection of these attacks is a highly valued domain. To meet these expectations, the algorithms are adapted in order to respond more efficiently to the issue encountered. Here, we will highlight the use of the different approaches mentioned above for the detection of Sybil, black-hole or jamming attacks. With detection approaches able to adapt to different attacks, the literature features a large number of examples of attack detection using these approaches. We will mention here just a few examples for the attack cases mentioned.

13.5.3.1. *AI for Sybil attack detection*

Firstly, let us consider the case of Sybil attacks. In the literature, several algorithms are deployed to prevent these attacks on large navigation networks or vehicular networks. We find detection algorithms using nearby nodes, using communication characteristics such as the RSSI criterion but, increasingly, learning approaches are deployed. A few examples can be used to illustrate the usage of learning approaches.

In James (2020), the authors show the usefulness of unsupervised and supervised approaches to identify attacking sources in the network. The authors propose using a Bayesian model that is self-supervised by the use of a recurrent autoencoder. An autoencoder is a neural network used for unsupervised learning of discriminative characteristics. The objective of an autoencoder is to learn a representation of a set of data, generally with the aim of reducing the size of this set. The objective of the model is to detect adversarial trajectories of Sybil attacks targeting communicating-navigation systems based on the synchronization of a vast number of sources. The proposed deep learning model is able to exploit the chronological characteristics of vehicle trajectories with sampling and displacement uncertainties. By adopting an encoder/reconstructor architecture (robust reconstruction of motion vectors applied to the encoder), the trajectories are embedded in a latent trajectory distribution space as multivariate random variables, which are then used to reconstruct trajectories considered authentic by the model. Next, by comparing the entry trajectory with those obtained, a credibility score can be computed statistically. Lastly, the training program is designed on the basis of trajectory gathering. To assess the efficiency of this proposed model, a series of comprehensive case studies are conducted on different sets of vehicle trajectory data in the real world.

We can also cite the works of Gu et al. (2017) that propose to detect Sybil attacks through support vector machine (SVM) learning using a core to transform communications. Therefore, they present three classifiers that distinguish malicious nodes from benign nodes by assessing the variance of their driving pattern matrices (DPM). We also cite the works of Quevedo et al. (2020), in which the authors describe a detection mechanism called SyDVELM, a mechanism based on an extreme machine learning approach to detect the presence of Sybil attacks in vehicular ad hoc networks (VANETs).

13.5.3.2. *AI for the detection of black-hole attacks*

We will now focus on the detection of black-hole attacks. In the literature, we also find numerous works aiming to detect this type of attack. Here, we will focus particularly on detection systems combining several machine learning approaches.

The work by Acharya et al. (2021) is of particular interest in the usage of learning classification approaches. In this article, the authors present a model that applies an SVM approach, as well as a gradient-boosting approach, a naive Bayesian classification approach, a k-NN approach and a logistic regression approach to prevent black-hole attacks in a VANET environment. To measure the quality of these models, the authors use an Akaike information criterion (AIC) maximum likelihood estimation as well as the ROC (Receiver Operating Characteristic) curve, each of which are measurements to assess the performance of a classifier.

Other articles also use more complex learning approaches to detect black-hole attacks in a VANET environment, such as the works of Malyshev et al. (2019) which deploy neural network learning to detect different types of attacks including black-hole attacks.

13.5.3.3. *AI for the detection of jamming attacks*

As regards jamming attacks, the case most often considered in the literature is that of detection. The mobile aspect of the jamming signal is rarely taken into account. Among the most widely used methods for interference signal detection, the signal-to-interference ratio (SIR) is the criterion most often put forward for this kind of detection. Through statistical studies, a tolerance threshold is put in place to detect the presence of jamming. The detection approaches then seek to highlight an anomaly in the behavior of this criterion. The example of the threshold is the most frequent case but we can also find other statistical methods called "novelty detection" to highlight the presence of jamming signals. For example, we can find approaches based on statistical tests, Bayesian approaches or learning-based approaches. We can cite the works of Strasser et al. (2010), Çakıroğlu and Özcerit (2021) and of Puñal et al. (2014) as examples on this subject.

We will take here as an example the works described in Villain et al. (2019) and Villain et al. (2021). In these works, the authors put in place a method for detecting a jamming signal received at different strength levels. In Villain et al. (2019), the work focuses on putting in place a communication channel monitoring system to detect the presence of a jamming signal measured at three different strength levels, ranging from very low strength (barely perceptible) to a strength level that prevents any communication. In order to detect these jamming signals, a support vector machine approach is deployed to learn the communication channel behavior in the presence of these signals. A temporal correction is then deployed on the detection to avoid the detection of a false positive. These works highlighted the difficulty that detection algorithms have in detecting the presence of very low-strength jamming signals. In Villain et al. (2021), an unsupervised approach is put in place on the measurement data collected for a very low jamming signal strength and a state without the presence of jamming signals. The groups resulting from this classification made it possible to highlight different profiles characteristic of the presence of a very low-intensity jamming signal.

13.6. Conclusion

With the evolution of communication systems and the desire to deploy increasingly connected vehicles, the ability to interact with vehicle systems has become increasingly critical. Attacks on these communication networks have evolved and can disrupt the road domain. The attacks presented in this chapter demonstrate the importance of securing or assessing the risk incurred by these attacks. It is therefore essential to detect these attacks in order to deploy efficient countermeasures. The advent of artificial intelligence is a significant point in the handling of this kind of issue. It makes it possible, through learning or putting in place a distinct profile, to detect characteristics on the networks and therefore to deploy a countermeasure adapted to the situation. The majority of research in the domain of detecting attacks in cybersecurity or telecommunications is seeing the development of new Artificial Intelligence approaches to meet the new security challenges of these networks.

13.7. References

3GPP (2022). TR 21.916 V16.2.0; 3rd Generation Partnership Project; Technical Specification Group Services and System Aspects; Release 16 Description; Summary of Rel-16 Work Items (Release 16). 3GPP Technical Report [Online]. Available at: https://www.3gpp.org/specifications-technologies/releases/release-16.

Abramson (1970). THE ALOHA SYSTEM: Another alternative for computer communications. In *Proceedings of the 17–19 November 1970, Fall Joint Computer Conference (AFIPS '70 (Fall))*, 281–285. ACM, New York.

Acharya, A. and Oluoch, J. (2021). A dual approach for preventing blackhole attacks in vehicular ad hoc networks using statistical techniques and supervised machine learning. In *2021 IEEE International Conference on Electro Information Technology* (EIT), Mt. Pleasant, 230–235.

Afanasyev, A., Mahadevan, P., Moiseenko, I., Uzun, E., Zhang, L. (2013). Interest flooding attack and countermeasures in Named Data Networking. In *Proceedings of the 2013 IFIP Networking Conference, IFIP Networking 2013*, Brooklyn, 1–9.

Çakiroglu, M. and Özcerit, A.T. (2010). Jamming detection mechanisms for wireless sensor networks. In *3rd International ICST Conference on Scalable Information Systems*. Springer Berlin, Heidelberg.

Charu, C.A. (2018). *Neural Networks and Deep Learning: A Textbook*. Springer, New York.

Clarke, B., Fokoue, E., Zhang, H.H. (2009). *Principles and Theory for Data Mining and Machine Learning*. Springer, New York.

ETSI (2022). ETSI TR 121 916 V16.2.0; Digital cellular telecommunications system (Phase 2+) (GSM); Universal Mobile Telecommunications System (UMTS); LTE; 5G; Release 16 Description; Summary of Rel-16 Work Items (3GPP TR 21.916 version 16.2.0 Release 16). ETSI Technical Report [Online]. Available at: https://www.etsi.org/deliver/etsi_tr/121900_121999/121916/16.02.00_60/tr_121916v160200p.pdf.

Gu, P., Khatoun, R., Begriche, Y., Serhrouchni, A. (2017). Support vector machine (SVM) based sybil attack detection in vehicular networks. In *2017 IEEE Wireless communications and networking conference. (WCNC)*, San Francisco, 1–6.

James, J.Q. (2020). Sybil attack identification for crowdsourced navigation: A self-supervised deep learning approach. *IEEE Transactions on Intelligent Transportation Systems*, 22(7), 4622–4634.

James, G., Witten, D., Hastie, T., Tibshirani, R. (2013). *An Introduction to Statistical Learning*. Springer, New York.

Malyshev, E.V., Moskvin, D.A., Zegzhda, D.P. (2019). Application of an artificial neural network for detection of attacks in VANETs. *Automatic Control and Computer Sciences*, 53(8), 889–894.

Murphy, K.P. (2022). *Probabilistic Machine Learning: An Introduction*. MIT Press, Cambridge, MA.

Pirayesh, H., Zeng, H. (2022). Jamming attacks and anti-jamming strategies in wireless networks: A comprehensive survey. *IEEE Communications Surveys & Tutorials*, 24(2), 767–809.

Puñal, O., Aktaş, I., Schnelke, C.J., Abidin, G., Wehrle, K., Gross, J. (2014). Machine learning-based jamming detection for IEEE 802.11: Design and experimental evaluation. In *Proceeding of IEEE International Symposium on a World of Wireless, Mobile and Multimedia Networks 2014*, Sydney, 1–10.

Quevedo, C.H., Quevedo, A.M., Campos, G.A., Gomes, R.L., Celestino, J., Serhrouchni, A. (2020). An intelligent mechanism for sybil attacks detection in vanets. In *ICC 2020–2020 IEEE International Conference on Communications*, (ICC), Dublin, 1–6.

Sakiz, F. and Sen, S. (2017). A survey of attacks and detection mechanisms on intelligent transportation systems: VANETs and IoV. *Ad Hoc Networks Journal*, 61, 33–50.

Storck, C.R. and Duarte-Figueiredo, F. (2020). A survey of 5G technology evolution, standards, and infrastructure associated with vehicle-to-everything communications by internet of vehicles. *IEEE Access*, 8, 117593–117614. doi: 10.1109/ACCESS.2020.3004779.

Strasser, M., Danev, B., Capkun, S. (2010). Detection of reactive jamming in sensor networks. *ACM Transactions on Sensor Networks* (TOSN), 7(2), 1–29.

Villain, J., Deniau, V., Fleury, A., Simon, E.P., Gransart, C., Kousri, R. (2019). EM monitoring and classification of IEMI and protocol-based attacks on IEEE 802.11 n communication networks. *IEEE Transactions on Electromagnetic Compatibility*, 61(6), 1771–1781.

Villain, J., Deniau, V., Gransart, C., Fleury, A., Simon, E.P. (2021). Characterization of IEEE 802.11 communications and detection of low-power jamming attacks in noncontrolled environment based on a clustering study. *IEEE Systems Journal*, 16(1), 683–692.

List of Authors

Toufik AHMED
LaBRI
Bordeaux INP
France

Lylia ALOUACHE
ETIS
CY Cergy Paris Université
France

Hasnaâ ANISS
ERENA
COSYS
Gustave Eiffel University
Marne-la-Vallée
France

Marwane AYAIDA
IEMN
Université Polytechnique
Hauts-de-France
Valenciennes
France

Sabrine BELMEKKI
COSYS-LEOST
Gustave Eiffel University
Villeneuve-d'Ascq
France

Mohamed BENZAGOUTA
ERENA
COSYS
Gustave Eiffel University
Marne-la-Vallée
France

Marion BERBINEAU
COSYS-LEOST
Gustave Eiffel University
IFSTTAR
Villeneuve-d'Ascq
France

Jean-Marie BONNIN
IMT Atlantique
Rennes
France

Ramzi BOUTAHALA
Université de Reims
Champagne-Ardenne
France

Mohamed Aymen CHALOUF
IRISA
Université de Rennes
Lannion
France

Christophe COUTURIER
YoGoKo
Rennes
France

Sébastien DEMMEL
Queensland University of Technology
Brisbane
Australia

Virginie DENIAU
COSYS-LEOST
Gustave Eiffel University
Villeneuve-d'Ascq
France

Romain DULOUT
ERENA
COSYS
Gustave Eiffel University
Marne-la-Vallée
France

Secil ERCAN
ERENA
COSYS
Gustave Eiffel University
Marne-la-Vallée
France

Hacène FOUCHAL
CReSTIC
Université de Reims
Champagne-Ardenne
France

David GÉRAULT
Technology Innovation Institute
Abu Dhabi
UAE

Imed GHNAYA
LaBRI
Bordeaux INP
France

Sébastien GLASER
Queensland University of Technology
Brisbane
Australia

Christophe GRANSART
COSYS-LEOST
Gustave Eiffel University
Villeneuve-d'Ascq
France

Dominique GRUYER
PICS-L
Gustave Eiffel University
Marne-la-Vallée
France

Sio-Song IENG
PICS-L
Gustave Eiffel University
Marne-la-Vallée
France

Sabri KHAMARI
LaBRI
Bordeaux INP
France

Francine KRIEF
LaBRI
Bordeaux INP
France

Pascal LAFOURCADE
LIMOS
Université Clermont-Auvergne
Clermont-Ferrand
France

Sassi MAALOUL
ERENA
COSYS
Gustave Eiffel University
Marne-la-Vallée
and
JUNIA
École d'ingénieurs
Lille
France

Léo MENDIBOURE
ERENA
COSYS
Université Gustave Eiffel
Marne-la-Vallée
France

Mohamed MOSBAH
LaBRI
Bordeaux INP
France

Yannis POUSSET
XLIM
Université de Poitiers
Chasseneuil-du-Poitou
France

Abdennour RACHEDI
LaBRI
Bordeaux INP
France

Léo ROBERT
MIS
Université de Picardie Jules Verne
Amiens
France

Tidiane SYLLA
LaMIA
Université des Sciences
des Techniques et des Technologies de
Bamako (USTT-B)
Mali

Charles TATKEU
COSYS-LEOST
Gustave Eiffel University
Villeneuve-d'Ascq
France

Jonathan VILLAIN
COSYS-LEOST
IMT Nord Europe
Lille
France

Geoffrey WILHEM
CReSTIC
Université de Reims
Champagne-Ardenne
France

Badreddine Yacine YACHEUR
LaBRI
Bordeaux INP
France

Index

E, F

G, H

I, J

L, M

N, O

P, Q

R, S

T, U, V

Printed and bound by CPI Group (UK) Ltd, Croydon, CR0 4YY

13/09/2026

14961070-0008